Study Guide and Solutions Manual

for
Seager and Slabaugh's

Chemistry for Today: General, Organic, and Biochemistry

Fifth Edition

Jennifer P. Harris
Portland Community College

Australia • Canada • Mexico • Singapore • Spain • United Kingdom • United States

Cover Image: © Terje Rakke/Getty Images

COPYRIGHT © 2005 Brooks/Cole, a division of Thomson Learning, Inc. Thomson Learning™ is a trademark used herein under license.

ALL RIGHTS RESERVED. No part of this work covered by the copyright hereon may be reproduced or used in any form or by any means—graphic, electronic, or mechanical, including but not limited to photocopying, recording, taping, Web distribution, information networks, or information storage and retrieval systems—without the written permission of the publisher.

Printed in the United States of America
1 2 3 4 5 6 7 07 06 05 04 03

Printer: West Group

ISBN: 0-534-39973-8

For more information about our products, contact us at:
Thomson Learning Academic Resource Center
1-800-423-0563

For permission to use material from this text, contact us by:
Phone: 1-800-730-2214
Fax: 1-800-730-2215
Web: http://www.thomsonrights.com

Brooks/Cole—Thomson Learning, Inc.
10 Davis Drive
Belmont, CA 94002-3098
USA

Asia
Thomson Learning
5 Shenton Way #01-01
UIC Building
Singapore 068808

Australia/New Zealand
Thomson Learning
102 Dodds Street
Southbank, Victoria 3006
Australia

Canada
Nelson
1120 Birchmount Road
Toronto, Ontario M1K 5G4
Canada

Europe/Middle East/South Africa
Thomson Learning
High Holborn House
50/51 Bedford Row
London WC1R 4LR
United Kingdom

Latin America
Thomson Learning
Seneca, 53
Colonia Polanco
11560 Mexico D.F.
Mexico

Spain/Portugal
Paraninfo
Calle/Magallanes, 25
28015 Madrid, Spain

Table of Contents

General Chemistry

Chapter 1	Matter, Measurements, and Calculations	1
Chapter 2	Atoms and Molecules	19
Chapter 3	Electronic Structure and Periodic Law	33
Chapter 4	Forces Between Particles	43
Chapter 5	Chemical Reactions	59
Chapter 6	States of Matter	73
Chapter 7	Solutions and Colloids	89
Chapter 8	Reaction Rates and Equilibrium	109
Chapter 9	Acids, Bases, and Salts	123
Chapter 10	Radioactivity and Nuclear Processes	143

Organic Chemistry

Chapter 11	Organic Compounds: Alkanes	153
Chapter 12	Unsaturated Hydrocarbons	169
Chapter 13	Alcohols, Phenols, and Ethers	185
Chapter 14	Aldehydes and Ketones	201
Chapter 15	Carboxylic Acids and Esters	213
Chapter 16	Amines and Amides	225

Biochemistry

Chapter 17	Carbohydrates	239
Chapter 18	Lipids	251
Chapter 19	Proteins	263
Chapter 20	Enzymes	277
Chapter 21	Nucleic Acids and Protein Synthesis	287
Chapter 22	Nutrition and Energy for Life	299
Chapter 23	Carbohydrate Metabolism	309
Chapter 24	Lipid and Amino Acid Metabolism	321
Chapter 25	Body Fluids	333
	Answers to the Chapter Self-Tests	343

Preface

This book is designed to serve as both a reference and a study tool. Each chapter includes a Chapter Outline, Learning Objectives/Assessment statements, Solutions for the Even-numbered End of Chapter (EOC) Exercises, Additional Activities, Answers to the Additional Activities, and Self-Test Questions. The Additional Activities include specific questions for each section as well as a summary activity entitled, "Tying It All Together with a Laboratory Application." The Answers to the Self-Test Questions are at the end of the book.

The EOC Exercises marked with the ☑ symbol are questions identified as "quick checks." Your ability to answer the "quick check" questions correctly indicates how well you have met the learning objectives for the chapter. It is important to remember that your instructor may have additional learning objectives that are not identified in this text.

I hope this book is useful to you and that your experience studying chemistry is positive. The subject matter is fascinating (ask any chemist) and applies to daily life. As with any meaningful scholastic experience, you will need to study. Ideally, using this book will focus that process. If you have questions, comments, or concerns about the book, please, e-mail me at jpharris@pcc.edu.

Thank you to all of the people who helped me with this book, especially my students, Lauren Raike, and Lynn Carlson. I also want to thank a few of the many notable chemists in my life: Fred R. Harris, my husband and best friend; Dr. Barinder Deu, my first high school chemistry teacher; Dr. Debra Feakes, my mentor at Southwest Texas State University (now, Texas State University-San Marcos); and my colleagues at Portland Community College. Additionally, I want to thank my mom and dad, COL (Ret.) and Mrs. David L. Pointer, and my sister, Colleen E. Pointer, MS, RD, LD (dietician extraordinaire), for their unfailing support. Last, but certainly not least, thank you to my dog, Candy Girl Harris, for both keeping my feet warm as I typed and taking me for walks in the Oregon sunshine throughout this process.

Chapter 1: Matter, Measurements, and Calculations

CHAPTER OUTLINE
1.1 What Is Matter?
1.2 Properties and Changes
1.3 A Model of Matter
1.4 Classifying Matter
1.5 Measurement Units
1.6 The Metric System
1.7 Large and Small Numbers
1.8 Significant Figures
1.9 Using Units in Calculations
1.10 Calculating Percentages
1.11 Density

LEARNING OBJECTIVES/ASSESSMENT
When you have completed your study of this chapter, you should be able to:
1. Explain what matter is. (Section 1.1; Exercise 1.2)
2. Explain differences between the terms *physical* and *chemical* as applied to:
    a. Properties of matter (Section 1.2; Exercises 1.10 b & c)
    b. Changes in matter (Section 1.2; Exercises 1.8 a & b)
3. Describe matter in terms of the accepted scientific model. (Section 1.3; Exercise 1.12)
4. On the basis of observation or information given to you, classify matter into the correct category based on the following pairs:
    a. Heterogeneous or homogeneous (Section 1.4; Exercise 1.22)
    b. Solution or pure substance (Section 1.4; Exercise 1.24)
    c. Element or compound (Section 1.4; Exercise 1.18)
5. Recognize the units of the metric system. (Section 1.6; Exercise 1.30)
6. Convert metric-system measurement units into related units. (Section 1.6; Exercise 1.40)
7. Express numbers using scientific notation. (Section 1.7; Exercise 1.48)
8. Do calculations involving numbers expressed in scientific notation. (Section 1.7; Exercise 1.60)
9. Express the results of measurements and calculations using the correct number of significant figures. (Section 1.8; Exercises 1.64 and 1.66)
10. Use the factor-unit method to solve numerical problems. (Section 1.9; Exercise 1.82)
11. Do calculations involving percentages. (Section 1.10; Exercise 1.92)
12. Do calculations involving densities. (Section 1.11; Exercise 1.98)

SOLUTIONS FOR THE END OF CHAPTER EXERCISES
WHAT IS MATTER? (SECTION 1.1)

☑1.2 All matter occupies space and has mass. Mass is a measurement of the amount of matter in an object. The mass of an object is constant regardless of where the mass is measured. Weight is a measurement of the gravitational force acting on an object. The weight of an object will change with gravity; therefore, the weight of an object will be different at different altitudes and on different planets.

1.4 The distance you can throw a bowling ball through the air will change more than the distance you can roll a bowling ball on a flat, smooth surface. When throwing a ball through the air, gravity pulls the ball toward the ground. The gravitational force on the moon is approximately 1/6th the gravitational force that is present on the earth; therefore, when throwing a ball on the moon, you should be able to throw it further than you can on earth. When rolling a ball, friction helps to slow down the ball. If the flat, smooth surface is the same on the earth and the moon, the amount of friction should remain constant.

2   Chapter 1

1.6   The attractive force of gravity for objects near the earth's surface increases as you get closer to the center of the earth (Exercise 1.5). If the earth bulges at the equator, the people at the equator are further from the center of the earth than people at the North Pole. If two people with the same mass were weighed at the equator and at the North Pole, the person at the equator would weigh less than the person at the North Pole because the gravitational force at the North Pole is stronger than the gravitational force at the equator.

PROPERTIES AND CHANGES (SECTION 1.2)

1.8
- ☑ a. The two pieces of the stick still have the same chemical composition as the original stick. This was a change that did not involve composition; therefore, it is a **physical change**.
- ☑ b. As the candle burns, it produces carbon dioxide, water, soot, and other products. This is a change that involves composition; therefore, it is a **chemical change**.
- c. The pieces of rock salt have the same chemical composition as the original larger piece of rock salt. This was a change that did not involve composition; therefore, it is a **physical change**.
- d. Many tree leaves are green in the spring and summer because of the green chlorophyll that is used in photosynthesis to produce energy for the tree. During these seasons, the tree stores the extra energy so that in autumn when the days grow shorter, the chlorophyll is no longer needed. As the leaves in the cell stop producing chlorophyll, the other colors present in the leaves become more visible. This change involves composition; therefore, it is a **chemical change**.

1.10
- a. This is a **physical property** because the composition of mercury does not change while making this observation.
- ☑ b. This is a **chemical property** because the composition of the sodium metal and water molecules changes while making this observation. The products for this reaction are sodium hydroxide and hydrogen gas. (Note: Predicting the products for this type of chemical reaction is covered in Section 9.6.)
- ☑ c. This is a **physical property** because the composition of water does not change while making this observation.
- d. This is a **chemical property** because rust would have a different chemical composition than gold. Attempting to change the chemical composition of a material is a test of a chemical property regardless of whether the attempt is successful.
- e. This is a **physical property** because the composition of chlorophyll does not change while making this observation.

A MODEL OF MATTER (SECTION 1.3)

☑1.12
- a. Yes, the succinic acid molecules have been changed by the process. The molecules of succinic acid released at least one atom each in the form of a gas. Without those atoms, the molecules cannot be succinic acid molecules. Also, if they were still succinic acid molecules, the melting point of the remaining solid would still be 182°C.
- b. No, the white solid that remains after heating is not succinic acid. The melting point of succinic acid is 182°C, but the melting point of this new solid is not 182°C.
- c. The succinic acid molecules contain more atoms than the molecules of the white solid produced by this process. Some of the atoms that were originally part of the succinic acid molecules are given off as a gas. That leaves fewer atoms to be a part of the molecules of new white solid.
- d. Succinic acid is **heteroatomic**. Some of the atoms were able to leave the succinic acid molecule in the form of a gas. Other atoms remained as part of a new molecule. If all of

Matter, Measurements, and Calculations 3

the atoms were the same type, they would have all turned into a gas or they would have all remained as a solid.

1.14   Carbon dioxide is **heteroatomic**. If oxygen and carbon atoms react to form one product, then carbon dioxide must contain these two types of atoms.

1.16   Water is **heteroatomic**. If breaking water apart into its components produces both hydrogen gas and oxygen gas, then water must contain two types of atoms.

CLASSIFYING MATTER (SECTION 1.4)

☑1.18   a. Substance A is a **compound** because it is a pure substance that contains at least two elements.
        b. Substance D is an **element** because it is a pure substance that contains only one element.
        c. Substance E is a **compound** because it is a pure substance that can break down into at least two different materials. Substances G and J **cannot be classified** because no tests were performed on them.

1.20   a. Substance R might appear to be an element based on the tests performed. It has not decomposed into any simpler substances based on these tests; however, this is not an exhaustive list of tests that could be performed on substance R. Substance R **cannot be classified** as an element or a compound based on the information given.
        b. Substance T is a **compound**. It is composed of at least two different elements because it produced two different substances on heating.
        c. The solid left in part b **cannot be classified** as an element or a compound. No tests have been performed on it.

☑1.22 | | | |
        |---|---|---|
        | a. | A gold chain | It is **homogeneous** because it is the same composition throughout. |
        | b. | Liquid eye drops | It is **homogeneous** because it is the same composition throughout. |
        | c. | Chunky peanut butter | It is **heterogeneous** because it is not the same composition throughout (peanut chunk vs. smooth regions). |
        | d. | A slice of watermelon | It is **heterogeneous** because it is not the same composition throughout (rind, meat, and seeds). |
        | e. | Cooking oil | It is **homogeneous** because it is the same composition throughout. |
        | f. | Italian salad dressing | It is **heterogeneous** because it is not the same composition throughout (oil, vinegar, and seasonings). |
        | g. | Window glass | It is **homogeneous** because it is the same composition throughout. |

☑1.24 | | | |
        |---|---|---|
        | a. | A gold chain | If the chain is only made of gold, then it is a **pure substance**. Most jewelry is made of gold alloyed with another metal for strength. If this chain is made of a gold alloy, then it is a **solution**. |
        | b. | Liquid eye drops | This is a **solution** because it contains saline (water and sodium chloride) as well as other substances. |
        | e. | Cooking oil | It can be a **solution** or a **pure substance** depending on the oil. Some oils only contain a single compound (pure substances); however, most oils are a mixture of several different compounds (solution). |
        | g. | Window glass | It is a **solution** because it contains a mixture of silicon dioxide and other components like iron. |

Chapter 1

MEASUREMENT UNITS (SECTION 1.5)

1.26 Modern society is complex and interdependent. Accomplishing projects like building a bridge, constructing a house, or machining an engine may require many different people to participate. Some people design the project, others supply the necessary materials, and yet another group does the construction. In order for the project to be successful, all of these people need a common language of measurement. Measurement is also important for giving directions, keeping track of the time people work, and keeping indoor environments at a comfortable temperature and pressure.

1.28 The amount of weight that a horse could carry or drag might have been measured in stones. It could also be used to measure people or other items in the 50-500 pound range. It is likely that a large stone was picked as the standard weight for the "stone" unit.

THE METRIC SYSTEM (SECTION 1.6)

☑1.30 Degrees Celsius (a), liters (b), milligrams (d), and seconds (f) are metric units. Feet (c) and quarts (e) are English units.

1.32 Meters are a metric unit that could replace the English unit feet in the measurement of the ceiling height. Liters are a metric unit that could replace the English unit quarts in the measurement of the volume of a cooking pot.

1.34
a. $1.00 \, \text{L} \left(\dfrac{1 \, \mu\text{L}}{10^{-6} \, \text{L}} \right) = 1.00 \times 10^6 \, \mu\text{L}$

b. $75 \, \text{kilowatts} \left(\dfrac{1000 \, \text{watts}}{1 \, \text{kilowatt}} \right) = 7.5 \times 10^4 \, \text{watts}$

c. $15 \, \text{megahertz} \left(\dfrac{10^6 \, \text{hertz}}{1 \, \text{megahertz}} \right) = 1.5 \times 10^7 \, \text{hertz}$

d. $200 \, \text{picometers} \left(\dfrac{10^{-12} \, \text{meters}}{1 \, \text{picometer}} \right) = 2 \times 10^{-10} \, \text{meters}$

1.36 $1 \, \text{cup} \left(\dfrac{240 \, \text{mL}}{1 \, \text{cup}} \right) \left(\dfrac{1 \, \text{L}}{1000 \, \text{mL}} \right) = 0.24 \, \text{L}$ or $1 \, \text{cup} \left(\dfrac{240 \, \text{mL}}{1 \, \text{cup}} \right) \left(\dfrac{1 \, \text{cm}^3}{1 \, \text{mL}} \right) = 240 \, \text{cm}^3$

1.38 $4.0 \, \text{kg} \left(\dfrac{2.20 \, \text{lbs.}}{1 \, \text{kg}} \right) = 8.8 \, \text{lbs.}$

☑1.40
a. $1 \, \text{m} = 1.094 \, \text{yd}$, so: $1 \, \text{m} - 1 \, \text{yd} = 1.094 \, \text{yd} - 1 \, \text{yd} = 0.094 \, \text{yd} \left(\dfrac{3 \, \text{ft}}{1 \, \text{yd}} \right) \left(\dfrac{12 \, \text{in}}{1 \, \text{ft}} \right) = 3.4 \, \text{in}$

b. $1°\text{C} = 1 \, \text{K}$, so $65°\text{C} = 65 \, \text{K}$.

c. $5 \, \text{lbs.} \left(\dfrac{1 \, \text{kg}}{2.20 \, \text{lbs.}} \right) = 2.27 \, \text{kg} = 2 \, \text{kg}$ with significant figures

Matter, Measurements, and Calculations 5

$$1.0 \text{ cm}^3 \left(\frac{1 \text{ dm}^3}{1000 \text{ cm}^3}\right)\left(\frac{1 \text{ kg}}{1.0 \text{ dm}^3}\right) = 1.0 \times 10^{-3} \text{ kg} \quad \text{or}$$

1.42 a.
$$1.0 \text{ cm}^3 \left(\frac{1 \text{ dm}^3}{1000 \text{ cm}^3}\right)\left(\frac{1 \text{ kg}}{1.0 \text{ dm}^3}\right)\left(\frac{1000 \text{ g}}{1 \text{ kg}}\right) = 1.0 \text{ g}$$

b. $$2.0 \text{ L}\left(\frac{1.057 \text{ qt}}{1 \text{ L}}\right)\left(\frac{32 \text{ fl oz}}{1 \text{ qt}}\right) = 68 \text{ fl oz}$$

c. $$5 \text{ grain}\left(\frac{1 \text{ mg}}{0.015 \text{ grain}}\right) = 333 \text{ mg} = 300 \text{ mg with significant figures}$$

1.44 $°F = \frac{9}{5}(°C) + 32$ $°F = \frac{9}{5}(36.1°C) + 32 = 97.0°F$ $°F = \frac{9}{5}(37.2°C) + 32 = 99.0°F$

LARGE AND SMALL NUMBERS (SECTION 1.7)

1.46
- a. 02.7×10^{-3} — Improper form because no leading zero is necessary. (2.7×10^{-3})
- b. 4.1×10^2 — Correct.
- c. 71.9×10^{-6} — Improper form because only one digit should be to the left of the decimal point. (7.19×10^{-5})
- d. 10^3 — Improper form because a nonexponential term should be written before the exponential term. (1×10^3)
- e. 0.0405×10^{-2} — Improper form because one nonzero digit should be to the left of the decimal point. (4.05×10^{-4})
- f. 0.119 — Improper form because one nonzero digit should be to the left of the decimal point and an exponential term should be to the right of the nonexponential term. (1.19×10^{-1})

☑1.48
- a. 14 thousand   = 14,000 = 1.4×10^4
- b. 365 3.65×10^2
- c. 0.00204 2.04×10^{-3}
- d. 461.8 4.618×10^2
- e. 0.00100 1.00×10^{-3}
- f. 9.11 hundred = 9.11 × 100 = 9.11×10^2

1.50
- a. 186 thousand mi/s $186 \times 1000 = 1.86 \times 10^5$ mi/s
- b. 1100 million km/h $1100 \times 1{,}000{,}000 = 1.1 \times 10^9$ km/h

1.52 0.000 000 000 000 000 000 000 0299
The decimal point has been moved 23 places to the left. This places 22 zeros to the right of the decimal point and before the numbers 299.

1.54
- a. $(8.2 \times 10^{-3})(1.1 \times 10^{-2})$ $= 9.02 \times 10^{-5}$ $= 9.0 \times 10^{-5}$ with significant figures
- b. $(2.7 \times 10^2)(5.1 \times 10^4)$ $= 1.377 \times 10^7$ $= 1.4 \times 10^7$ with significant figures
- c. $(3.3 \times 10^{-4})(2.3 \times 10^2)$ $= 7.59 \times 10^{-2}$ $= 7.6 \times 10^{-2}$ with significant figures
- d. $(9.2 \times 10^{-4})(2.1 \times 10^4)$ $= 1.932 \times 10^1$ $= 1.9 \times 10^1$ with significant figures
- e. $(4.3 \times 10^6)(6.1 \times 10^5)$ $= 2.623 \times 10^{12}$ $= 2.6 \times 10^{12}$ with significant figures

1.56
- a. $(144)(0.0876)$ $= (1.44 \times 10^2)(8.76 \times 10^{-2})$ $= 1.26144 \times 10^1$ $= 1.26 \times 10^1$ with SF
- b. $(751)(106)$ $= (7.51 \times 10^2)(1.06 \times 10^2)$ $= 7.9606 \times 10^4$ $= 7.96 \times 10^4$ with SF

6   Chapter 1

        c. (0.0422)(0.00119)   = $(4.22 \times 10^{-2})(1.19 \times 10^{-3})$ = 5.0218×10^{-5} = 5.02×10^{-5} with SF
        d. (128,000)(0.0000316)   = $(1.28 \times 10^{5})(3.16 \times 10^{-5})$ = 4.0448×10^{0} = 4.04×10^{0} with SF

1.58   a. $\dfrac{3.1 \times 10^{-3}}{1.2 \times 10^{2}}$ = 2.583×10^{-5} = 2.6×10^{-5} with SF

        b. $\dfrac{7.9 \times 10^{4}}{3.6 \times 10^{2}}$ = 2.194×10^{2} = 2.2×10^{2} with SF

        c. $\dfrac{4.7 \times 10^{-1}}{7.4 \times 10^{2}}$ = 6.35135×10^{-4} = 6.4×10^{-4} with SF

        d. $\dfrac{0.00229}{3.16}$ = 7.2468354×10^{-4} = 7.25×10^{-4} with SF

        e. $\dfrac{119}{3.8 \times 10^{3}}$ = $3.131578947 \times 10^{-2}$ = 3.1×10^{-2} with SF

☑1.60   a. $\dfrac{(5.3)(0.22)}{(6.1)(1.1)}$ = 1.7377×10^{-1} = 1.7×10^{-1} with SF

        b. $\dfrac{(3.8 \times 10^{-4})(1.7 \times 10^{-2})}{6.3 \times 10^{3}}$ = 1.025×10^{-9} = 1.0×10^{-9} with SF

        c. $\dfrac{4.8 \times 10^{6}}{(7.4 \times 10^{3})(2.5 \times 10^{-4})}$ = 2.59459×10^{6} = 2.6×10^{6} with SF

        d. $\dfrac{5.6}{(0.022)(109)}$ = 2.335279×10^{0} = 2.3×10^{0} with SF

        e. $\dfrac{(4.6 \times 10^{-3})(2.3 \times 10^{2})}{(7.4 \times 10^{-4})(9.4 \times 10^{-5})}$ = 1.520989×10^{7} = 1.5×10^{7} with SF

SIGNIFICANT FIGURES (SECTION 1.8)

1.62   a.   A ruler with smallest scale marking of 0.1 cm   0.01 cm
         b.   A measuring telescope with smallest scale marking of 0.1 mm   0.01 mm
         c.   A protractor with smallest scale marking of 1°   0.1°
         d.   A tire pressure with smallest scale marking of 1 lb/in²   0.1 lb/in²

☑1.64   a.   Exactly 6 mL of water measured with a graduated cylinder
             that has smallest markings of 0.1 mL.   6.00 mL
         b.   A temperature that appears to be exactly 37 degrees using a
             thermometer with smallest markings of 1°C.   37.0°C
         c.   A time of exactly nine seconds measured with a stopwatch
             that has smallest markings of 0.1 second.   9.00 seconds
         d.   Fifteen and one-half degrees measured with a protractor that
             has 1 degree scale markings.   15.5°

☑1.66   a.   Measured = 5.06 lbs. $\dfrac{5.06 \text{ lb.}}{16 \text{ potatoes}} = 0.31625 \dfrac{\text{lb.}}{\text{potato}} = 0.316 \dfrac{\text{lb.}}{\text{potato}}$ with SF
               Exact = 16 potatoes
         b.   Measured = percentages $\dfrac{71.2\% + 66.9\% + 74.1\% + 80.9\% + 63.6\%}{5 \text{ players}}$ = 71.34% with SF
               Exact = 5 players

1.68 a. 0.0400   3 SF (0.0<u>400</u>)   d. 4.4 × 10⁻³   2 SF
     b. 309   3 SF   e. 1.002   4 SF
     c. 4.006   4 SF   f. 255.02   5 SF

1.70 a. $(3.71)(1.4)$ 5.194   = 5.2 with significant figures

     b. $(0.0851)(1.2262)$ 0.10434962   = 0.104 with significant figures

     c. $\dfrac{(0.1432)(2.81)}{(0.7762)}$ 0.518412780211   = 0.518 with significant figures

     d. $(3.3 \times 10^4)(3.09 \times 10^{-3})$ 101.97   = 1.0 × 10² with significant figures

     e. $\dfrac{(760)(2.00)}{6.02 \times 10^{20}}$ 2.52491694352 × 10⁻¹⁸   = 2.5 × 10⁻¹⁸ with significant figures
                                                                         (assuming 0 in 760 is not significant)

1.72 a. $0.208 + 4.9 + 1.11$ = 6.218   = 6.2 with significant figures
     b. $228 + 0.999 + 1.02$ = 230.019   = 2.30 × 10² with significant figures
     c. $8.543 - 7.954$ = 0.589   = 0.589 with significant figures
     d. $(3.2 \times 10^{-2}) + (5.5 \times 10^{-1})$
        (Hint: Write in decimal   = 0.582   = 0.58 with significant figures
        form first, then add.)
     e. $336.86 - 309.11$ = 27.75   = 27.75 with significant figures
     f. $21.66 - 0.02387$ = 21.63613   = 21.64 with significant figures

1.74 a. $\dfrac{(0.0267 + 0.00119)(4.626)}{28.7794}$ = 0.004483037867   = 0.00448 with significant figures

     b. $\dfrac{212.6 - 21.88}{86.37}$ = 2.20817413454   = 2.208 with significant figures

     c. $\dfrac{27.99 - 18.07}{4.63 - 0.88}$ = 2.6453333   = 2.65 with significant figures

     d. $\dfrac{18.87}{2.46} - \dfrac{18.07}{0.88}$
        (HINT: Do divisions first,   = -12.8633592018   = -13 with significant figures
        then subtract.)

     e. $\dfrac{(8.46 - 2.09)(0.51 + 0.22)}{(3.74 + 0.07)(0.16 + 0.2)}$ = 3.3902741324   = 3 with significant figures

     f. $\dfrac{12.06 - 11.84}{0.271}$ = 0.811808118081   = 0.81 with significant figures

1.76 a. **Area** $(A = l \times w)$ **Perimeter** $(P = 2(l) + 2(w))$

   Black $A = 12.00 \text{ cm} \times 10.40 \text{ cm} = 124.8 \text{ cm}^2$ $P = 2(12.00 \text{ cm}) + 2(10.40 \text{ cm}) = 44.80 \text{ cm}$
   Red $A = 20.20 \text{ cm} \times 2.42 \text{ cm} = 48.884 \text{ cm}^2 = 48.9 \text{ cm}^2$ $P = 2(20.20 \text{ cm}) + 2(2.42 \text{ cm}) = 45.24 \text{ cm}$
   Green $A = 3.18 \text{ cm} \times 2.55 \text{ cm} = 8.109 \text{ cm}^2 = 8.11 \text{ cm}^2$ $P = 2(3.18 \text{ cm}) + 2(2.55 \text{ cm}) = 11.46 \text{ cm}$
   Orange $A = 13.22 \text{ cm} \times 0.68 \text{ cm} = 8.9896 \text{ cm}^2 = 9.0 \text{ cm}^2$ $P = 2(13.22 \text{ cm}) + 2(0.68 \text{ cm}) = 27.80 \text{ cm}$

   b. **Length**   **Width**

   Black $12.00 \text{ cm} \left(\dfrac{1 \text{ m}}{100 \text{ cm}}\right) = 0.1200 \text{ m}$ $10.40 \text{ cm} \left(\dfrac{1 \text{ m}}{100 \text{ cm}}\right) = 0.1040 \text{ m}$

8   Chapter 1

Red $20.20 \, \cancel{cm} \left(\dfrac{1 \, m}{100 \, \cancel{cm}} \right) = 0.2020 \, m$ $2.42 \, \cancel{cm} \left(\dfrac{1 \, m}{100 \, \cancel{cm}} \right) = 0.0242 \, m$

Green $3.18 \, \cancel{cm} \left(\dfrac{1 \, m}{100 \, \cancel{cm}} \right) = 0.0318 \, m$ $2.55 \, \cancel{cm} \left(\dfrac{1 \, m}{100 \, \cancel{cm}} \right) = 0.0255 \, m$

Orange $13.22 \, \cancel{cm} \left(\dfrac{1 \, m}{100 \, \cancel{cm}} \right) = 0.1322 \, m$ $0.68 \, \cancel{cm} \left(\dfrac{1 \, m}{100 \, \cancel{cm}} \right) = 0.0068 \, m$

Area $(A = l \times w)$ **Perimeter** $(P = 2(l) + 2(w))$

Black $A = 0.1200 \, m \times 0.1040 \, m = 0.01248 \, m^2$ $P = 2(0.1200 \, m) + 2(0.1040 \, m) = 0.4480 \, m$

Red $A = 0.2020 \, m \times 0.0242 \, m = 0.0048884 \, m^2$ $P = 2(0.2020 \, m) + 2(0.0242 \, m) = 0.4524 \, m$
$= 0.00489 \, m^2$

Green $A = 0.0318 \, m \times 0.0255 \, m = 8.109 \times 10^{-4} \, m^2$ $P = 2(0.0318 \, m) + 2(0.0255 \, m) = 0.1146 \, m$
$= 8.11 \times 10^{-4} \, m^2$

Orange $A = 0.1322 \, m \times 0.0068 \, m = 8.9896 \times 10^{-4} \, m^2$ $P = 2(0.1322 \, m) + 2(0.0068 \, m) = 0.2780 \, m$
$= 9.0 \times 10^{-4} \, m^2$

c.   No, the number of significant figures in the answers remains constant. The numbers of places past the decimal are different; however, that could be fixed by rewriting all of the answers in scientific notation.

USING UNITS IN CALCULATIONS (SECTION 1.9)

1.78   a.   20 mg to grains $\dfrac{0.015 \, grains}{1 \, mg}$ c.   4 qt to liters $\dfrac{1 \, L}{1.057 \, qt}$

       b.   350 mL to fl oz $\dfrac{0.0338 \, fl \, oz}{1 \, mL}$ d.   5 yd to meters $\dfrac{1 \, m}{1.094 \, yd}$

1.80 $26 \, \cancel{miles} \left(\dfrac{1 \, km}{0.621 \, \cancel{miles}} \right) = 41.8679549114 \, km = 42 \, km$

☑1.82 $250 \, \cancel{mL} \left(\dfrac{0.0338 \, \cancel{fl \, oz}}{1 \, \cancel{mL}} \right) \left(\dfrac{1 \, cup}{8 \, \cancel{fl \, oz}} \right) = 1.05625 \, cups = 1.1 \, cups$

(Note: Cups are not measured in 0.1 increments.)

1.84 $18.0 \, \cancel{kg} \left(\dfrac{2.20 \, lbs.}{1 \, \cancel{kg}} \right) = 39.6 \, lbs$ The bag is not overweight.

1.86 $131 \dfrac{\cancel{mg}}{\cancel{dL}} \left(\dfrac{1 \, g}{1000 \, \cancel{mg}} \right) \left(\dfrac{10 \, \cancel{dL}}{1 \, L} \right) = 1.31 \dfrac{g}{L}$

CALCULATING PERCENTAGES (SECTION 1.10)

1.88 $\dfrac{\$25.73}{\$467.80} \times 100 = 5.500\%$

1.90 $\dfrac{1.0\,\frac{mg}{day}}{1.4\,\frac{mg}{day}} \times 100 = 71\%$

☑1.92 Total = 987.1 mg + 213.3 mg + 99.7 mg + 14.4 mg + 0.1 mg = 1314.6 mg

IgG = $\dfrac{987.1\,mg}{1314.6\,mg} \times 100 = 75.09\%$; IgA = $\dfrac{213.3\,mg}{1314.6\,mg} \times 100 = 16.23\%$; IgM = $\dfrac{99.7\,mg}{1314.6\,mg} \times 100 = 7.58\%$;

IgD = $\dfrac{14.4\,mg}{1314.6\,mg} \times 100 = 1.10\%$; IgE = $\dfrac{0.1\,mg}{1314.6\,mg} \times 100 = 0.008\%$

DENSITY (SECTION 1.11)

1.94 a. $\dfrac{39.6\,g}{50.0\,mL} = 0.792\,\frac{g}{mL}$ c. $\dfrac{39.54\,g}{20.0\,L} = 1.98\,\frac{g}{L}$

b. $\dfrac{243\,g}{236\,mL} = 1.03\,\frac{g}{mL}$ d. $\dfrac{222.5\,g}{25.0\,cm^3} = 8.90\,\frac{g}{cm^3}$

1.96 Volume = $(3.98\,cm)^3 = 63.0\,cm^3$

Density = $\dfrac{mass}{volume} = \dfrac{718.3\,g}{(3.98\,cm)^3} = 11.4\,\frac{g}{cm^3}$

☑1.98 $280\,g\left(\dfrac{1\,mL}{0.736\,g}\right) = 380\,mL$

ADDITIONAL ACTIVITIES

Section 1.1 Review: Is air matter? Describe how you might test your answer.

Which of these devices will give the same readings for an object at the top of a mountain, at the bottom of a valley, and on another planet? Explain.

Section 1.2 Review: Identify the following as chemical or physical AND as properties or changes:

		Chemical or Physical	Property or Change
a.	the boiling point of water		
b.	mixing baking soda and vinegar to produce carbon dioxide gas		
c.	freezing vinegar		
d.	milk reacts with lemon juice		

10 Chapter 1

Section 1.3 Review: Prefix review. Scientists make use of many prefixes. Complete the table below.

Prefix	Meaning	Meaning when combined with "–atomic molecule"	
mono-		monoatomic molecule*	
di-		diatomic molecule	
tri-		triatomic molecule	
poly-		polyatomic molecule	
homo-		homoatomic molecule	
hetero-		heteroatomic molecule	

* This term is not actually used. Can you think of a reason why that might be?

Draw examples of a:
  a. heteroatomic diatomic molecule
  b. homoatomic triatomic molecule
  c. polyatomic molecule

Section 1.4 Review: Complete the following organizational chart to visualize the relationships between the definitions related to the classification of matter.

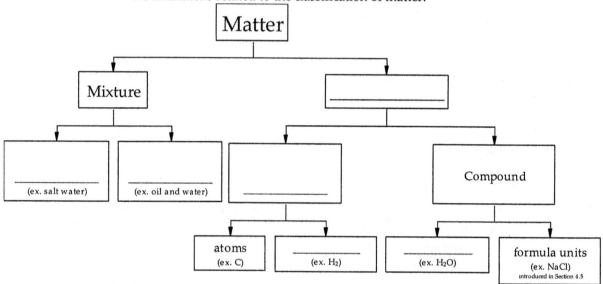

Section 1.6 Review: Two nutrition labels are shown below. Answer the following questions based on those labels.

(1) How many nutritional calories are in 355 mL of fat free milk?

(2) How many kilocalories are in 355 mL of both beverages?

(3) How many Joules are in 355 mL of both beverages?

(4) Which beverage provides more energy per 355 mL serving?

(5) Is energy the only consideration when selecting foods? (Preview of Chapter 22)

Matter, Measurements, and Calculations 11

Section 1.7 Review: How many pennies are needed to have $58.39? Write this number in both expanded form and scientific notation.

If you have 6.4×10^{-2} pennies, is this more, less, or the same as having $58.39?

Section 1.8 Review: Record your measurements based on each of the following instruments. Pay careful attention to significant figures and units. When appropriate, read from the bottom of the meniscus. Underline the digit in your measurement that contains uncertainty.

10-mL graduated cylinder

50-mL buret

−10°C – 110°C thermometer

60-minute stopwatch
(when reset, both hands point to 60)

top-loading balance

100-mL beaker

125-mL Erlenmeyer flask

5-cm ruler*

*Many rulers have the centimeter scale marked with mm. On a ruler that has both inches and centimeters, you can double check whether the markings are millimeters or centimeters by remembering: 2.54 cm = 1 inch.

12   Chapter 1

Section 1.9 Review: Convert the units in the Section 1.8 review as shown below:
- volumes to cubic centimeters and liters
- temperature to degrees Fahrenheit and Kelvin
- length measurement to inches, feet, millimeters, meters, and kilometers
- mass to kilograms and milligrams
- time to hours, minutes, and microseconds.

Report your answers in both expanded form and scientific notation.

Section 1.10 Review: A solution of ammonia in water has a concentration of 27%(w/w).

$$\%(w/w) = \frac{\text{mass of ammonia}}{\text{mass of solution}} \times 100$$

How many grams of ammonia would be needed to make 575 g of solution?
How many grams of water would be needed to make 575 g of solution?
What is the weight percentage of water in this solution?

(Solution concentrations are covered in more detail in Section 7.4.)

Section 1.11 Review: The density of 27%(w/w) ammonia in water is 0.90 g/mL.
What is the mass in grams of 75 mL of this solution?
What is the volume in milliliters of 8.4 g of this solution?

Tying It All Together with a Laboratory Application:

A chemistry student performs an experiment to separate a 2.503 g mixture of sand, salt, and calcium carbonate. The sample has the same consistency throughout; it is a (1) _____ mixture. The student adds 20.8 mL of water to the mixture and stirs. The salt dissolves in the water, but the sand and calcium carbonate do not. Salt dissolving in water is a (2) _____ process. The student allows the mixture to stand and the sand and calcium carbonate fall to the bottom of the container. This is a (3) _____ mixture. The student filters the mixture. The salt water goes through the filter paper, but the sand and calcium carbonate do not. The salt water is a (4) _____ mixture. The student pours the salt water into a beaker with a mass of 25.842 g, then places the beaker on a hotplate. The water boils and leaves salt crystals in the beaker. The student used a (5) _____ change to isolate the salt from the water. The mass of the salt crystals and the beaker are 26.048 g. The mass of salt in the original mixture is (6) _____, which is (7) _____ % of the original mixture. In order to isolate the sand from the sand and calcium carbonate, the student adds 20.0 mL of 3.00 M hydrochloric acid (HCl). The student notices that the addition of HCl, produces a slight fizzing. Once the fizzing ends, the student observes that the white calcium carbonate has disappeared, but the sand remains in the container. The reaction of hydrochloric acid with calcium carbonate to produce carbon dioxide, water, and calcium chloride is a (8) _____ property. The student decants (pours off) the liquid from the (9) _____ mixture of sand, water, and calcium chloride. The student washes the sand with distilled water and decants it into the same container as the water and calcium chloride. The student dries the sand in a container on the hotplate. The mass of the sand is 1.942×10^{-3} kg, which is (10) _____ g, or (11) _____ % of the original mixture. The student adds 20.00 mL of 1.25 M potassium carbonate to the clear calcium chloride and water mixture. A white precipitate of calcium carbonate forms immediately. This was a (12) _____ change. The student filters the mixture and the calcium carbonate remains on the 425 mg piece of filter paper. After allowing the calcium carbonate and filter paper to dry, the student finds their combined mass is 765 mg. The mass of the calcium carbonate is (13) _____ mg or

(14) _____ g, which is (15) _____% of the original mixture. The total mass of the recovered substances, sand, salt, and calcium carbonate, is (16) _____ which is (17) _____ (greater than, less than, or the same as) the mass of original mixture. Which of the following might have occurred (18)? _____

- A. All of the components were completely recovered.
- B. Some of the material was lost during the many transferring steps of this procedure.
- C. One of the components was not dried thoroughly.

SOLUTIONS FOR THE ADDITIONAL ACTIVITIES

Section 1.1 Review:

Air is matter because it has mass and occupies space. The mass of air can be verified by finding the mass of an empty balloon, blowing up the balloon, and taking the mass of the filled balloon. The difference in the mass of the full balloon and the empty balloon is the mass of the air in the balloon. The volume of the air is visible because the balloon is filled.

The lab balance will give the same readings for an object regardless of where the measurement is taken because it measures mass. The bathroom scale will not be consistent because it measures weight. Weight is the measurement of the gravitational force on an object, not the mass of the object.

Section 1.2 Review:

		Chemical or Physical	Property or Change
a.	the boiling point of water	physical	property
b.	mixing baking soda and vinegar to produce carbon dioxide gas	chemical	change
c.	freezing vinegar	physical	change
d.	milk reacts with lemon juice	chemical	property

Section 1.3 Review:

Prefix	Meaning	Meaning when combined with "–atomic molecule"	
mono-	1	monoatomic molecule*	a molecule containing 1 atom
di-	2	diatomic molecule	a molecule containing 2 atoms
tri-	3	triatomic molecule	a molecule containing 3 atoms
poly-	many	polyatomic molecule	a molecule containing many atoms
homo-	same	homoatomic molecule	a molecule containing only one type of atom
hetero-	different	heteroatomic molecule	a molecule containing at least 2 types of atoms

* A molecule must contain at least two atoms, otherwise it is called an atom.

Draw examples of a:

a.	heteroatomic diatomic molecule		This could be any molecule containing 2 different types of atoms.
b.	homoatomic triatomic molecule		This could be any molecule containing 3 of the same type of atoms.
c.	polyatomic molecule		This could be any molecule containing 4 or more atoms.

14 Chapter 1

Section 1.4 Review:

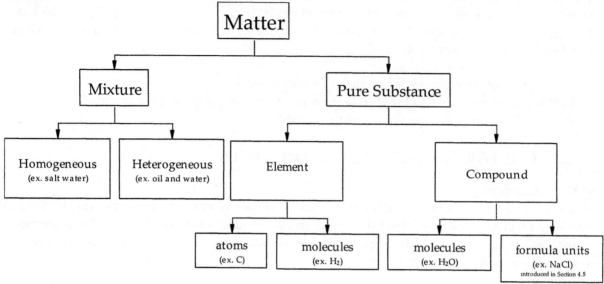

Section 1.6 Review:

(1) $355 \text{ mL milk} \times \dfrac{90 \text{ Cal}}{240 \text{ mL}} = 133.125 \text{ Cal} = 130 \text{ Calories}$

(2)
$355 \text{ mL soda} \times \dfrac{160 \text{ Calories}}{355 \text{ mL}} \times \dfrac{1 \text{ kilocalorie}}{1 \text{ Calorie}} = 160 \text{ kcal}$

$355 \text{ mL milk} \times \dfrac{90 \text{ Cal}}{240 \text{ mL}} \times \dfrac{1 \text{ kilocalorie}}{1 \text{ Calorie}} = 130 \text{ kcal}$

(3)
$355 \text{ mL soda} \times \dfrac{160 \text{ Cal}}{355 \text{ mL}} \times \dfrac{1 \text{ kilocalorie}}{1 \text{ Calorie}} \times \dfrac{1000 \text{ cal}}{1 \text{ kcal}} \times \dfrac{4.184 \text{ J}}{1 \text{ cal}} = 669440 \text{ J} = 6.7 \times 10^5 \text{ J}$

$355 \text{ mL milk} \times \dfrac{90 \text{ Cal}}{240 \text{ mL}} \times \dfrac{1 \text{ kilocalorie}}{1 \text{ Calorie}} \times \dfrac{1000 \text{ cal}}{1 \text{ kcal}} \times \dfrac{4.184 \text{ J}}{1 \text{ cal}} = 556995 \text{ J} = 5.6 \times 10^5 \text{ J}$

(4) The soda provides more energy per 355 mL serving.

(5) Energy is not the only consideration when selecting food. Nutritional value is also important. The milk has more nutritional value than the soda.

Section 1.7 Review:

$\$58.39 \times \dfrac{100 \text{ pennies}}{\$1.00} = 5839 \text{ pennies} = 5.839 \times 10^3 \text{ pennies}$

$6.4 \times 10^{-2} = 0.064$ pennies; This value is less than one penny and much less than $58.39.

Section 1.8 and 1.9 Review:

Instrument	Section 1.8	Section 1.9
graduated cylinder	7.7$\underline{4}$ mL	$7.74 \text{ mL} \left(\dfrac{1 \text{ cm}^3}{1 \text{ mL}}\right) = 7.74 \text{ cm}^3 = 7.74 \times 10^0 \text{ cm}^3$ $7.74 \text{ mL} \left(\dfrac{1 \text{ L}}{1000 \text{ mL}}\right) = 0.00774 \text{ L} = 7.74 \times 10^{-3} \text{ L}$

buret	29.2 mL	$29.2 \text{ mL} \left(\dfrac{1 \text{ cm}^3}{1 \text{ mL}}\right) = 29.2 \text{ cm}^3 = 2.92 \times 10^1 \text{ cm}^3$ $29.2 \text{ mL} \left(\dfrac{1 \text{ L}}{1000 \text{ mL}}\right) = 0.0292 \text{ L} = 2.92 \times 10^{-2} \text{ L}$
thermometer	73.3°C	$°F = \tfrac{9}{5}(73.3°C) + 32 = 164°F$ $73.3°C + 273 = 346.3 \text{ K}$
stopwatch	14.2 sec	$14.2 \text{ sec} \left(\dfrac{1 \text{ min}}{60 \text{ sec}}\right)\left(\dfrac{1 \text{ hour}}{60 \text{ min}}\right) = 0.00394 \text{ hr.} = 3.94 \times 10^{-3} \text{ hr.}$ $14.2 \text{ sec} \left(\dfrac{1 \text{ min.}}{60 \text{ sec}}\right) = 0.237 \text{ min.} = 2.37 \times 10^{-1} \text{ min.}$ $14.2 \text{ sec} \left(\dfrac{10^6 \text{ μs}}{1 \text{ sec}}\right) = 14200000 \text{ μs} = 1.42 \times 10^7 \text{ μs}$
balance	42.381 g	$42.381 \text{ g} \left(\dfrac{1 \text{ kg}}{1000 \text{ g}}\right) = 0.042381 \text{ kg} = 4.2381 \times 10^{-2} \text{ kg}$ $42.381 \text{ g} \left(\dfrac{1000 \text{ mg}}{1 \text{ g}}\right) = 42381 \text{ mg} = 4.2381 \times 10^4 \text{ mg}$
beaker	47 mL	$47 \text{ mL} \left(\dfrac{1 \text{ cm}^3}{1 \text{ mL}}\right) = 47 \text{ cm}^3 = 4.7 \times 10^1 \text{ cm}^3$ $47 \text{ mL} \left(\dfrac{1 \text{ L}}{1000 \text{ mL}}\right) = 0.047 \text{ L} = 4.7 \times 10^{-2} \text{ L}$
Erlenmeyer flask	40 mL	$40 \text{ mL} \left(\dfrac{1 \text{ cm}^3}{1 \text{ mL}}\right) = 40 \text{ cm}^3 = 4.0 \times 10^1 \text{ cm}^3$ $40 \text{ mL} \left(\dfrac{1 \text{ L}}{1000 \text{ mL}}\right) = 0.040 \text{ L} = 4.0 \times 10^{-2} \text{ L}$
ruler	3.66 cm	$3.66 \text{ cm} \left(\dfrac{0.394 \text{ in.}}{1 \text{ cm}}\right) = 1.44 \text{ in.} = 1.44 \times 10^0 \text{ in.}$ $3.66 \text{ cm} \left(\dfrac{0.394 \text{ in.}}{1 \text{ cm}}\right)\left(\dfrac{1 \text{ ft.}}{12 \text{ in.}}\right) = 0.120 \text{ ft.} = 1.20 \times 10^{-1} \text{ ft.}$ $3.66 \text{ cm} \left(\dfrac{10 \text{ mm}}{1 \text{ cm}}\right) = 36.6 \text{ mm} = 3.66 \times 10^1 \text{ mm}$ $3.66 \text{ cm} \left(\dfrac{1 \text{ m}}{100 \text{ cm}}\right) = 0.0366 \text{ m} = 3.66 \times 10^{-2} \text{ m}$ $3.66 \text{ cm} \left(\dfrac{1 \text{ m}}{100 \text{ cm}}\right)\left(\dfrac{1 \text{ km}}{1000 \text{ m}}\right) = 0.0000366 \text{ km} = 3.66 \times 10^{-5} \text{ km}$

Section 1.10 Review: $575 \text{ g} \times 27\% = 155.25 \text{ g} = 160 \text{ g}$ ammonia; $575 \text{ g} - 155.25 \text{ g} = 419.75 \text{ g} = 420 \text{ g}$ water

$\dfrac{419.75 \text{ g}}{575 \text{ g}} \times 100 = 73\%(w/w)$ or $100\% - 27\% = 73\%(w/w)$

16   Chapter 1

Section 1.11 Review: $\quad 75 \text{ mL} \times \dfrac{0.90 \text{ g}}{1 \text{ mL}} = 67.5 \text{ g} = 68 \text{ g} \qquad 8.4 \text{ g} \times \dfrac{1 \text{ mL}}{0.90 \text{ g}} = 9.\overline{3} \text{ mL} = 9.3 \text{ mL}$

Tying It All Together with a Laboratory Application:

(1) homogeneous   (6) 0.206 g   (11) 77.59%   (16) 2.488 g
(2) physical   (7) 8.23%   (12) chemical   (17) less than
(3) heterogeneous   (8) chemical   (13) 340 mg   (18) B
(4) homogeneous   (9) heterogeneous   (14) 0.340 g
(5) physical   (10) 1.942 g   (15) 13.6%

SELF-TEST QUESTIONS

Multiple Choice

1. Which of the following involves a chemical change?
   a. stretching a rubber band
   b. breaking a stick
   c. lighting a candle
   d. melting an ice cube

2. A solid substance is subjected to a number of test and observations. Which of the following would be classified as a chemical property of the substance?
   a. It is gray in color.
   b. It has a density of 2.04 g/mL.
   c. It dissolves in acid and a gas is liberated.
   d. It is not attracted to either pole of a magnet.

3. Which of the following terms could not be properly used in the description of a compound?
   a. solution   b. polyatomic   c. pure substance   d. heteroatomic

4. Which of the following is an example of heterogeneous matter?
   a. water containing sand
   b. a sample of salt water
   c. a pure sample of iron
   d. a sample of pure table salt

5. When a substance undergoes a physical change, which of the following is always true?
   a. It melts.
   b. A new substance is produced.
   c. Heat is given off.
   d. The molecular composition is unchanged.

6. Which of the following is not a chemical change?
   a. burning magnesium
   b. pulverizing sulfur
   c. exploding nitroglycerine
   d. rusting iron

7. Which of the following is the basic unit of length in the metric system?
   a. centimeter   b. meter   c. millimeter   d. kilometer

8. Which of the following is a derived unit?
   a. calorie   b. cubic decimeter   c. Joule   d. kilogram

9. In the number 3.91×10^{-3}, the original decimal position is located _____ from its current position.
   a. 3 places to the right
   b. 3 places to the left
   c. 2 places to the right
   d. 2 places to the left

10. How many significant figures are included in the number 0.02102?
    a. two   b. three   c. four   d. five

11. Twenty-one (21) students in a class of 116 got a B grade on an exam. What percent of the students in the class got B's?
    a. 21.0   b. 22.1   c. 15.3   d. 18.1

12. What single factor derived from Table 1.3 would allow you to calculate the number of quarts in a 2.0 L bottle of soft drink?
    a. $\dfrac{1.057 \text{ quarts}}{1 \text{ L}}$ b. $\dfrac{0.0338 \text{ fl oz}}{1 \text{ mL}}$ c. $\dfrac{1 \text{ L}}{1.057 \text{ quarts}}$ d. $\dfrac{1 \text{ mL}}{0.0338 \text{ fl oz}}$

13. On a hot day, a Fahrenheit thermometer reads 97.3°F. What would this reading be on a Celsius thermometer?
    a. 118°C   b. 22.1°C   c. 36.3°C   d. 143°C

14. The density of a 1 mL sample of a patient's blood is 1.08 g/mL. The density of a pint of blood taken at the same time from the same patient would be _____ 1.08 g/mL.
    a. greater than   c. equal to
    b. less than   d. more than one possible answer

15. A 125 mL urine specimen weighs 136.0 g. The density of the specimen is:
    a. $1.09 \frac{g}{mL}$ b. $0.919 \frac{g}{mL}$ c. $261 \frac{g}{mL}$ d. $11.0 \frac{g}{mL}$

Matching
Match the type of measurement given on the right as responses to the measurement units given on the left.
16. Kelvin   a. mass
17. milliliter   b. volume
18. gram   c. length
19. centimeter   d. temperature
20. cubic decimeter   e. density
21. kilometer
22. pounds per cubic foot

True-False
23. The mass of an object is the same as its weight.
24. A physical property can be observed without attempting any composition changes.
25. The cooking of food involves chemical changes.
26. The smallest piece of water that has the properties of water is called an atom.
27. Carbon monoxide molecules are diatomic and heteroatomic.
28. The prefix milli- means one thousand times.
29. One meter is shorter than one yard.
30. A pure substance containing sulfur and oxygen atoms must be classified as a compound.
31. The calorie and Joule are both units of energy.
32. In scientific notation, the exponent on the 10 cannot be larger than 15.
33. The correctly rounded sum resulting from adding 13.0, 1.094, and 0.132 will contain five significant figures.
34. If an object floats in water, it must have a higher density than water.
35. Most gases are less dense than liquids.

Chapter 2: Atoms and Molecules

CHAPTER OUTLINE
2.1 Symbols and Formulas
2.2 Inside the Atom
2.3 Isotopes
2.4 Relative Masses of Atoms and Molecules
2.5 Isotopes and Atomic Weights
2.6 Avogadro's Number: The Mole
2.7 The Mole and Chemical Formulas

LEARNING OBJECTIVES/ASSESSMENT
When you have completed your study of this chapter, you should be able to:
1. Use chemical element symbols to write formulas for compounds. (Section 2.1; Exercise 2.4)
2. Identify the characteristics of protons, neutrons, and electrons. (Section 2.2; Exercises 2.10 and 2.12)
3. Use the concepts of atomic number and mass number to determine the number of subatomic particles in isotopes and to write correct symbols for isotopes. (Section 2.3; Exercises 2.16 and 2.22)
4. Use atomic weights of the elements to calculate molecular weights of compounds. (Section 2.4; Exercise 2.32)
5. Use isotope percent abundances and masses to calculate atomic weights of the elements. (Section 2.5; Exercise 2.38)
6. Use the mole concept to obtain relationships between number of moles, number of grams, and number of atoms for elements, and use those relationships to obtain factors for use in factor-unit calculations. (Section 2.6; Exercises 2.44 a & b and 2.46 a & b)
7. Use the mole concept and molecular formulas to obtain relationships between number of moles, number of grams, and number of atoms or molecules for compounds, and use those relationships to obtain factors for use in factor-unit calculations. (Section 2.7; Exercise 2.50 b and 2.52 b)

SOLUTIONS FOR THE END OF CHAPTER EXERCISES
SYMBOLS AND FORMULAS (SECTION 2.1)

2.2   a. A triatomic molecule of a compound*

b. A molecule of a compound containing two atoms of one element and two atoms of a second element*

c. A molecule of a compound containing two atoms of one element, one atom of a second element, and four atoms of a third element*

d. A molecule containing two atoms of one element, six atoms of a second element, and one atom of a third element*

*Note: Each of these structures could be drawn in many different ways.

2.4   a. A molecule of water (two hydrogen atoms and one oxygen atom) — H_2O; like Exercise 2.2 a*

b. A molecule of hydrogen peroxide (two hydrogen atoms and two oxygen atoms) — H_2O_2; like Exercise 2.2 b*

c. A molecule of sulfuric acid (two hydrogen atoms, one sulfur atom, and four oxygen atoms) — H_2SO_4; like Exercise 2.2 c*

*The number and variety of atoms are alike. The actual structures of the molecules are different.

20   Chapter 2

      d. A molecule of ethyl alcohol (two carbon atoms, six hydrogen atoms, and one oxygen atom) C_2H_6O; like Exercise 2.2 d*

*The number and variety of atoms are alike. The actual structures of the molecules are different.

2.6
  a. methane (CH_4)   1 carbon atom; 4 hydrogen atoms
  b. perchloric acid ($HClO_4$)   1 hydrogen atom; 1 chlorine atom; 4 oxygen atoms
  c. aluminum chloride ($AlCl_3$)   1 aluminum atom; 3 chlorine atoms
  d. propane (C_3H_8)   3 carbon atoms; 8 hydrogen atoms

2.8
  a. HSH (hydrogen sulfide)   More than one H is part of the compound; a subscript should be used: H_2S
  b. $HCLO_2$ (chlorous acid)   The elemental symbol for chlorine is Cl (the second letter of a symbol must be lowercase): $HClO_2$
  c. $2HN_2$ (hydrazine – two hydrogen atoms and four nitrogen atoms)   The subscripts should reflect the number of each type of atom in the compound: H_2N_4
  d. C2H6 (ethane)   The numbers should be subscripted: C_2H_6

INSIDE THE ATOM (SECTION 2.2)

☑2.10

	Charge	Mass (u)
a. 4 protons and 5 neutrons	4	9
b. 9 protons and 10 neutrons	9	19
c. 20 protons and 23 neutrons	20	43
d. 47 protons and 60 neutrons	47	107

☑2.12   The number of protons and electrons are equal in a neutral atom.
    a. 4 electrons   b. 9 electrons   c. 20 electrons   d. 47 electrons

ISOTOPES (SECTION 2.3)

2.14

	Electrons	Protons
a. potassium	19	19
b. Cd	48	48
c. element number 51	51	51

☑2.16

	Protons	Neutrons	Electrons
a. $^{34}_{16}S$	16	18	16
b. $^{91}_{40}Zr$	40	51	40
c. $^{131}_{54}Xe$	54	77	54

2.18
  a. silicon-28 $^{28}_{14}Si$
  b. argon-40 $^{40}_{18}Ar$
  c. strontium-88 $^{88}_{38}Sr$

2.20

	Mass Number	Atomic Number	Symbol
a. 4 protons and 5 neutrons	9	4	$^{9}_{4}Be$
b. 9 protons and 10 neutrons	19	9	$^{19}_{9}F$
c. 20 protons and 23 neutrons	43	20	$^{43}_{20}Ca$
d. 47 protons and 60 neutrons	107	47	$^{107}_{47}Ag$

Atoms and Molecules   21

☑2.22   a.   contains 17 electrons and 20 neutrons $^{37}_{17}Cl$

            b.   a copper atom with a mass number of 65 $^{65}_{29}Cu$

            c.   a zinc atom that contains 36 neutrons $^{66}_{30}Zn$

RELATIVE MASSES OF ATOMS AND MOLECULES (SECTION 2.4)

2.24 $12.01 \text{ u} \left(\dfrac{1 \text{ atom He}}{4.00 \text{ u He}} \right) = 3.00 \text{ atoms He}$

2.26 $77.1\% \times 52.00 \text{ u} = 0.771 \times 52.00 \text{u} = 40.1 \text{ u}$; Ca; calcium

2.28 $\dfrac{1}{2} \times 28.09 \text{ u} = 14.045 \text{ u}$; N; nitrogen

2.30   a.   sulfur trioxide (SO_3) $(1 \times 32.07 \text{ u}) + (3 \times 16.00 \text{ u}) = 80.07 \text{ u}$

           b.   glycerin ($C_3H_8O_3$) $(3 \times 12.01 \text{ u}) + (8 \times 1.01 \text{ u}) + (3 \times 16.00 \text{ u}) = 92.11 \text{ u}$

           c.   sulfuric acid (H_2SO_4) $(2 \times 1.01 \text{ u}) + (1 \times 32.07 \text{ u}) + (4 \times 16.00 \text{ u}) = 98.09 \text{ u}$

           d.   nitrogen (N_2) $2 \times 14.01 \text{ u} = 28.02 \text{ u}$

           e.   propane (C_3H_8) $(3 \times 12.01 \text{ u}) + (8 \times 1.01 \text{ u}) = 44.11 \text{ u}$

☑2.32   The gas is most likely to be ethylene based on the following calculations:
$$\text{acetylene}: (2 \times 12.01 \text{ u}) + (2 \times 1.01 \text{ u}) = 26.04 \text{ u}$$
$$\text{ethylene}: (2 \times 12.01 \text{ u}) + (4 \times 1.01 \text{ u}) = 28.06 \text{ u}$$
$$\text{ethane}: (2 \times 12.01 \text{ u}) + (6 \times 1.01 \text{ u}) = 30.08 \text{ u}$$

2.34   The y in the formula for serine stands for 3, the nearest whole number from the following calculation:
$$(y \times 12.01 \text{ u}) + (7 \times 1.01 \text{ u}) + (1 \times 14.01 \text{ u}) + (3 \times 16.00 \text{ u}) = 105.10 \text{ u}$$
$$y \times 12.01 \text{ u} + 69.08 \text{ u} = 105.10 \text{ u}$$
$$y \times 12.01 \text{ u} = 36.02 \text{ u}$$
$$y = 2.999$$

ISOTOPES AND ATOMIC WEIGHTS (SECTION 2.5)

2.36   a.   The number of neutrons in the nucleus $26.982 - 13 = 13.982 \approx 14$ neutrons

           b.   The mass (in u) of the nucleus (to three   27.0 u
                significant figures)

☑2.38 $19.78\% \times 10.0129 \text{ u} + 80.22\% \times 11.0093 \text{ u} =$
             $0.1978 \times 10.0129 \text{ u} + 0.8022 \times 11.0093 \text{ u} = 10.81221208 \text{ u}$; 10.812 u with SF

                                            or

$$\dfrac{(19.78 \times 10.0129 \text{ u}) + (80.22 \times 11.0093 \text{ u})}{100} = 10.81221208 \text{ u}; 10.812 \text{ u with SF}$$

The atomic weight listed for boron in the periodic table is 10.81 u. The two values are close to one another.

22   Chapter 2

2.40 $69.09\% \times 62.9298 \text{ u} + 30.91\% \times 64.9278 \text{ u} =$

$0.6909 \times 62.9298 \text{ u} + 0.3091 \times 64.9278 \text{ u} = 63.5473818 \text{ u}; 63.55 \text{ u with SF}$

or

$$\frac{(69.09 \times 62.9298 \text{ u}) + (30.91 \times 64.9278 \text{ u})}{100} = 63.5473818 \text{ u}; 63.55 \text{ u with SF}$$

The atomic weight listed for copper in the periodic table is 63.55 u. The two values are the same.

AVOGADRO'S NUMBER: THE MOLE (SECTION 2.6)

2.42 $1.60 \text{ g O} \left(\dfrac{1 \text{ mole O}}{16.00 \text{ g O}} \right) \left(\dfrac{6.02 \times 10^{23} \text{ atoms O}}{1 \text{ mole O}} \right) = 6.02 \times 10^{22} \text{ atoms O}$

$6.02 \times 10^{22} \text{ atoms F} \left(\dfrac{1 \text{ mole F}}{6.02 \times 10^{23} \text{ atoms F}} \right) \left(\dfrac{19.00 \text{ g F}}{1 \text{ mole F}} \right) = 1.90 \text{ g F}$

2.44 ☑a. silicon

1 mol Si atoms = 6.02×10^{23} Si atoms
6.02×10^{23} Si atoms = 28.09 g Si
_____
1 mol Si atoms = 28.09 g Si

☑b. calcium

1 mol Ca atoms = 6.02×10^{23} Ca atoms
6.02×10^{23} Ca atoms = 40.08 g Ca
_____
1 mol Ca atoms = 40.08 g Ca

c. argon

1 mol Ar atoms = 6.02×10^{23} Ar atoms
6.02×10^{23} Ar atoms = 39.95 g Ar
1 mol Ar atoms = 39.95 g Ar

2.46 ☑a. The number of grams of silicon in 1.25 mol of silicon

1 mol Si atoms = 28.09 g Si; $\dfrac{28.09 \text{ g Si}}{1 \text{ mol Si}}$

$1.25 \text{ mol Si} \left(\dfrac{28.09 \text{ g Si}}{1 \text{ mol Si}} \right) = 35.1 \text{ g Si}$

☑b. The mass in grams of one calcium atom

6.02×10^{23} Ca atoms = 40.08 g Ca; $\dfrac{40.08 \text{ g Ca}}{6.02 \times 10^{23} \text{ Ca atoms}}$

$1 \text{ atom Ca} \left(\dfrac{40.08 \text{ g Ca}}{6.02 \times 10^{23} \text{ Ca atoms}} \right) = 6.658 \times 10^{-23} \text{ g Ca}$

c. The number of argon atoms in a 20.5-g sample of argon

6.02×10^{23} Ar atoms = 39.95 g Ar; $\dfrac{6.02 \times 10^{23} \text{ Ar atoms}}{39.95 \text{ g Ar}}$

$20.5 \text{ g Ar} \left(\dfrac{6.02 \times 10^{23} \text{ Ar atoms}}{39.95 \text{ g Ar}} \right) = 3.09 \times 10^{23} \text{ atoms Ar}$

THE MOLE AND CHEMICAL FORMULAS (SECTION 2.7)

2.48 $(1 \times 10.81 \text{ u}) + (3 \times 19.00 \text{ u}) = 67.81 \text{ u}$; 1 mole BF_3 = 67.81 g BF_3

$(2 \times 1.01 \text{ u}) + (1 \times 32.07 \text{ u}) = 34.09 \text{ u}$; 1 mole H_2S = 34.09 g H_2S

$$0.34 \text{ g H}_2\text{S} \left(\frac{1 \text{ mole H}_2\text{S}}{34.09 \text{ g H}_2\text{S}} \right) \left(\frac{6.02 \times 10^{23} \text{ molecules H}_2\text{S}}{1 \text{ mole H}_2\text{S}} \right) = 6.0 \times 10^{21} \text{ molecules H}_2\text{S}$$

$$6.0 \times 10^{21} \text{ molecules BF}_3 \left(\frac{1 \text{ mole BF}_3}{6.02 \times 10^{23} \text{ molecules BF}_3} \right) \left(\frac{67.81 \text{ g BF}_3}{1 \text{ mole BF}_3} \right) = 0.68 \text{ g BF}_3$$

2.50   a. benzene (C_6H_6)
1. 2 C_6H_6 molecules contain 12 C atoms and 12 H atoms.
2. 10 C_6H_6 molecules contain 60 C atoms and 60 H atoms.
3. 100 C_6H_6 molecules contain 600 C atoms and 600 H atoms.
4. 6.02×10^{23} C_6H_6 molecules contain 36.12×10^{23} C atoms and 36.12×10^{23} H atoms.
5. 1 mol of C_6H_6 molecules contain 6 moles of C atoms and 6 moles of H atoms.
6. 78.12 g of benzene contains 72.06 g of C and 6.06 g of H.

☑b. nitrogen dioxide (NO_2)
1. 2 NO_2 molecules contain 2 N atoms and 4 O atoms.
2. 10 NO_2 molecules contain 10 N atoms and 20 O atoms.
3. 100 NO_2 molecules contain 100 N atoms and 200 O atoms.
4. 6.02×10^{23} NO_2 molecules contain 6.02×10^{23} N atoms and 12.04×10^{23} O atoms.
5. 1 mol of NO_2 molecules contain 1 mole of N atoms and 2 moles of O atoms.
6. 46.01 g of nitrogen dioxide contains 14.01 g of N and 32.00 g of O.

c. hydrogen chloride (HCl)
1. 2 HCl molecules contain 2 H atoms and 2 Cl atoms.
2. 10 HCl molecules contain 10 H atoms and 10 Cl atoms.
3. 100 HCl molecules contain 100 H atoms and 100 Cl atoms.
4. 6.02×10^{23} HCl molecules contain 6.02×10^{23} H atoms and 6.02×10^{23} Cl atoms.
5. 1 mol of HCl molecules contain 1 mole of H atoms and 1 mole of Cl atoms.
6. 36.46 g of hydrogen chloride contains 1.01 g of H and 35.45 g of Cl.

2.52   a. **Statement 5.** 1 mol of C_6H_6 molecules contain 6 moles of C atoms and 6 moles of H atoms.

$$\text{Factor}: \left(\frac{6 \text{ moles H atoms}}{1 \text{ mole } C_6H_6} \right)$$

$$0.75 \text{ mol } C_6H_6 \left(\frac{6 \text{ moles H atoms}}{1 \text{ mole } C_6H_6} \right) = 4.5 \text{ moles H atoms}$$

24   Chapter 2

☑ b.   **Statement 5.** 1 mol of NO_2 molecules contain 1 moles of N atoms and 2 moles of O atoms.

$$\text{Factor}: \left(\frac{2 \text{ moles O atoms}}{1 \text{ mole } NO_2}\right)$$

$$0.50 \text{ mole } NO_2 \left(\frac{2 \text{ moles O atoms}}{1 \text{ mole } NO_2}\right)\left(\frac{6.02 \times 10^{23} \text{ O atoms}}{1 \text{ mole O atoms}}\right) = 6.0 \times 10^{23} \text{ O atoms}$$

c.   **Statement 6.** 36.46 g of hydrogen chloride contains 1.01 g of H and 35.45 g of Cl.

$$\frac{35.45 \text{ g Cl}}{36.46 \text{ g HCl}} \times 100 = 97.23\% \text{ Cl in HCl}$$

2.54

$$0.75 \text{ mole } H_2O \left(\frac{1 \text{ mole O atoms}}{1 \text{ mole } H_2O}\right)\left(\frac{6.02 \times 10^{23} \text{ O atoms}}{1 \text{ mole O atoms}}\right) = 4.515 \times 10^{23} \text{ O atoms}$$

$$4.515 \times 10^{23} \text{ O atoms} \left(\frac{1 \text{ mole O atoms}}{6.02 \times 10^{23} \text{ O atoms}}\right)\left(\frac{1 \text{ mole } C_2H_6O}{1 \text{ mole O atoms}}\right)\left(\frac{46.08 \text{ g } C_2H_6O}{1 \text{ mole } C_2H_6O}\right)$$

$$= 34.56 \text{ g } C_2H_6O \approx 35 \text{ g with SF}$$

2.56 $\dfrac{4.04 \text{ g H}}{16.05 \text{ g } CH_4} \times 100 = 25.2\% \text{ H in } CH_4$ $\dfrac{6.06 \text{ g H}}{30.08 \text{ g } C_2H_6} \times 100 = 20.1\% \text{ H in } C_2H_6$

2.58   **Statement 4.** 6.02×10^{23} H_3PO_4 molecules contain 18.06×10^{23} H atoms, 6.02×10^{23} P atoms, and 24.08×10^{23} O atoms.

Statement 5. 1 mol H_3PO_4 molecules contain 3 moles of H atoms, 1 mole of P atoms, and 4 moles of O atoms.

Statement 6. 98.00 g of phosphoric acid contains 3.03 g of H, 30.97 g of P, and 64.00 g of O.

a.   **Statement 6.** 98.00 g of phosphoric acid contains 3.03 g of H, 30.97 g of P, and 64.00 g of O.

$$\text{Factor}: \left(\frac{3.03 \text{ g H}}{98.00 \text{ g } H_3PO_4}\right)$$

$$46.8 \text{ g } H_3PO_4 \left(\frac{3.03 \text{ g H}}{98.00 \text{ g } H_3PO_4}\right) = 1.45 \text{ g H}$$

b.   **Statement 5.** 1 mol H_3PO_4 molecules contain 3 moles of H atoms, 1 mole of P atoms, and 4 moles of O atoms.

$$\text{Factor}: \left(\frac{4 \text{ moles of O atoms}}{1 \text{ mole } H_3PO_4}\right)$$

$$1.25 \text{ moles } H_3PO_4 \left(\frac{4 \text{ moles of O atoms}}{1 \text{ mole } H_3PO_4}\right) = 5.00 \text{ moles of O atoms}$$

c.   **Statement 4.** 6.02×10^{23} H_3PO_4 molecules contain 18.06×10^{23} H atoms, 6.02×10^{23} P atoms, and 24.08×10^{23} O atoms.

$$\text{Factor}: \left(\frac{6.02 \times 10^{23} \text{ P atoms}}{6.02 \times 10^{23} \text{ } H_3PO_4 \text{ molecules}}\right)$$

$$8.42 \times 10^{21} \text{ molecules } H_3PO_4 \left(\frac{6.02 \times 10^{23} \text{ P atoms}}{6.02 \times 10^{23} \text{ } H_3PO_4 \text{ molecules}}\right) = 8.42 \times 10^{21} \text{ P atoms}$$

Atoms and Molecules 25

2.60 Magnetite contains the higher mass percentage of iron as shown in the calculation below:

$$\frac{167.55 \text{ g Fe}}{231.55 \text{ g Fe}_3\text{O}_4} \times 100 = 72.360\% \text{ Fe in Fe}_3\text{O}_4 \qquad \frac{111.70 \text{ g Fe}}{159.7 \text{ g Fe}_2\text{O}_3} \times 100 = 69.944\% \text{ Fe in Fe}_2\text{O}_3$$

ADDITIONAL ACTIVITIES

Section 2.1 Review: The crossword game pieces are chemical symbols combined to spell the name of different substances. The molecular formulas for these named substances (or one of their components) are written to the right of the pieces. Write out the element names for the all of the elemental symbols.

| Ca | F | Fe | I | Ne | | $C_8H_{10}N_4O_2$ |

| As | P | Ir | In | | $C_9H_8O_4$ |

| S | U | Cr | O | Se | | $C_{12}H_{22}O_{11}$ |

| W | At | Er | | H_2O |

| C | Ho | Co | La | Te | | $C_7H_8N_4O_2$ (theobromine – one of the components of chocolate) |

Section 2.2 Review: Assume the following picture represents an atom.

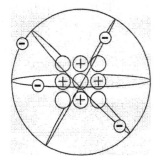

(1) Circle the nucleus.
(2) Which particle(s) is(are) found in the nucleus?
(3) Which particle(s) is(are) found outside the nucleus?
(4) What is the charge on the nucleus?
(5) What is the charge on the whole atom?
(6) What is the mass of this atom in amu?
(7) What is the mass of this atom in grams?

Section 2.3 Review:

(1) Add the atomic numbers to each of the game pieces from the Section 2.1 Review.
(2) What is the atomic number for the atom represented in the Section 2.2 Review?
(3) What is the mass number for the atom represented in the Section 2.2 Review?
(4) What are the name and symbol for the isotope represented in the Section 2.2 Review?
(5) Write the symbol for carbon-14.
(6) What is the atomic number for carbon-14?
(7) What is the mass number for carbon-14?
(8) How many protons are in carbon-14?
(9) How many neutrons are in carbon-14?
(10) How many electrons are in carbon-14?
(11) Sketch an atom of carbon-14 (like the drawing in the Section 2.2 Review).

Section 2.4 Review:

(1) Calculate the molecular weights for the crossword game "words" and their molecular formulas from the Section 2.1 Review.

A scientist claims to have discovered a new element, jupiterium. Its atomic mass is 5.994 times the mass of sulfur.

(2) What is the atomic weight of this element?
(3) Add the atomic weight of this element to the box from the periodic table.
(4) Does any known element have this atomic weight? If so, what are the name and symbol for this element?
(5) What information in addition to the atomic mass could confirm the real identity of this element?

Section 2.5 Review:

Gallium has two naturally occurring isotopes as shown in the table below:

Isotope symbol	Natural Abundance	Isotope Mass
^{69}Ga	60.11%	68.925580 amu
^{71}Ga	39.89%	70.9247005 amu

(1) In a sample of 10,000 gallium atoms, how many are gallium-69?
(2) If you were able to pull out individual gallium atoms from sample of 10,000 gallium atom, what is the likelihood of the atom being a gallium-71 isotope?
(3) Calculate the average atomic weight of gallium. Does this match the periodic table?
(4) Will any individual gallium atom have this average atomic weight?
(5) How many atoms are in 5.00 g of gallium?
(6) How many gallium-69 atoms are in a 5.00 g sample of gallium?

Section 2.6 Review:

(1) Write Avogadro's number in expanded form.

This number is enormous! To gain perspective on this number, consider the following information. The surface area of the earth is approximately 196,935,000 square miles. The dimensions of a crossword game box are as follows: 7.5 inches wide, 15.75 inches long, and 1.515 inches deep. The box can hold 1566 game pieces (although each game is sold with only 100 pieces).

(2) Write the surface area of the earth in scientific notation.
(3) Calculate the surface area (length x width) of the top of the game box in square inches.
(4) Convert the surface area of the box into square miles. (12 inches = 1 foot, 144 in² = 1 ft², 5280 feet = 1 mile, 2.78784×10^7 ft² = 1 mi²)

(5) How many boxes could be placed on the earth's surface?
(6) How many game pieces would all of these boxes contain?
(7) Divide Avogadro's number by the number of game pieces in the boxes that would cover the earth's surface. This will tell you how many game boxes deep each stack would need to be in order to cover the earth with one mole of game pieces.
(8) Convert the depth of the stack into miles by multiplying the number of boxes in a stack by the depth of each box in miles.

As a point of comparison of the size of atoms to the game pieces:
(9) What is the mass of one mole of silicon atoms?
(10) If the density of silicon is 2.33 g/cm³, what is the volume (in cm³) of 1 mole of silicon atoms?
(11) Convert this volume to Tablespoons. (15 mL = 1 Tablespoon)
(12) Could you hold one mole of silicon atoms in your hand?
(13) Assuming the atoms occupy all of the space in the 1 mole sample of silicon*, what is the volume (in cm³) of a silicon atom?
(14) How many silicon atoms will fit into the volume of a game piece (0.715 in. x 0.795 in. x 0.175 in.).

*Note: This assumption is not valid. The solid will contain empty space and the atoms are actually smaller than this calculation implies.

Section 2.7 Review:
(1) Calculate the molecular weight for all of the molecular formulas given in the right column of the table in the Section 2.1 Review.
(2) How many molecules are in 1 mole of each of those compounds?
(3) What is the mass in grams for 1 mole of each of those compounds?
(4) How many moles of carbon are in 1 mole of each of those compounds?
(5) How many atoms of hydrogen are in 1 mole of each of those compounds?
(6) How many grams of oxygen are in 1 mole of each of those compounds?
(7) What is the mass percentage of nitrogen in each of those compounds?

Tying It All Together with a Laboratory Application:
In 1911 and 1914, Ernest Rutherford published papers in *Philosophical Magazine* about the structure of the atom. At this time, the charge and mass of an electron were known; however, the proton and the neutron had not been discovered. The experimental setup is shown below:

The gold foil is 4×10^{-5} cm thick. If the gold foil was 4×10^{-5} cm wide by 4×10^{-5} cm high, the volume of the gold foil would be (1) _____ cm³. The density of gold is 19.3 g/cm³; therefore the mass of this gold foil would be (2) _____ g or (3) _____ u. The number of moles of gold atoms in this sample of gold foil would be (4) _____ moles. The number of atoms in this sample of gold foil would be (5) _____ atoms. The gold foil would be approximately (6) _____ atoms thick. The actual width and the height of the gold foil were greater than 4×10^{-5} cm for ease of experimental setup.

The alpha particles are produced by radioactive decay, which is covered in Chapter 10. One of the accepted chemical symbols for an alpha particle is $_{2}^{4}\alpha$. The mass number of an alpha particle is (7) _____. The atomic number of an alpha particle is (8) _____. The element with the same atomic number as an alpha particle is (9) _____. The relative mass of gold to an alpha particle is (10) _____. An alpha particle contains (11) _____ protons and (12) _____ neutrons. The alpha particle has a 2+ charge, which means it contains (13) _____ electrons. The alpha particles travel at 2.09×10^9 cm/sec. Light travels at 3.00×10^8 m/s. The speed of an alpha particle is (14) _____ percentage the speed of light.

The alpha particles were aimed at the gold foil which was surrounded by a circular zinc sulfide (ZnS) screen. ZnS is a phosphor. It emits light after being struck by an energetic material (like an alpha particle). In this experiment, the angle was measured between the initial path of the alpha particle to the gold foil and the location on the screen that the alpha particle struck after interacting with the gold foil. Most of the alpha particles in this experiment went straight through the gold foil (~98%) or were deflected only a small amount (~2%). Approximately 1 in 20,000 alpha particles, though, were turned back through an average angle of 90°. Rutherford's explanation for this phenomenon was that an atom is mainly empty space with an exceedingly small, dense, positively charged (15) _____ at the center of the atom and (16) _____ distributed around the outside of the nucleus that maintain a neutral atom. If an alpha particle passed close to the nucleus of one of the gold atoms it would be (pick one: attracted or repelled) (17) _____ because the alpha particle is positive and the nucleus is positive.

SOLUTIONS FOR THE ADDITIONAL ACTIVITIES
Section 2.1 Review:
Caffeine – calcium, fluorine, iron, iodine, neon, carbon, hydrogen, nitrogen, oxygen
Aspirin – arsenic, phosphorus, iridium, indium, carbon, hydrogen, oxygen
Sucrose – sulfur, uranium, chromium, oxygen, selenium, carbon, hydrogen, oxygen
Water – tungsten, astatine, erbium, hydrogen, oxygen
Chocolate – carbon, holmium, cobalt, lanthanum, tellurium, carbon, hydrogen, nitrogen, oxygen

Section 2.2 Review:

(1) _____, (2) protons and neutrons, (3) electrons, (4) +4, (5) 0,

(6) $4 \text{ protons} \left(\dfrac{1 \text{ u}}{1 \text{ proton}} \right) + 5 \text{ neutrons} \left(\dfrac{1 \text{ u}}{1 \text{ neutron}} \right) + 4 \text{ electrons} \left(\dfrac{\frac{1}{1836} \text{ u}}{1 \text{ electron}} \right) = 9.0021786 \text{ u} = 9 \text{ u}$,

(7) $4 \text{ protons} \left(\dfrac{1.67 \times 10^{-24} \text{ g}}{1 \text{ proton}} \right) + 5 \text{ neutrons} \left(\dfrac{1.67 \times 10^{-24} \text{ g}}{1 \text{ neutron}} \right) + 4 \text{ electrons} \left(\dfrac{9.07 \times 10^{-28} \text{ g}}{1 \text{ electron}} \right) = 1.50 \times 10^{-23} \text{ g}$

Section 2.3 Review:

(1)

(2) 4, (3) 9, (4) beryllium-9, $^{9}_{4}Be$,
(5) $^{14}_{6}C$, (6) 6, (7) 14, (8) 6, (9) 8, (10) 6,
(11)

Section 2.4 Review:

(1) Caffeine – $40.08\ u + 19.00\ u + 55.85\ u + 126.90\ u + 20.18\ u = 262.01\ u$;

$C_8H_{10}N_4O_2$ – $8\ C\left(\dfrac{12.01\ u}{1\ C}\right) + 10\ H\left(\dfrac{1.0079\ u}{1\ H}\right) + 4\ N\left(\dfrac{14.01\ u}{1\ N}\right) + 2\ O\left(\dfrac{16.00\ u}{1\ O}\right) = 194.20\ u$;

Aspirin – $74.92\ u + 30.97\ u + 192.22\ u + 114.82\ u = 412.93\ u$;

$C_9H_8O_4$ – $9\ C\left(\dfrac{12.01\ u}{1\ C}\right) + 8\ H\left(\dfrac{1.0079\ u}{1\ H}\right) + 4\ O\left(\dfrac{16.00\ u}{1\ O}\right) = 180.15\ u$;

Sucrose – $32.07\ u + 238.03\ u + 52.00\ u + 16.00\ u + 78.96\ u = 417.06\ u$;

$C_{12}H_{22}O_{11}$ – $12\ C\left(\dfrac{12.01\ u}{1\ C}\right) + 22\ H\left(\dfrac{1.0079\ u}{1\ H}\right) + 11\ O\left(\dfrac{16.00\ u}{1\ O}\right) = 342.29\ u$;

Water – $183.9\ u + 210\ u + 167.26\ u = 561\ u$;

H_2O – $2\ H\left(\dfrac{1.0079\ u}{1\ H}\right) + 1\ O\left(\dfrac{16.00\ u}{1\ O}\right) = 18.02\ u$;

Chocolate – $12.01\ u + 164.93\ u + 58.93\ u + 138.91\ u + 127.60\ u = 502.38\ u$;

$C_7H_8N_4O_2$ – $7\ C\left(\dfrac{12.01\ u}{1\ C}\right) + 8\ H\left(\dfrac{1.0079\ u}{1\ H}\right) + 4\ N\left(\dfrac{14.01\ u}{1\ N}\right) + 2\ O\left(\dfrac{16.00\ u}{1\ O}\right) = 180.17\ u$;

(2) $5.994 \times 32.07\ u = 192.23\ u$; (3) ; (4) iridium, Ir; (5) If the atomic number of jupiterium is 77, then this element is really iridium. The atomic number is the number of protons in the nucleus of an atom. The atomic number is unique for each element.

Section 2.5 Review:

(1) $60.11\% \times 10{,}000 = 0.6011 \times 10{,}000 = 6{,}011$; (2) $\dfrac{3989}{10000} \times 100 = 39.89\%$;

(3) $60.11\%(68.925580\ u) + 39.89\%(70.9247005\ u) = 0.6011(68.925580\ u) + 0.3989(70.9247005\ u) = 69.72\ u$,

yes; (4) no; (5) $5.00\ \cancel{g}\left(\dfrac{1\ \cancel{u}}{1.67 \times 10^{-24}\ \cancel{g}}\right)\left(\dfrac{1\ atom}{69.72\ \cancel{u}}\right) = 4.29 \times 10^{22}\ atoms$;

(6) $5.00\ \cancel{g}\left(\dfrac{1\ \cancel{u}}{1.67 \times 10^{-24}\ \cancel{g}}\right)\left(\dfrac{1\ atom}{69.72\ \cancel{u}}\right)60.11\% = 5.00\ \cancel{g}\left(\dfrac{1\ \cancel{u}}{1.67 \times 10^{-24}\ \cancel{g}}\right)\left(\dfrac{1\ atom}{69.72\ \cancel{u}}\right)0.6011 =$

$= 2.58 \times 10^{22}$ gallium – 69 atoms

30   Chapter 2

Section 2.6 Review:
(1) 602 200 000 000 000 000 000 000; (2) 1.96935×10^8 square miles;
(3) 7.5 inches $\times$ 15.75 inches = 118.125 in^2 = 120 in^2;

(4) $118.125 \text{ in}^2 \left(\dfrac{1 \text{ ft}}{12 \text{ in}}\right)^2 \left(\dfrac{1 \text{ mi}}{5280 \text{ ft}}\right)^2 = 2.9424662104 \times 10^{-8} \text{ mi}^2 = 2.9 \times 10^{-8} \text{ mi}^2$;

(5) $196935000 \text{ mi}^2 \left(\dfrac{1 \text{ box}}{2.9424662104 \times 10^{-8} \text{ mi}^2}\right) = 6.69285510583 \times 10^{15}$ boxes = 6.7×10^{15} boxes;

(6) $6.69285510583 \times 10^{15}$ boxes $\left(\dfrac{1566 \text{ pieces}}{1 \text{ box}}\right) = 1.04810110957 \times 10^{19}$ pieces = 1.0×10^{19} pieces;

(7) 6.022×10^{23} pieces $\left(\dfrac{\text{boxes to cover earth's surface}}{1.04810110957 \times 10^{19} \text{ pieces}}\right) = 57456.2888$ layers of boxes = 5.7×10^4 boxes deep;

(8) 57456.2888 layers of boxes $\left(\dfrac{1.515 \text{ inches}}{1 \text{ box}}\right)\left(\dfrac{1 \text{ foot}}{12 \text{ inches}}\right)\left(\dfrac{1 \text{ mile}}{5280 \text{ feet}}\right) = 1.37383644996$ miles = 1.4 miles;

(9) 28.09 g; (10) $28.09 \text{ g} \left(\dfrac{1 \text{ cm}^3}{2.33 \text{ g}}\right) = 12.0557939914 \text{ cm}^3 = 12.1 \text{ cm}^3$;

(11) $12.0557939914 \text{ cm}^3 \left(\dfrac{1 \text{ mL}}{1 \text{ cm}^3}\right)\left(\dfrac{1 \text{ Tablespoon}}{15 \text{ mL}}\right) = 0.803719599427$ Tablespoons = 0.80 Tablespoons;

(12) yes; (13) $\dfrac{12.0557939914 \text{ cm}^3}{1 \text{ mole of atoms}} \left(\dfrac{1 \text{ mole}}{6.022 \times 10^{23} \text{ atoms}}\right) = 2.00195848413 \times 10^{-23} \dfrac{\text{cm}^3}{\text{atoms}} = 2.0 \times 10^{-23} \dfrac{\text{cm}^3}{\text{atoms}}$;

(14) $0.715 \text{ in} \times 0.795 \text{ in} \times 0.175 \text{ in} \left(\dfrac{2.54 \text{ cm}}{1 \text{ in}}\right)^3 \left(\dfrac{2.33 \text{ g}}{1 \text{ cm}^3}\right)\left(\dfrac{1 \text{ mole}}{28.09 \text{ g}}\right)\left(\dfrac{6.022 \times 10^{23} \text{ atoms}}{1 \text{ mole}}\right) = 8.14 \times 10^{22}$ atoms

Section 2.7 Review:
(1) $C_8H_{10}N_4O_2$ = 194.20 u; $C_9H_8O_4$ = 180.15 u; $C_{12}H_{22}O_{11}$ = 342.29 u; H_2O = 18.02 u; $C_7H_8N_4O_{12}$ = 180.17 u;
(2) 1 mole of any compound contains 6.022×10^{23} molecules; (3) $C_8H_{10}N_4O_2$ = 194.20 g; $C_9H_8O_4$ = 180.15 g; $C_{12}H_{22}O_{11}$ = 342.29 g; H_2O = 18.02 g; $C_7H_8N_4O_{12}$ = 180.17 g; (4) $C_8H_{10}N_4O_2$ = 8 moles C; $C_9H_8O_4$ = 9 moles C; $C_{12}H_{22}O_{11}$ = 12 moles C; H_2O = 0 moles C; $C_7H_8N_4O_{12}$ = 7 moles C; (5) $C_8H_{10}N_4O_2$ =

$1 \text{ mole } C_8H_{10}N_4O_2 \left(\dfrac{6.022 \times 10^{23} \text{ molecules}}{1 \text{ mole } C_8H_{10}N_4O_2}\right)\left(\dfrac{10 \text{ atoms H}}{1 \text{ molecule } C_8H_{10}N_4O_2}\right) = 6.022 \times 10^{24}$ atoms H; $C_9H_8O_4$ =

4.818×10^{24} atoms H; $C_{12}H_{22}O_{11}$ = 1.32×10^{25} atoms H; H_2O = 1.204×10^{24} atoms H; $C_7H_8N_4O_{12}$ = 4.818×10^{24}

atoms H; (6) $C_8H_{10}N_4O_2$ = $1 \text{ mole } C_8H_{10}N_4O_2 \left(\dfrac{2 \text{ moles O}}{1 \text{ mole } C_8H_{10}N_4O_2}\right)\left(\dfrac{16.00 \text{ g O}}{1 \text{ mole O}}\right) = 32.00$ g O; $C_9H_8O_4$ =

64.00 g O; $C_{12}H_{22}O_{11}$ = 176.00 g O; H_2O = 16.00 g O; $C_7H_8N_4O_{12}$ = 192.00 g O; (7) $C_8H_{10}N_4O_2$ =
$\dfrac{56.04 \text{ g N}}{194.20 \text{ g } C_8H_{10}N_4O_2} \times 100 = 28.86\%$ N; $C_9H_8O_4$ = 0% N; $C_{12}H_{22}O_{11}$ = 0% N; H_2O = 0% N; $C_7H_8N_4O_{12}$ =
31.10% N

Tying It All Together with a Laboratory Application:
(1) 6.4×10^{-14} cm^3 = 6×10^{-14} cm^3
(2) 1.2352×10^{-12} g = 1×10^{-12} g
(3) 7.3964×10^{11} u = 7×10^{11} u
(4) 6.2710×10^{-15} moles = 6×10^{-15} moles
(5) 3.7764×10^9 atoms = 4×10^9 atoms
(6) 1557.25 atoms = 2000 atoms = 2×10^3 atoms
(7) 4
(8) 2

(9) helium   (11) 2   (13) 0   (15) nucleus   (17) repelled
(10) 49   (12) 2   (14) 6.97%   (16) electrons

SELF-TEST QUESTIONS
Multiple Choice
1. Which of the following is an incorrect symbol for an element?
   a. Ce   b. Au   c. K   d. CR

2. Which of the following is an incorrect formula for a compound?
   a. CO_2 b. CO_1 c. N_2O d. NO_2

3. Two objects have masses of 3.2 g and 0.80 g. What is the relative mass of the 3.2 g object compared to the other?
   a. 4.0 to 1   b. 2.0 to 1   c. 0.50 to 1   d. 0.25 to 1

4. Suppose the atomic weights of the elements were assigned in such a way that the atomic weight of helium, He, was 1.00 u. What would be the atomic weight of oxygen, O, in u, on this scale?
   a. 16.0   b. 8.00   c. 4.00   d. 0.250

5. What is the molecular weight of phosphoric acid, H_3PO_4, in u?
   a. 48.0   b. 50.0   c. 96.0   d. 98.0

6. How many neutrons are there in the nucleus of a potassium-39 atom?
   a. 1   b. 19   c. 20   d. 39

7. What is the mass in grams of 1.00 mole of chlorine molecules, Cl_2?
   a. 6.02×10^{23} b. 71.0   c. 35.5   d. 1.18×10^{-22}

8. Calculate the weight percent of sulfur, S, in SO_2.
   a. 50.1   b. 33.3   c. 66.7   d. 25.0

Matching
Match the molecules represented on the left with the terms on the right to the correct classification given.

9.   a. homoatomic and polyatomic
10.   b. homoatomic and not polyatomic
11.   c. heteroatomic and polyatomic
12.   d. heteroatomic and not polyatomic
13.

32 Chapter 2

Match the number given as responses to the following:
14. The number of moles of oxygen atoms in 2 moles of NO_2.   a. 1
15. The number of moles of NH_3 that contain 3 moles of nitrogen atoms.   b. 2
16. The number of moles of nitrogen atoms in one-half mole of N_2O_5.   c. 3
17. The number of moles of electrons in one mole of helium atoms.   d. 4
18. The number of moles of neutrons in one mole of ^3_1H.

True-False
19. In some instances, two different elements are represented by the same symbol.
20. The mass of a single atom of silicon, Si, is 28.1 g.
21. All isotopes of a specific element have the same atomic number.
22. All atoms of a specific element have the same number of protons and electrons.
23. One mole of water molecules, H_2O, contain two moles of hydrogen atoms, H.
24. 1.00 mol of sulfur, S, contains the same number of atoms as 14.0 g of nitrogen, N.
25. 6.02×10^{23} molecules of methane, CH_4, contains 6.02×10^{23} atoms of hydrogen.

Chapter 3: Electronic Structure and the Periodic Law

CHAPTER OUTLINE
3.1 The Periodic Law and Table
3.2 Electronic Arrangements in Atoms
3.3 The Shell Model and Chemical Properties
3.4 Electronic Configurations
3.5 Another Look at the Periodic Table
3.6 Property Trends Within the Periodic Table

LEARNING OBJECTIVES/ASSESSMENT
When you have completed your study of this chapter, you should be able to:
1. Locate elements in the periodic table on the basis of group and period designations. (Section 3.1; Exercise 3.4)
2. Determine numerical relationships such as the number of electrons in designated orbitals, shells, or subshells. (Section 3.2; Exercise 3.12)
3. Determine the number of electrons in the valence shell. (Section 3.3; Exercise 3.18)
4. Relate electronic structure to the location of elements in the periodic table. (Section 3.3; Exercise 3.22)
5. Write electronic configurations for elements. (Section 3.4; Exercise 3.24)
6. Determine the number of unpaired electrons for elements. (Section 3.4; Exercise 3.24)
7. Identify element from their electronic configurations. (Section 3.4; Exercise 3.28)
8. Determine the shell and subshell location of the distinguishing electron in elements. (Section 3.5; Exercise 3.34)
9. Based on their location in the periodic table, correctly classify elements into the following categories: representative element, transition element, inner-transition element, noble gas, metal, nonmetal, and metalloid. (Section 3.5; Exercise 3.36)
10. Recognize property trends within the periodic table. (Section 3.6; Exercises 3.40 and 3.42)
11. Use property trends in the periodic table to predict selected properties of elements. (Section 3.6; Exercises 3.40 and 3.42)

SOLUTIONS FOR THE END OF CHAPTER EXERCISES
THE PERIODIC LAW AND TABLE (SECTION 3.1)

3.2
		Group	Period
a.	Sb	V A (15)	5
b.	cadmium	II B (12)	5
c.	element number 31	III A (13)	4
d.	Zr	IV B (4)	5

☑3.4
		Symbol	Name
a.	The noble gas belonging to period 4	Kr	krypton
b.	The fourth element (reading down) in group IV A (14)	Sn	tin
c.	Belongs to group VI B (6) and period 5	Mo	molybdenum
d.	The sixth element (reading left to right) in period 6	W	tungsten

3.6
a.	How many elements are located in group VII B (7) of the periodic table?	4
b.	How many total elements are found in periods 1 and 2 of the periodic table?	10
c.	How many elements are found in period 5 of the periodic table?	18

3.8
a.	This is a horizontal arrangement of elements in the periodic table.	period
b.	Element 11 begins this arrangement in the periodic table.	period

34 Chapter 3

    c. The element nitrogen is the first member of this arrangement.   group
    d. Elements 9, 17, 35, and 53 belong to this arrangement.   group

ELECTRONIC ARRANGEMENTS IN ATOMS (SECTION 3.2)

3.10   Protons are subatomic particles with a positive charge that are located in the nucleus.

☑3.12   a. A $2p$ orbital   2 electrons
    b. A $2p$ subshell   6 electrons
    c. The second shell   8 electrons

3.14   Four (4) orbitals are found in the second shell: one $2s$ orbital and three $2p$ orbitals.

3.16   Seven (7) orbitals are found in a $4f$ subshell. The maximum number of electrons that can be located in this subshell is 14 because each of the seven orbitals can hold two electrons.

THE SHELL MODEL AND CHEMICAL PROPERTIES (SECTION 3.3)

☑3.18   a. element number 54   8 electrons
    b. The first element (reading down) in group V A (15)   5 electrons
    c. Sn   4 electrons
    d. The fourth element (reading left to right) in period 3   4 electrons

3.20   Cesium is the period 6 element with chemical properties most like sodium. Cesium has 1 valence-shell electron. Sodium also has only 1 valence-shell electron.

☑3.22   I would expect to find silver and gold in addition to the copper because these elements are all in the same group on the periodic table. Elements that are in the same group have similar chemical properties; therefore, if copper is part of this ore, then the other elements that are most similar to it, are also likely to be part of the ore.

ELECTRONIC CONFIGURATIONS (SECTION 3.4)

☑3.24

		Electron Configuration	**Unpaired Electrons**
a.	element number 37	$1s^22s^22p^63s^23p^64s^23d^{10}4p^65s^1$	1
b.	Si	$1s^22s^22p^63s^23p^2$	2
c.	titanium	$1s^22s^22p^63s^23p^64s^23d^2$	2
d.	Ar	$1s^22s^22p^63s^23p^6$	0

3.26

		Electron Configuration	**Solutions**
a.	s electrons in potassium	$1s^22s^22p^63s^23p^64s^1$	7
b.	unpaired electrons in aluminum	$1s^22s^22p^63s^23p^1$	1
c.	filled subshells in magnesium	$1s^22s^22p^63s^2$	4

☑3.28

		Symbol	**Name**
a.	Contains only two $2p$ electrons	C	carbon
b.	Contains an unpaired $3s$ electron	Na	sodium
c.	Contains two unpaired $3p$ electrons	Si or S	silicon or sulfur
d.	Contains three $4d$ electrons	Nb	niobium
e.	Contains three unpaired $3d$ electrons	V or Co	vanadium or cobalt

Electronic Structure and the Periodic Law 35

3.30   a. arsenic $[Ar]4s^23d^{10}4p^3$
       b. An element that contains 25 electrons $[Ar]4s^23d^5$
       c. silicon $[Ne]3s^23p^2$
       d. element number 53 $[Kr]5s^24d^{10}5p^5$

3.32   a. sodium $[Ne]3s^1$ e. phosphorus $[Ne]3s^23p^3$
       b. magnesium $[Ne]3s^2$ f. sulfur $[Ne]3s^23p^4$
       c. aluminum $[Ne]3s^23p^1$ g. chlorine $[Ne]3s^23p^5$
       d. silicon $[Ne]3s^23p^2$ h. argon $[Ne]3s^23p^6$

ANOTHER LOOK AT THE PERIODIC TABLE (SECTION 3.5)

☑3.34 a. lead p area
       b. element 27 d area
       c. Tb f area
       d. Rb s area

☑3.36 a. iron   transition   3.38 a. argon   nonmetal
       b. element 15   representative   b. element 3   metal
       c. U   inner-transition   c. Ge   metalloid
       d. xenon   noble gas   d. boron   metalloid
       e. tin   representative   e. Pm   metal

PROPERTY TRENDS WITHIN THE PERIODIC TABLE (SECTION 3.6)

☑3.40 a. K or Ti   K more metallic   ☑3.42 a. Ga or Se   Ga larger radius
       b. As or Bi   Bi   b. N or Sb   Sb
       c. Mg or Sr   Sr   c. O or C   C
       d. Sn or Ge   Sn   d. Te or S   Te

3.44   a. Si or Cl   Si loses e⁻ more easily
       b. Mg or Ba   Ba
       c. F or Br   Br
       d. Ca or S   Ca

ADDITIONAL ACTIVITIES

Section 3.1 Review: Does this periodic table of shapes have the same orientation of periods and groups as the periodic table of elements? Explain why or why not.

36 Chapter 3

Section 3.2 Review: The periodic table can be used to help remember the relationships between shells, subshells, orbitals, and electrons.

(1) Number the periods on the left side of the periodic table. The period numbers are the same as the shell numbers. How many shells are known?

(2) The blocks on the periodic table represent the subshells. Color the blocks of the periodic table. Use the following instructions and/or Figure 3.9 as a guide.
  (a) The s block consists of the first 2 columns of the periodic table and the box that represents He.
  (b) The p block consists of the last 6 columns of the periodic table (excluding the box that represents He).
  (c) The d block consists of 10 columns in the middle of the periodic table. The first of these ten columns is next to the s block. The last nine columns are immediately to the right of the p block.
  (d) The f block consists of 14 columns in the middle of the periodic table. The f block splits the d block into 2 groups and is usually placed below the periodic table.

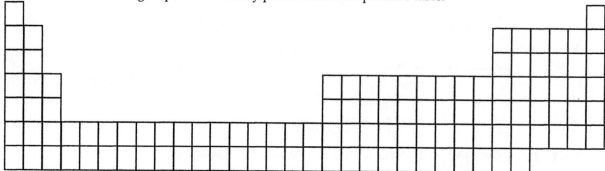

(3) The shell number not only describes the location and energy of the electrons around a nucleus, but it also gives the number of subshells within the shell. How many subshells should each of the labeled periods have? Name these subshells. (Only the first four subshells are given names in the textbook. After f the lettering becomes alphabetical (except letter *j* is excluded).) Based on your colored periodic table, why does the textbook exclude the higher energy subshells?

(4) Trace across each period and count the number of subshells in each period. Does this match your prediction from (3)? Propose a reason for any discrepancies.

(5) The maximum number of electrons in a subshell can be obtained by counting the number of boxes in one row of the subshell. What is the maximum number of electrons in the...
    (a) s subshell?   (b) p subshell?   (c) d subshell?   (d) f subshell?

(6) The number of orbitals in a subshell can be obtained by dividing the maximum number of electrons in a subshell by two (because two electrons can be placed in each orbital). How many orbitals are in the...
    (a) s subshell?   (b) p subshell?   (c) d subshell?   (d) f subshell?

Section 3.3 Review:

(1) How many valence electrons are in cesium (#55)?
(2) How many valence electrons are in selenium (#34)?
(3) How many valence electrons are in xenon (#54)?
(4) Which subshells must be filled in order to have 8 electrons in the valence shell?
(5) Can the first shell have 8 electrons in its valence shell? Explain why or why not.
(6) Which elements in the periodic table contain valence shells with filled s and p orbitals? (These elements are called **noble gases**. Helium is normally included as a noble gas because its valence shell is filled with only 2 electrons.)
(7) Add a "Noble Gases" label to the appropriate column of the periodic table in the Section 3.2 Review.

Section 3.4 Review:
(1) Rank the three subshells in the third shell in order of increasing energy.
(2) Based on Figure 3.7 in the textbook, how does the energy level of the 4s subshell compare to the energy levels of each of the subshells in the third shell?
(3) Referring back to the coloring exercise from the Section 3.2 Review, label the start of each subshell on the periodic table (ex. 1s).

For questions 4-7, you may refer to the Section 3.2 Review or Figure 3.8 in the textbook.
(4) Which subshell is completely filled immediately before the 4f subshell is filled?
(5) Which subshell is completely filled immediately before the 5f subshell is filled?
(6) Which subshell is completely filled immediately before the 5d subshell is filled?
(7) Which subshell is completely filled immediately before the 6d subshell is filled?

Section 3.5 Review:
(1) How many electrons do each of the following symbols imply?
    (a) [He]   (b) [Ne]   (c) [Kr]   (d) [Rn]
(2) What are the electron configurations associated with each of the following symbols?
    (a) [He]   (b) [Ne]   (c) [Kr]   (d) [Rn]
(3) Which subshell would be filled immediately after each of these symbols?
    (a) [He]   (b) [Ne]   (c) [Kr]   (d) [Rn]
(4) What do each of these symbols have in common?
    (a) [He]   (b) [Ne]   (c) [Kr]   (d) [Rn]
(5) Would [Ir] ever be used in an electron configuration? Explain why or why not.
(6) Add the following labels to the appropriate columns from the Section 3.2 Review.
    (a) representative elements   (b) inner-transition elements   (c) transition elements

(7) In which subshell(s) can the distinguishing electron be found for each of the following classes?
    (a) metal   (b) nonmetal   (c) metalloid
(8) How many electrons are in the valence shell of each of the following elements?
    (a) Sr   (b) W   (c) U   (d) As
(9) In which subshell is the distinguishing electron for each of the following elements found?
    (a) Sr   (b) W   (c) U   (d) As

Section 3.6 Review:
(1) On the periodic table below, draw and label the trend for increasing metallic character in a period and a group.
(2) On the periodic table below, draw and label the trend for increasing atomic size in a period and a group.
(3) On the periodic table below, draw and label the trend for increasing ionization energy in a period and a group.
(4) What is the relationship between the trends for metallic character and atomic size?
(5) What is the relationship between the trends for metallic character and ionization energy?
(6) What is the relationship between the trends for atomic size and ionization energy?

38 Chapter 3

Tying It All Together with a Laboratory Application:
When sodium metal is placed in water, it reacts to produce sodium hydroxide and hydrogen gas. When potassium is placed in water, it will have a (pick one: different or similar) (1) _____ chemical reaction because sodium and potassium are part of the same (2) _____. The number of electrons in the (3) _____ shell of sodium is 1. The number of electrons in the valence shell of potassium is (4) _____. The distinguishing electron for sodium is in the (5) _____ subshell. The distinguishing electron for potassium is in shell (6) _____. Sodium is a (pick one: representative element, noble gas, inner-transition element, transition element) (7) _____. Potassium is a (pick one: metal, nonmetal, metalloid) (8) _____. Potassium atoms are (pick one: larger or smaller) (9) _____ than sodium atoms. Potassium atoms are (10) _____ metallic than sodium atoms. In terms of chemical behavior, potassium atoms will likely react (11) _____ vigorously with water because they lose electrons (12) _____ easily than sodium atoms.

Magnesium is in group (13) _____ and period (14) _____. The chemical behavior of magnesium in water (15) _____ be predicted based on the knowledge of the chemical behavior of sodium because elements in the same (16) _____ do not necessarily share the same chemical properties. Based on periodic trends, though, magnesium atoms are (pick one: larger or smaller) (17) _____ than sodium atoms. Magnesium atoms are (18) _____ metallic than sodium atoms. Magnesium atoms lose electrons (19) _____ easily than sodium atoms. In Section 9.6, we will find that magnesium atoms will not react with cold water, but they will react with steam.
In terms of physical properties, magnesium will have a (20) _____ thermal conductivity, a (21) _____ electrical conductivity, and a (22) _____ luster. Magnesium can be drawn into wires because it is (23) _____ and hammered into sheets because it is (24) _____. Magnesium has these properties because it is a (25) _____.

SOLUTIONS FOR THE ADDITIONAL ACTIVITIES

Section 3.1 Review: The periodic table of shapes does not have the same orientation as the periodic table of elements. The groups in the periodic table of shapes are circles, squares, and triangles. The groups in the periodic table of shapes are along rows instead of along columns as the groups are in the periodic table of elements. The periods in the periodic table of shapes are in columns instead of rows like the periods in the periodic table of elements. The orientation of the periodic table of shapes can be changed to match the orientation of the periodic table of elements as shown to the right.

Corrected periodic table of shapes.

Section 3.2 Review:

Electronic Structure and the Periodic Law   39

(1) 7 known shells

(3)
shell number	1	2	3	4	5	6	7
number of subshells	1	2	3	4	5	6	7
name of subshells	s	s, p	s, p, d	s, p, d, f	s, p, d, f, g	s, p, d, f, g, h	s, p, d, f, g, h, i

The higher energy subshells (g, h, and i) are not actually occupied in the elements on the periodic table.

(4)
period number	1	2	3	4	5	6	7
number of subshells	1	2	2	3	3	4	4

Periods 1 and 2 match the prediction from (3); however, periods 3, 4, 5, 6, and 7 do not. Periods 5, 6, and 7 could not match the proposed values from (3) because the higher energy subshells (g, h, and i) are not represented on the periodic table. Period 3 does not contain a d subshell and period 4 has a d subshell before the p subshell. Perhaps, the 3d subshell is in period 4 (which means that the 4d subshell is in period 5, the 5d subshell is in period 6, and the 6d subshell is in period 7). Periods 4 and 5 do not contain f subshells. Perhaps the 4f subshell is in period 6 and the 5f subshell is in period 7.
(5) (a) 2; (b) 6; (c) 10; (d) 14; (6) (a) 1; (b) 3; (c) 5; (d) 7

Section 3.3 Review:
(1) 1; (2) 6; (3) 8; (4) s and p; (5) No, the first shell can have a maximum of 2 electrons in its valence shell.
(6) neon, argon, krypton, xenon, radon

Section 3.4 Review:
(1) 3s < 3p < 3d; (2) The 4s subshell is higher in energy than the 3s and 3p subshells, but lower in energy than the 3d subshell. 3s < 3p < 4s < 3d; (4) 6s; (5) 7s; (6) 4f; (7) 5f

Section 3.5 Review:
(1) (a) 2; (b) 10; (c) 36; (d) 86; (2) (a) $1s^2$; (b) $1s^2 2s^2 2p^6$; (c) $1s^2 2s^2 2p^6 3s^2 3p^6 4s^2 3d^{10} 4p^6$; (d) $1s^2 2s^2 2p^6 3s^2 3p^6 4s^2 3d^{10} 4p^6 5s^2 4d^{10} 5p^6 6s^2 4f^{14} 5d^{10} 6p^6$; (3) (a) 2s; (b) 3s; (c) 5s; (d) 7s; (4) All of these elements are noble gases.
(5) [Ir] would never be used in an abbreviated electron configuration because it is not a noble gas. The abbreviated electron configurations must start with a noble gas in brackets. (7) (a) s, p, d, f; (b) s, p; (c) p;
(8) (a) 2; (b) 2; (c) 2; (d) 5; (9) (a) 5s; (b) 5d; (c) 5f; (d) 4p

Section 3.6 Review:

(4) the trends are the same; (5) the trends are opposite; (6) the trends are opposite

Tying It All Together with a Laboratory Application:
(1) similar
(2) group or family
(3) valence
(4) 1
(5) 3s
(6) 4
(7) representative element
(8) metal
(9) larger
(10) more
(11) more
(12) more
(13) II A or (2)
(14) 3
(15) cannot
(16) period
(17) smaller
(18) less
(19) less
(20) high
(21) high
(22) metallic
(23) ductile
(24) malleable
(25) metal

SELF-TEST QUESTIONS
Multiple Choice

1. Elements in the same group
   a. have similar chemical properties
   b. have consecutive atomic numbers
   c. are called isotopes
   d. constitute a period of elements

2. Which element of those with the following atomic numbers should have properties similar to those of oxygen (element number 8)?
   a. 15
   b. 4
   c. 2
   d. 34

3. Which of the following elements is found in period 3 of the periodic table?
   a. Al
   b. Ga
   c. B
   d. more than one response is correct

4. The electronic configuration for an element containing 15 protons would be
   a. $1s^2 2s^2 2p^6 3s^2 3p^6$
   b. $1s^2 2s^2 2p^6 3s^2 4p^3$
   c. $1s^2 2s^2 2p^6 3s^2 3p^6 4s^2$
   d. $1s^2 2s^2 2p^6 3s^2 3p^3$

5. Which of the following is a true statement for an electronic configuration of $1s^2 2s^2 2p^6 3s^2 3p^6$?
   a. there are 6 electrons in the 3p orbital
   b. there are 6 electrons in the 3p shell
   c. there are 6 electrons in the 3p subshell
   d. more than one response is correct

6. The maximum number of electrons that may occupy a 4d orbital is
   a. 10
   b. 4
   c. 2
   d. 8

7. How many unpaired electrons are found in titanium (element number 22)?
   a. 1
   b. 2
   c. 3
   d. 4

8. Which element has the electronic configuration $1s^2 2s^2 2p^6 3s^2 3p^6$?
   a. Ne
   b. Ar
   c. K
   d. Kr

9. The element with unpaired electrons in a d subshell is element number
   a. 33
   b. 27
   c. 20
   d. 53

10. Which of the following elements has an electronic configuration ending in $4d^7$?
    a. Co
    b. Ir
    c. Rh
    d. more than one response is correct

11. How many electrons are found in the valence shell of oxygen (element number 8)?
    a. 3
    b. 4
    c. 5
    d. 6

12. The distinguishing electronic configuration of np^4 is characteristic of which group in the periodic table?
    a. II A (2)
    b. IV A (14)
    c. VI A (16)
    d. noble gases

13. The element with the electronic configuration $1s^2 2s^2 2p^3$ will be found in
    a. period 1, group VI A (16)
    b. period V A (15), group 2
    c. period 2, group V A (15)
    d. period III A (13), group V A (15)

14. How many electrons would be contained in the valence shell of an atom in group VI A (16)?
    a. 6   b. 2   c. 4   d. 16

15. Which of the following atoms is the most metallic?
    a. Sr   b. Ge   c. Br   d. Rb

16. Which of the following atoms has the largest radius?
    a. Si   b. C   c. Sn   d. Ge

17. Which of the following atoms would have the lowest ionization potential?
    a. Ca   b. Na   c. K   d. Rb

True-False
18. Elements 20 and 21 are in the same period of the periodic table.
19. Hund's rule states that electrons within a subshell remain unpaired if possible.
20. A 2p and a 3p subshell contain the same number of orbitals.
21. The maximum number of electrons an orbital may contain does not vary with the type of orbital.
22. The distinguishing electron in Br is found in a p orbital.

Matching
From the list on the right, choose a response that is consistent with the description on the left as far as electronic configurations are concerned. You may use a response more than once.

23. a noble gas   a. $1s^2 2s^2 2p^6 3s^2$
24. Mg   b. $1s^2 2s^2 2p^6$
25. a transition element   c. $1s^2 2s^2 2p^6 3s^2 3p^6 4s^2 3d^3$
26. an element just completing the filling of the n=2 shell   d. $1s^2 2s^2 2p^4$

Match each of the categories on the right with an element on the left.

27. neon (Ne)   a. a representative metal
28. phosphorus (P)   b. a noble gas nonmetal
29. calcium (Ca)   c. a nonmetal but not a noble gas
30. element number 47   d. a transition metal
31. element number 82   e. an inner-transition metal
32. top element of group V A (15)
33. the element containing 66 protons
34. the element with $5s^2$ as its distinguishing electron
35. the element in period 3 with 8 valence electrons

Chapter 4: Forces Between Particles

CHAPTER OUTLINE
4.1 Noble Gas Configurations
4.2 Ionic Bonding
4.3 Ionic Compounds
4.4 Naming Binary Ionic Compounds
4.5 The Smallest Unit of Ionic Compounds
4.6 Covalent Bonding
4.7 Polyatomic Ions
4.8 Shapes of Molecules and Polyatomic Ions
4.9 The Polarity of Covalent Molecules
4.10 More About Naming Compounds
4.11 Other Interparticle Forces

LEARNING OBJECTIVES/ASSESSMENT
When you have completed your study of this chapter, you should be able to:
1. Draw correct Lewis structures for atoms of representative elements. (Section 4.1; Exercise 4.2)
2. Use electronic configurations to determine the number of electrons gained or lost by atoms as they achieve noble gas configurations. (Section 4.2; Exercise 4.12)
3. Use the octet rule to correctly predict the ions formed during the formation of ionic compounds. (Section 4.3; Exercise 4.20)
4. Write correct formulas for the following types of ionic compounds:
   a. Binary compounds containing a representative metal and representative nonmetal. (Section 4.3; Exercise 4.22)
   b. Compounds containing a representative metal and a polyatomic ion. (Section 4.10; Exercise 4.70)
5. Correctly name the following:
   a. binary ionic compounds. (Section 4.4; Exercise 4.30)
   b. binary covalent compounds. (Section 4.10; Exercise 4.66)
   c. ionic compounds containing a representative metal and a polyatomic ion. (Section 4.10; Exercise 4.72)
6. Determine formula weights for ionic compounds. (Section 4.5; Exercise 4.38)
7. Draw Lewis structures for molecules and polyatomic ions. (Section 4.8; Exercise 4.52)
8. Use the VSEPR theory to predict the shapes of molecules and polyatomic ions (Section 4.8; Exercise 4.52)
9. Use electronegativities to determine the type of bonding that is likely to occur between pairs of representative elements. (Section 4.9; Exercise 4.58)
10. Determine whether a covalent molecule is polar or nonpolar. (Section 4.9; Exercise 4.64)
11. Relate melting points and boiling points of pure substances to the strength and types of interparticle forces present in the substances. (Section 4.11; Exercise 4.78)

SOLUTIONS FOR THE END OF CHAPTER EXERCISES
NOBLE GAS CONFIGURATIONS (SECTION 4.1)

☑4.2 a. iodine   ·Ï: c. tin   ·S̈n· or ·S̈n·
       b. strontium S̈r or ·Sr·   d. sulfur   ·S̈: or :S̈:

4.4   a. antimony $[Kr]5s^2 4d^{10} 5p^3$
      b. Rb $[Kr]5s^1$
      c. element number 52 $[Kr]5s^2 4d^{10} 5p^4$
      d. cesium $[Xe]6s^1$

44 Chapter 4

4.6 a. antimony $\cdot\overset{..}{\underset{.}{Sb}}\cdot$ or $\overset{..}{\underset{.}{Sb}}:$ c. element number 52 $\cdot\overset{..}{\underset{..}{Te}}:$ or $\overset{..}{\underset{..}{Te}}:$

   b. Rb $\overset{}{\underset{}{Rb}}\cdot$ d. cesium $\overset{}{\underset{}{Cs}}\cdot$

4.8 a. Any group III A (13) element $\overset{}{\underset{}{E}}\cdot$ or $\cdot\overset{}{\underset{}{E}}\cdot$

   b. Any group VI A (16) element $\cdot\overset{..}{\underset{..}{E}}:$ or $\overset{..}{\underset{..}{E}}:$

IONIC BONDING (SECTION 4.2)

4.10

		Added electrons	Removed electrons
a.	antimony	3	15 (5 not including d electrons)
b.	Rb	17 (7 not including d electrons)	1
c.	element number 52	2	16 (6 not including d electrons)
d.	cesium	31 (7 not including d or f electrons)	1

☑4.12

		Number of electrons lost/gained	Equation
a.	Cs	1 electron lost	$Cs \rightarrow Cs^+ + e^-$
b.	oxygen	2 electrons gained	$O + 2e^- \rightarrow O^{2-}$
c.	element number 7	3 electrons gained	$N + 3e^- \rightarrow N^{3-}$
d.	iodine	1 electron gained	$I + e^- \rightarrow I^-$

4.14

		Equation	Ion Symbol
a.	A selenium ion that has gained two electrons	$Se + 2e^- \rightarrow Se^{2-}$	Se^{2-}
b.	A rubidium atom that has lost one electron	$Rb \rightarrow Rb^+ + e^-$	Rb^+
c.	An aluminum atom that has lost three electrons	$Al \rightarrow Al^{3+} + 3e^-$	Al^{3+}

4.16 a. E^{2-} sulfur   c. E^+ sodium
      b. E^{3+} aluminum d. E^- chlorine

4.18 a. Rb^+ krypton   c. Se^{2-} krypton
      b. P^{3-} argon   d. Mg^{2+} neon

IONIC COMPOUNDS (SECTION 4.3)

☑4.20

		Cation formation	Anion formation	Formula
a.	Ca and Cl	$Ca \rightarrow Ca^{2+} + 2e^-$	$Cl + e^- \rightarrow Cl^-$	$CaCl_2$
b.	lithium and bromine	$Li \rightarrow Li^+ + e^-$	$Br + e^- \rightarrow Br^-$	$LiBr$
c.	elements number 12 and 16	$Mg \rightarrow Mg^{2+} + 2e^-$	$S + 2e^- \rightarrow S^{2-}$	MgS

☑4.22 a. Te^{2-} $BaTe$ c. F^- BaF_2
       b. N^{3-} Ba_3N_2 d. P^{3-} Ba_3P_2

4.24 a. CaS   binary   d. $NaHCO_3$ not binary
      b. Na_3P binary   e. P_2O_5 binary
      c. $MgSO_4$ not binary

Forces Between Particles 45

NAMING BINARY IONIC COMPOUNDS (SECTION 4.4)

4.26 a. Li^+ lithium ion   c. Ba^{2+} barium ion
      b. Mg^{2+} magnesium ion   d. Cs^+ cesium ion

4.28 a. Br^- bromide ion   c. P^{3-} phosphide ion
      b. O^{2-} oxide ion   d. Te^{2-} telluride ion

☑4.30 a. $CaCl_2$ calcium chloride   d. Al_2S_3 aluminum sulfide
       b. BaS barium sulfide   e. SrF_2 strontium fluoride
       c. $ZnBr_2$ zinc bromide

4.32 a. SnS and SnS_2 tin (II) sulfide and tin (IV) sulfide
      b. $FeCl_2$ and $FeCl_3$ iron (II) chloride and iron (III) chloride
      c. Cu_2O and CuO copper (I) oxide and copper (II) oxide
      d. $AuCl$ and $AuCl_3$ gold (I) chloride and gold (III) chloride

4.34 a. SnS and SnS_2 stannous sulfide and stannic sulfide
      b. $FeCl_2$ and $FeCl_3$ ferrous chloride and ferric chloride
      c. Cu_2O and CuO cuprous oxide and cupric oxide
      d. $AuCl$ and $AuCl_3$ aurous chloride and auric chloride

4.36 a. lead (IV) oxide PbO_2 d. nickel (III) nitride NiN
      b. cobalt (II) chloride $CoCl_2$ e. platinum (II) phosphide Pt_3P_2
      c. copper (II) sulfide CuS

THE SMALLEST UNIT OF IONIC COMPOUNDS (SECTION 4.5)

☑4.38 a. $NaBr$ $(1 \times 22.99 \text{ u}) + (1 \times 79.90 \text{ u}) = 102.89 \text{ u}$
       b. CaF_2 $(1 \times 40.08 \text{ u}) + (2 \times 19.00 \text{ u}) = 78.08 \text{ u}$
       c. Cu_2S $(2 \times 63.55 \text{ u}) + (1 \times 32.07 \text{ u}) = 159.17 \text{ u}$
       d. Li_3N $(3 \times 6.94 \text{ u}) + (1 \times 14.01 \text{ u}) = 34.83 \text{ u}$

4.40 a. $NaBr$ Na^+, Br^- c. Cu_2S Cu^+, S^{2-}
      b. CaF_2 Ca^{2+}, F^- d. Li_3N Li^+, N^{3-}

4.42 a. $NaBr$ $1 \times 22.99 \text{ g} = 22.99 \text{ g } Na^+$ $1 \times 79.90 \text{ g} = 79.90 \text{ g } Br^-$
      b. CaF_2 $1 \times 40.08 \text{ g} = 40.08 \text{ g } Ca^{2+}$ $2 \times 19.00 \text{ g} = 38.00 \text{ g } F^-$
      c. Cu_2S $2 \times 63.55 \text{ g} = 127.10 \text{ g } Cu^+$ $1 \times 32.07 \text{ g} = 32.07 \text{ g } S^{2-}$
      d. Li_3N $3 \times 6.94 \text{ g} = 20.82 \text{ g } Li^+$ $1 \times 14.01 \text{ g} = 14.01 \text{ g } N^{3-}$

4.44 a. $NaBr$
$$1 \text{ mole} \times \frac{6.02 \times 10^{23} \text{ ions}}{1 \text{ mole}} = 6.02 \times 10^{23} \text{ ions } Na^+$$

$$1 \text{ mole} \times \frac{6.02 \times 10^{23} \text{ ions}}{1 \text{ mole}} = 6.02 \times 10^{23} \text{ ions } Br^-$$

46   Chapter 4

b. CaF₂

$$1 \text{ mole} \times \frac{6.02 \times 10^{23} \text{ ions}}{1 \text{ mole}} = 6.02 \times 10^{23} \text{ ions Ca}^{2+}$$

$$2 \text{ moles} \times \frac{6.02 \times 10^{23} \text{ ions}}{1 \text{ mole}} = 1.204 \times 10^{24} \text{ ions F}^{-}$$

c. Cu₂S

$$2 \text{ mole} \times \frac{6.02 \times 10^{23} \text{ ions}}{1 \text{ mole}} = 1.204 \times 10^{24} \text{ ions Cu}^{+}$$

$$1 \text{ mole} \times \frac{6.02 \times 10^{23} \text{ ions}}{1 \text{ mole}} = 6.02 \times 10^{23} \text{ ions S}^{2-}$$

d. Li₃N

$$3 \text{ mole} \times \frac{6.02 \times 10^{23} \text{ ions}}{1 \text{ mole}} = 1.806 \times 10^{24} \text{ ions Li}^{+}$$

$$1 \text{ mole} \times \frac{6.02 \times 10^{23} \text{ ions}}{1 \text{ mole}} = 6.02 \times 10^{23} \text{ ions N}^{3-}$$

COVALENT BONDING (SECTION 4.6)

4.46

4.48   a. H₂S (each H atom is bonded to the S atom)   H—S̈—H

b. ClF   :C̈l—F̈:

c. HBr   H—B̈r:

d. HClO (the H and Cl are each bonded to O)   H—Ö—C̈l:

POLYATOMIC IONS (SECTION 4.7)

4.50   a. PH₄⁺ (each H atom is bonded to the P atom)

b. HPO₄²⁻ (each O atom is bonded to the P atom, and the H atom is boned to an O atom)

c. HSO₄⁻ (each O atom is bonded to the S atom, and the H atom is boned to an O atom)

SHAPES OF MOLECULES AND POLYATOMIC IONS (SECTION 4.8)

4.52		Lewis structure	3D Structure	VSEPR
a.	H_2S (each H atom is bonded to the S atom)	H—S̈—H	(H above, S with lone pairs, H below)	bent or angular
b.	PCl_3 (each Cl atom is bonded to the P atom)	:C̈l—P—C̈l: with :C̈l: below	(P with three Cl)	pyramid
c.	OF_2 (each F atom is bonded to the O atom)	:F̈—Ö—F̈:	(O with two F)	bent or angular
d.	SnF_4 (each F atom is bonded to the Sn atom)	:F̈— with F above and below, Sn center, :F̈:	(Sn with four F)	tetrahedral

4.54		Lewis structure	3D Structure	VSEPR
a.	NO_2^- (each O is bonded to N)	$[:\ddot{O}=N—\ddot{O}:]^-$		bent or angular
b.	ClO_3^- (each O is bonded to Cl)	$[:\ddot{O}—\ddot{C}l—\ddot{O}:\;\;:\ddot{O}:]^-$		pyramid
c.	CO_3^{2-} (each O is bonded to C)	$[:\ddot{O}=C—\ddot{O}:\;\;:\ddot{O}:]^{2-}$		triangle
d.	H_3O^+ (each H is bonded to O) Note the positive charge; compare with NH_4^+.	$[H—\ddot{O}—H\;\;H]^+$		pyramid

THE POLARITY OF COVALENT MOLECULES (SECTION 4.9)

4.56 a. H—I
  b. (S=O with O below)
  c. O—O—O The bonds in this molecule are nonpolar.

4.58		Calculation	Classification
a.	LiBr	2.8 − 1.0 = 1.8	polar covalent
b.	HCl	3.0 − 2.1 = 0.9	polar covalent

Forces Between Particles 47

48   Chapter 4

    c. PH$_3$ (each H is bonded to P) $2.1 - 2.1 = 0$ nonpolar covalent

    d. SO$_2$ (each O is bonded to S) $3.5 - 2.5 = 1.0$ polar covalent

    e. CsF $4.0 - 0.7 = 3.3$ ionic

4.60 a. H——I $\delta+$ H——I $\delta-$ This is a polar molecule.

    b. (S=O with O below)   (polarized structure with $\delta-$ O)   This is a polar molecule.

    c. (O—O with O below)   The bonds in this molecule are nonpolar.   This is a nonpolar molecule.

4.62   **Calculation**   **Classification**

    a. nitrogen and oxygen $3.5 - 3.0 = 0.5$ polar covalent

    b. magnesium and oxygen $3.5 - 1.2 = 2.3$ ionic

    c. N and H $3.0 - 2.1 = 0.9$ polar covalent

☑4.64 a. C≡O $\delta+$ C≡O $\delta-$ polar molecule

    b. H—Se—H $\delta+$ H—Se $\delta-$, $\delta+$ H   polar molecule

    c. AlI$_3$ (Al with three $\delta-$ I atoms)   nonpolar molecule

MORE ABOUT NAMING COMPOUNDS (SECTION 4.10)

☑4.66 a. PCl$_3$ phosphorus trichloride   d. BF$_3$ boron trifluoride

    b. N$_2$O$_5$ dinitrogen pentoxide   e. CS$_2$ carbon disulfide

    c. CCl$_4$ carbon tetrachloride

4.68 a. dinitrogen tetroxide   N$_2$O$_4$ c. silicon dioxide   SiO$_2$

    b. sulfur hexachloride   SCl$_6$ d. oxygen difluoride   OF$_2$

☑4.70 a. calcium and the nitrite ion   Ca(NO$_2$)$_2$ calcium nitrite

    b. magnesium and the hypochlorite ion   Mg(ClO)$_2$ magnesium hypochlorite

    c. Cs and Cr$_2$O$_7^{2-}$ Cs$_2$Cr$_2$O$_7$ cesium dichromate

    d. K and SO$_3^{2-}$ K$_2$SO$_3$ potassium sulfite

☑4.72 a. barium hydroxide   Ba(OH)$_2$ d. ammonium sulfate   (NH$_4$)SO$_4$

    b. magnesium sulfite   MgSO$_3$ e. lithium hydrogen carbonate   LiHCO$_3$

    c. calcium carbonate   CaCO$_3$

4.74 a. Any group I A (1) element and SO$_3^{2-}$ M$_2$SO$_3$

    b. Any group I A (1) element and C$_2$H$_3$O$_2^-$ MC$_2$H$_3$O$_2$

    c. Any metal that forms M^{2+} ions and Cr$_2$O$_7^{2-}$ MCr$_2$O$_7$

Forces Between Particles   49

  d. Any metal that forms M^{3+} ions and PO_4^{3-} MPO_4
  e. Any metal that forms M^{3+} ions and NO_3^- $M(NO_3)_3$

OTHER INTERPARTICLE FORCES (SECTION 4.11)

4.76 The alcohol has higher melting and boiling points than the ether. The forces that hold the alcohol molecules together must be stronger and harder to break than the forces that hold the ether molecules together.

☑4.78 The noble gases are nonpolar; therefore, they only experience dispersion forces. These forces increase with the size of the particles; therefore, the order of increasing boiling point for the noble gases is: helium, neon, argon, krypton, xenon, and radon.

4.80 Dispersion forces and dipolar forces are unlikely to be the predominant ones in the lattice of solid sucrose because these forces are relatively weak. Ionic bonds, metallic bonds, and covalent bonds are unlikely to be the predominant ones in the lattice of solid sucrose because they are very strong.

ADDITIONAL ACTIVITIES

Section 4.1 Review:

(1) Complete the electron-dot formulas for each of the elements in the abbreviated periodic table below.

(1) I A	(2) II A		(13) III A	(14) IV A	(15) V A	(16) VI A	(17) VII A	(18) VIII A
H								He
Li	Be		B	C	N	O	F	Ne
Na	Mg		Al	Si	P	S	Cl	Ar

(2) What patterns do you notice after completing (1)?
(3) What is the relationship between the group numbers in parenthesis and the dots on the elements in the s block?
(4) What is the relationship between the group numbers in parenthesis and the dots on the elements in the p block?
(5) How many dots would be placed on an element from the d block?
(6) How many dots would be placed on an element from the f block?
(7) Is there a relationship between the number of dots and the group numbers (Roman numeral or parenthetical) for the elements in the d or f blocks? Explain your answer.

Section 4.2 Review:

(1) Referring to the abbreviated periodic table in the Section 4.1 Review, how many electrons must an element from groups I A (1) through VII A (17) gain in order to have the same valence shell electron configuration as the noble gas from the same period?

(2) For hydrogen and the elements in period 2 from groups IA (1) through VII A (17), write the symbols for the ions formed by gaining enough electrons to have the same electron configuration as the noble gas from the same period.

(3) Which ions from (2) are the most likely to exist? Why?

(4) Referring to the abbreviated periodic table in the Section 4.1 Review, how many electrons must an element from groups I A (1) through VII A (17) lose in order to have the same valence shell electron configuration as the noble gas from the previous period?

(5) For hydrogen and the elements in period 3 from groups IA (1) through VII A (17), write the symbols for the ions formed by losing enough electrons to have the same electron configuration as the noble gas from the previous period.

(6) Which ions from (5) are the most likely to exist? Why?

(7) Rb^+ is isoelectronic with krypton. Is the Rb^+ ion the same as a krypton atom? Explain your answer.

Section 4.3 Review:

(1) For the elements in groups I A (1) through IIV A (17), write the charge of the mostly likely ion for each group above the group numbers on the abbreviated periodic table from the Section 4.1 Review.

(2) Write formulas for the compounds formed between hydrogen and each of the elements in groups I A (1) through IIV A (17) from period 2.

(3) The elements in the d and f blocks usually form ions with a 2+ charge; however, many exceptions exist. All of the metal atoms in the p block except aluminum can also have charges other than the ones recorded in (1). Determine the number of positive and negative ions in the following compounds. Include the charges for each ion.

| $IrCl_4$ | Pr_2O_3 | BiF_5 | PbO_2 | $SnBr_2$ | InN | CuI | U_3N_2 | VSe_2 | $YbTe$ | Au_2S | GaP | WC |

Section 4.4 Review:

(1) Name the compounds from the Section 4.3 Review using a roman numeral in parenthesis to indicate the charge on the positive ion.

(2) Name the following compounds:

| Li_2S | BeO | NaF | $MgCl_2$ | K_3N | Ca_3P_2 | RbI | $SrBr$ | Cs_2Te | $BaSe$ | Fr_4C | Ra_3As_2 | AlF_3 |

(3) Do the compounds from (2) need roman numerals like the compounds from (1)? Explain why or why not.

Section 4.5 Review:

(1) Calculate the formula weight for the following compounds: $IrCl_4$, Pr_2O_3, Li_2S, BeO.

(2) Calculate the mass of 1 mole of the following compounds: BiF_5, PbO_2, NaF, $MgCl_2$.

(3) Calculate the number of formula units in 1 mole of the following compounds: $SnBr_2$, InN, K_3N, Ca_3P_2.

(4) Calculate the number and identify the types of ions in 1 mole of the following compounds: U_3N_2, RbI.

Section 4.6 Review:

(1) Based on the electron-dot symbols in the abbreviated periodic table from the Section 4.1 Review, what is the minimum number of electrons each nonmetal would need to share with other atoms in order to have a completed valence shell?

(2) What is wrong with each of these Lewis structures? Propose a better structure, if possible.

:P—P:
| |
:P—P:

:Br≡N: F=F   H—H—O:

Section 4.7 Review:
What charge should each of the following polyatomic ions have?

peroxide, hydronium, thiosulfate, succinate

hypochlorite, arsenite

Section 4.8 Review:
Predict the shapes of each of the "corrected" Lewis structures from the Section 4.6 Review as well as the polyatomic ions in the Section 4.7 Review. (For the succinate ion, predict the shape around each carbon atom.)

Section 4.9 Review:
Perform the following tasks for each of the "corrected" Lewis structures from the Section 4.6 Review as well as the polyatomic ions in the Section 4.7 Review.
(1) Draw the charge distribution on the Lewis Structure.
(2) Using Table 4.4 from the textbook, classify each bond as polar covalent, nonpolar covalent, or ionic.
(3) Classify the molecule or ion as polar or nonpolar.

Section 4.10 Review:
(1) Identify the following chemical formulas as covalent or ionic compounds or polyatomic ions.
(2) Name each of the chemical formulas.

Na_2O	N_2O	MnO_2	MnO_4^-	SCl_2	$MgCl_2$	P_2F_4	KF	NH_3	NH_4Br

Section 4.11 Review:
(1) What is the difference between the prefixes *inter-* and *intra-*?
(2) Identify the types of "particles" in each of the following types of materials.

network solids	ionic compounds	polar covalent compounds
metals	nonpolar covalent compounds	polar covalent compounds containing one of the following bonds: N—H, O—H, or F—H

(3) Match the materials from (2) to the major interparticle force that hold these materials together.

ionic bond	metallic bond	covalent bond
dipolar force	hydrogen bonding	dispersion force

(4) Match the following chemical substances with a material from (2) and an interparticle force from (3).

NH_3 (ammonia)	brass (an alloy of copper and zinc)	diamond (a form of pure C atoms)	Ne	KCl	SO_2

Tying It All Together with a Laboratory Application:
As discussed in the Chapter 3 Laboratory Application, sodium metal reacts with water to produce hydrogen gas and sodium hydroxide. The chemical equation for this reaction is:

$$2\ Na\ (s) + H_2O\ (l) \rightarrow H_2\ (g) + 2\ NaOH\ (aq)$$

52 Chapter 4

Rewrite this equation using Lewis Structures (1).

Sodium metal is held together by (2) _____ . During the reaction the sodium atoms (3) _____ electrons to form (4) _____ ions. The Na⁺ ion has the same electron configuration as the noble gas (5) _____.

Liquid water is held together primarily by (6) _____ interparticle forces. Water has a (7) _____ shape. The oxygen atom is (8) _____ electronegative than the hydrogen atom; consequently, the hydrogen atoms have a partially (9) _____ charge. Water is a(n) (10) _____ compound. If water were named as a (11) _____ covalent compound, it would be called (12) _____.

The hydrogen molecule has a (13) _____ shape. It has a (14) _____ bond. Hydrogen (15) _____ a compound. Hydrogen gas molecules experience (16) _____ interparticle forces.

Sodium hydroxide is a(n) (17) _____ compound. It is soluble in water; however, if all the water evaporated, it would form a(n) (18) _____. The sodium ion and the hydroxide ion would occupy different (19) _____. These ions would be held together by a(n) (20) _____ bond. The smallest unit of sodium hydroxide is called a(n) (21) _____ and has a formula weight of (22) _____.

The hydroxide ion is a(n) (23) _____ (simple or polyatomic) ion. It contains a (24) _____ bond. It has a (25) _____ shape.

SOLUTIONS FOR THE ADDITIONAL ACTIVITIES
Section 4.1 Review:

(1) Periodic table with electron-dot symbols: H (IA), Li, Be, Na, Mg in s-block; B, C, N, O, F, Ne, Al, Si, P, S, Cl, Ar in p-block; He in group VIIIA (18). Group labels: (2) IIA, (13) IIIA, (14) IVA, (15) VA, (16) VIA, (17) VIIA.

(2) The electron-dot symbols for elements in the same group have the same number of electrons. The number of electrons increases one per box from left to right across a period. (3) In the s block, the number in parenthesis is the same as the number of electrons in the electron-dot symbol. (4) In the p block, the number in parenthesis is 10 higher than the number of electrons in the electron-dot symbol. (5) 2; (6) 2; (7) No obvious connection exists between the Roman numerals or parenthetical numbers for the d block and the 2 valence-shell electrons these elements have. The f block elements do not have Roman numerals or parenthetical numbers, thus no connection can be established for these elements either.

Section 4.2 Review:

	Group	I A (1)	II A (2)	III A (13)	IV A (14)	V A (15)	VI A (16)	VII A (17)
(1)	e⁻ to gain	7	6	5	4	3	2	1
(2)	symbols	H⁻; Li⁷⁻	Be⁶⁻	B⁵⁻	C⁴⁻	N³⁻	O²⁻	F⁻

(3) The ions from (2) that are most likely to form are H⁻, C⁴⁻, N³⁻, O²⁻, and F⁻. These ions formed when the neutral atoms gained no more than half the total number of valence-shell electrons for the noble gas from the same period (≤1 for He, ≤4 Ne).

	Group	I A (1)	II A (2)	III A (13)	IV A (14)	V A (15)	VI A (16)	VII A (17)
(4)	e⁻ to lose	1	2	3	4	5	6	7
(5)	symbols	H⁺ *; Na⁺	Mg²⁺	Al³⁺	Si⁴⁺	P⁵⁺	S⁶⁺	Cl⁷⁺

*Hydrogen is in period 1; therefore, a noble gas from the previous period would have to be in period 0 which does not exist. The electron in the valence-shell of hydrogen was removed to form this ion.

(6) The ions from (5) that are most likely to form are H⁺, Na⁺, Mg²⁺, Al³⁺, and Si⁴⁺. These ions formed when the neutral atoms lost no more than half the total number of valence-shell electrons for the noble gas from the previous row (1 for H*, ≤4 Ne).

(7) No, Rb⁺ (the rubidium ion) is not the same as a krypton atom. The rubidium ion has one more proton than the krypton atom. Rb⁺ has the same electron configuration as a krypton atom as well as the same total number of electrons as a krypton atom. Consequently, the rubidium ion has a positive charge because it has one more proton than electron. The rubidium ion has chemical properties that more similar to a krypton atom than a rubidium atom because many chemical properties result from the electron configuration; however, the rubidium ion is not a krypton atom. The rubidium ion is smaller than a krypton atom because the additional proton pulls the electrons closer to the nucleus of the rubidium ion. The positive charge on the rubidium ion also causes it to be attracted to negatively charged ions. Krypton is electrically neutral and is not attracted to negative ions.

Section 4.3 Review:

(1)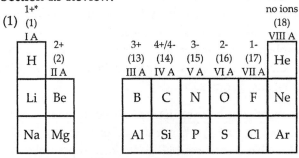

*Hydrogen can make both 1+ and 1- ions.

(2)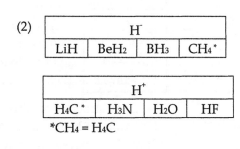

*CH₄ = H₄C

(3)	IrCl₄	Pr₂O₃	BiF₅	PbO₂	SnBr₂	InN	CuI	U₃N₂	VSe₂	YbTe	Au₂S	GaP	WC
−	4 Cl⁻	3 O²⁻	5 F⁻	2 O²⁻	2 Br⁻	N³⁻	I⁻	2 N³⁻	2 Se²⁻	Te²⁻	S²⁻	P³⁻	C⁴⁻
+	Ir⁴⁺	2 Pr³⁺	Bi⁵⁺	Pb⁴⁺	Sn²⁺	In³⁺	Cu⁺	3 U²⁺	V⁴⁺	Yb²⁺	2 Au⁺	Ga³⁺	W⁴⁺

Section 4.4 Review:

(1)

IrCl₄ iridium (IV) chloride	Pr₂O₃ praseodymium (III) oxide	BiF₅ bismuth (V) fluoride	PbO₂ lead (IV) oxide
SnBr₂ tin (II) bromide	InN indium (III) nitride	CuI copper (I) iodide	U₃N₂ uranium (II) nitride
VSe₂ vanadium (IV) selenide	YbTe ytterbium (II) telluride	Au₂S gold (I) sulfide	GaP gallium (III) phosphide
WC tungsten (IV) carbide			

(2)

Li₂S lithium sulfide	BeO beryllium oxide	NaF sodium fluoride	MgCl₂ magnesium chloride
K₃N potassium nitride	Ca₃P₂ calcium phosphide	RbI rubidium iodide	SrBr strontium bromide
Cs₂Te cesium telluride	BaSe barium selenide	Fr₄C francium carbide	Ra₃As₂ radium arsenide
AlF₃ aluminum fluoride			

(3) The compounds in (2) do not need roman numerals because the metals ions from group I A (1) always have a 1+ charge, the metals ions from group II A (2) always have a 2+ charge, and the aluminum ion always has a 3+ charge.

Section 4.5 Review:
(1) $IrCl_4$ = 334.02 u; Pr_2O_3 = 329.82 u; Li_2S = 45.95 u; BeO = 25.01 u; (2) BiF_5 = 303.98 g; PbO_2 = 239.2 g; NaF = 41.99 g; $MgCl_2$ = 95.21 g; (3) 6.022×10^{23} formula units $SnBr_2$; 6.022×10^{23} formula units InN; 6.022×10^{23} formula units K_3N; 6.022×10^{23} formula units Ca_3P_2; (4) 1.8066×10^{24} U^{2+} ions, 1.2044×10^{24} N^{3-} ions; 6.022×10^{23} Rb^+ ions, 6.022×10^{23} I^- ions.

Section 4.6 Review:
(1) H = 1; C = 4; N = 3; P = 3; O = 2; S = 2; F = 1; Cl = 1; He = 0; Ne = 0; Ar = 0; (2) The given structure for P_4 is incorrect because the picture shows 24 electrons (16 electrons in lone pairs and 8 electrons in bonds), but the structure should only contain 20 electrons (4 P atoms, each with 5 valence-shell electrons). Structure A is a viable alternative because this structure has only 20 electrons (8 electrons in lone pairs and 12 electrons in bonds). Structure B is not only a viable alternative to the given structure, but is also the actual structure. Like structure A, structure B contains 20 electrons (8 electrons in lone pairs and 12 electrons in bonds). Each phosphorus atom is bonded to 3 other phosphorus atoms in order to form a tetrahedron. This structure can be constructed with 4 paper clips (P atoms) and 6 straws (bonds).

The given structure for BrN is incorrect because the structure only contains 10 electrons (4 electrons in lone pairs and 6 electrons in bonds) when it should contain 12 electrons (Br has 7 valence-shell electrons and N has 5 valence-shell electrons). A possible alternative structure is structure C.

The structure for F_2 is incorrect because the structure only contains 12 electrons (8 electrons in lone pairs and 4 electrons in bonds) when it should contain 14 electrons (each F atom has 7 valence-shell electrons). A possible alternative structure is structure D.

The structure for H_2O is incorrect because the structure contains 10 electrons (6 electrons in lone pairs and 4 electrons in bonds) when it should contain 8 electrons (H has 1 valence-shell electrons and O has 6 valence-shell electrons). Also, the hydrogen atom in the middle of this structure has 2 bonds, but hydrogen is only capable of making 1 bond. A possible alternative structure is structure E.

Section 4.7 Review:
peroxide = 2-; hydronium = 1+; hypochlorite = 1-; arsenite = 1-; thiosulfate = 2-; succinate = 2-

Section 4.8 Review:
Structure A = bent or angular (based on each P atom); Structure B = pyramid with triangle base (based on each P atom); Structure C = linear; Structure D = linear; Structure E = bent or angular; peroxide = linear; hydronium = pyramid with triangle base; hypochlorite = linear; arsenite = bent or angular; thiosulfate = tetrahedron; succinate = triangle (both C atoms double bonded to an O atom and single bonded to an O atom and another C atom), tetrahedron (both C atoms bonded to 2 H and 2 other C atoms)

Section 4.9 Review:
Structures A & B   (1) no charge distribution because only P atoms, (2) nonpolar covalent bonds, (3) nonpolar molecule

Structure C

(1)   , (2) polar covalent bond, (3) polar molecule

Section 4.10 Review:

Na₂O	ionic compound	sodium oxide	MgCl₂	ionic compound	magnesium chloride
N₂O	covalent compound	dinitrogen monoxide	P₂F₄	covalent compound	diphosphorus tetrafluoride
MnO₂	ionic compound	manganese (IV) oxide	KF	ionic compound	potassium fluoride
MnO₄⁻	polyatomic ion	permanganate ion	NH₃	covalent compound	nitrogen trihydride*
SCl₂	covalent compound	sulfur dichloride	NH₄Br	ionic compound	ammonium bromide

*Note: NH₃ has the common name ammonia.

56 Chapter 4

Section 4.11 Review:
(1) *inter-* means between two different parts, *intra-* means within one part;
(2)

network solids **atoms**	ionic compounds **ions**	polar covalent compounds **polar molecules**
metals **metal atoms**	nonpolar covalent compounds **nonpolar covalent molecules**	polar covalent compounds containing one of the following bonds: N—H, O—H, or F—H **molecules with N—H, O—H, or F—H bonds**

(3)

ionic bond **ionic compounds**	metallic bond **metals**	covalent bond **network solids**
dipolar force **polar covalent compounds**	hydrogen bonding **polar covalent compounds containing one of the following bonds: N—H, O—H, or F—H**	dispersion force **nonpolar covalent compounds**

(4) NH₃ – polar covalent compounds containing one of the following bonds: N—H, O—H, or F—H and hydrogen bonding;
brass – metal and metallic bond;
diamond – covalent network solid and covalent bond;
Ne – nonpolar covalent molecules (atoms) and dispersion force;
KCl – ionic compound and ionic bond;
SO₂ – polar covalent compound and dipolar force

Tying It All Together with a Laboratory Application:

(1) 2 Na + 2 H—O—H → H—H + 2 Na⁺ [:Ö—H]⁻

(2) metallic bonding   (8) more   (13) linear   (17) ionic   (22) 40.00 u
(3) lose   (9) positive   (14) nonpolar   (18) crystal lattice (23) polyatomic
(4) sodium   (10) polar covalent   covalent   (19) lattice sites   (24) polar covalent
(5) neon   (11) binary   (15) is not   (20) ionic   (25) linear
(6) hydrogen bonding   (12) dihydrogen   (16) dispersion   (21) formula unit
(7) bent   monoxide

SELF-TEST QUESTIONS
Multiple Choice
1. Which of the following elements has the lowest electronegativity?
   a. As   b. P   c. Br   d. Cl

2. In describing the strength of interparticle forces we discover that the weakest forces or bonds are
   a. ionic   b. dipolar   c. dipolar   d. dispersion

3. The formula for the compound formed between the elements Ba and O would be
   a. BaO   b. Ba₂O   c. BaO₂   d. Ba₂O₃

4. The formula for the ionic compound containing Al^{3+} and SO_4^{2-} ions would be
   a. Al(SO₄)₂   b. AlSO₄   c. Al₃(SO₄)₂   d. Al₂(SO₄)₃

5. The expected formula of the molecule formed when nonmetals C and H combine in compliance with the octet rule is
   a. CH₄   b. CH₂   c. C₄H   d. CH₃

6. The name of the covalent compound PCl₃ is
   a. trichlorophosphide   c. phosphorus trichloride
   b. phosphorus trichlorine   d. phophorus chloride

7. The compound MgSO₄ is correctly named
   a. magnesium sulfur tetroxide   c. magnesium sulfide
   b. magnesium sulfoxide   d. magnesium sulfate

8. This bond is found in molecules such as HCl and H₂O.
   a. nonpolar covalent bond   c. ionic bond
   b. polar covalent bond   d. metallic bond

9. If the electronegativity difference between two elements A and B is 1.0, what type of bond is A-B?
   a. nonpolar covalent   c. ionic
   b. polar covalent   d. metallic

10. Which of the following is a correct electron dot formula for sulfur (element 16)?
    a. :S̈:   b. ·S̈:   c. ·S̈·   d. S̈

11. In the structures below, each bonding electron pair is denoted by a dash. A correct structure of SO₃²⁻ is

12. A covalent molecule forms between elements A and B. B is more electronegative. Which of the following molecules would be polar?

    a. :B̈=A=B̈:   b. :Ä—B̈:   c. (planar structure with A center, B's attached)   d. (structure with A center, 4 B's attached)

13. Which of the following contains polar bonds, but is a nonpolar molecule?
    a. H—S\
              \H
    b. O=C=O   c. H—Cl   d. F—F

14. According to the VSEPR theory, the shape of OF₂ will be
    a. linear   b. bent   c. triangle   d. tetrahedral

15. The name for OF₂ is
    a. monoxygen difluoride
    b. oxygen difluorine
    c. oxygen fluorine
    d. oxygen difluoride

True-False
16. No more than one pair of electrons can be shared to form covalent bonds between atoms.
17. Dispersion forces between particles are correctly classified as very strong.
18. A compound between elements with atomic numbers 7 and 9 will contain covalent bonds.
19. All covalent bonds are polar.
20. The interparticle forces in a solid noble gas would have to be polar in nature.
21. Neon (Ne) has a higher boiling point than krypton (Kr).
22. Nonpolar molecules only contain nonpolar bonds.

Matching
An ionic compound is formed from each of the pairs given on the left. For each pair, choose the correct formula for the resulting compound from the responses on the right.

23. X⁺ and Y²⁻
24. X⁺ and Y⁻
25. X³⁺ and Y³⁻
26. X is a group II A (2) ion and Y is a group VI A (16) ion

a. X₃Y₂
b. XY
c. XY₂
d. X₂Y

For each molecule given on the left, predict the molecular geometry based on VSEPR theory.
27. NH₃
28. BrCl
29. H₂S
30. CO₂
31. CH₄

a. linear
b. planar triangle
c. triangular pyramid
d. tetrahedral
e. bent

For each of the molecules on the left, choose the statement from the right that correctly gives the polarity of the bonds in the molecule and the polarity of the molecule as a whole.

32. H₂S, H—S—H
33. CO₂, O═C═O
34. N₂O, N≡N—O
35. O₃, O—O═O

a. a polar molecule containing all polar bonds
b. a nonpolar molecule containing all nonpolar bonds
c. a nonpolar molecule containing all polar bonds
d. a polar molecule containing polar bonds and nonpolar bonds

Chapter 5: Chemical Reactions

CHAPTER OUTLINE
5.1 Chemical Equations
5.2 Types of Reactions
5.3 Redox Reactions
5.4 Decomposition Reactions
5.5 Combination Reactions
5.6 Replacement Reactions
5.7 Ionic Equations
5.8 Energy and Reactions
5.9 The Mole and Chemical Equations
5.10 The Limiting Reactant
5.11 Reaction Yields

LEARNING OBJECTIVES/ASSESSMENT
When you have completed your study of this chapter, you should be able to:
1. Identify the reactants and products in written reaction equations. (Section 5.1; Exercise 5.2)
2. Balance simple reaction equations by inspection. (Section 5.1; Exercise 5.6)
3. Assign oxidation numbers to elements in chemical formulas. (Section 5.3; Exercise 5.10)
4. Identify the oxidizing and reducing agents in redox reactions. (Section 5.3; Exercise 5.18)
5. Classify reactions into the categories of redox or nonredox. (Sections 5.4; Exercise 5.20)
6. Classify reactions into the categories of decomposition, combination, single replacement, or double replacement. (Sections 5.4, 5.5, and 5.6; Exercise 5.20)
7. Write molecular equations in total ionic and net ionic forms. (Section 5.7; Exercise 5.30 a, b, & c)
8. Classify reactions as exothermic or endothermic. (Section 5.8; Exercise 5.34)
9. Use the mole concept to do calculations based on reaction equations. (Section 5.9; Exercise 5.42)
10. Use the mole concept to do calculations based on the limiting reactant principle. (Section 5.10; Exercise 5.52)
11. Use the mole concept to do percentage yield calculations. (Section 5.11; Exercise 5.56)

SOLUTIONS FOR THE END OF CHAPTER EXERCISES
CHEMICAL EQUATIONS (SECTION 5.1)

☑5.2

		Reactants	Products
a.	$2 H_2 (g) + O_2 (g) \rightarrow 2 H_2O (l)$	$H_2 (g), O_2 (g)$	$H_2O (l)$
b.	$CaCO_3 (s) \rightarrow CaO (s) + CO_2 (g)$	$CaCO_3 (s)$	$CaO (s), CO_2 (g)$
c.	sodium + water → hydrogen + sodium hydroxide	sodium, water	hydrogen, sodium hydroxide
d.	copper + silver nitrate → copper nitrate + silver	copper, silver nitrate	copper nitrate, silver

5.4
a. $Mg (s) + O_2 (g) \rightarrow MgO (s)$ is not consistent with the law of conservation of matter because two moles of oxygen atoms are on the reactant side of the equation, but only one mole of oxygen atoms is on the product side of the equation.
b. $2 SO_2 (g) + O_2 (g) \rightarrow 2 SO_3 (g)$ is consistent with the law of conservation of matter.
c. 1.60 g oxygen + 1.00 g hydrogen → 2.80 g water is not consistent with the law of conservation of matter because the mass of the reactants (2.60 g on the left side) does not equal the mass of the products (2.80 g on the right side).

Notice, the number of moles of oxygen and hydrogen are not equal on both sides of the equation either.

Reactants:

$$1.60 \text{ g } O_2 \left(\frac{1 \text{ mole } O_2}{32.00 \text{ g } O_2} \right) = 0.0500 \text{ moles } O_2; \quad 1.00 \text{ g } H_2 \left(\frac{1 \text{ mole } H_2}{2.0158 \text{ g } H_2} \right) = 0.496 \text{ moles } H_2$$

Products:

$$2.80 \text{ g H}_2\text{O} \left(\frac{1 \text{ mole H}_2\text{O}}{18.0158 \text{ g H}_2\text{O}}\right)\left(\frac{1 \text{ moles O}_2}{2 \text{ moles H}_2\text{O}}\right) = 0.0777 \text{ moles O}_2$$

$$2.80 \text{ g H}_2\text{O} \left(\frac{1 \text{ mole H}_2\text{O}}{18.0158 \text{ g H}_2\text{O}}\right)\left(\frac{1 \text{ moles H}_2}{1 \text{ mole H}_2\text{O}}\right) = 0.1554 \text{ moles H}_2$$

Balanced equation: 1.60 g oxygen + 0.20 g hydrogen → 1.80 g water

d. CuO (s) + H_2 (g) → H_2O (l) + Cu (s) is consistent with the law of conservation of matter.

☑5.6 a. Ag (s) + Cu(NO$_3$)$_2$ (aq) → Cu (s) + AgNO$_3$ (aq)

Elements	Ag	Cu	N	O
Reactant	1	1	2	6
Product	1	1	1	3

This equation is not balanced because the number of moles of nitrogen and oxygen are not balanced.

b. 2 N$_2$O (g) + 3 O$_2$ (g) → 4 NO$_2$ (g)

Elements	N	O
Reactant	4	8
Product	4	8

This equation is balanced.

c. Mg (s) + O$_2$ (g) → 2 MgO (s)

Elements	Mg	O
Reactant	1	2
Product	2	2

This equation is not balanced because the number of moles of magnesium are not balanced.

d. H$_2$SO$_4$ (aq) + Ca(OH)$_2$ (aq) → CaSO$_4$ (s) + 2 H$_2$O (l)

Elements	H	S	O	Ca
Reactant	4	1	6	1
Product	4	1	6	1

This equation is balanced.

5.8
a. KClO$_3$ (s) → KCl (s) + O$_2$ (g)　　　　　　　　　　2 KClO$_3$ (s) → 2 KCl (s) + 3 O$_2$ (g)
b. C$_2$H$_6$ (g) + O$_2$ (g) → CO$_2$(g) + H$_2$O (l)　　　　2 C$_2$H$_6$ (g) + 7 O$_2$ (g) → 4 CO$_2$(g) + 6 H$_2$O (l)
c. nitrogen + oxygen → dinitrogen pentoxide　　　　2 N$_2$ (g) + 5 O$_2$ (g) → 2 N$_2$O$_5$ (g)
d. MgCl$_2$ (s) + H$_2$O (g) → MgO (s) + HCl (g)　　　　MgCl$_2$ (s) + H$_2$O (g) → MgO (s) + 2 HCl (g)
e. CaH$_2$ (s) + H$_2$O (l) → Ca(OH)$_2$ (s) + H$_2$ (g)　　CaH$_2$ (s) + 2 H$_2$O (l) → Ca(OH)$_2$ (s) + 2 H$_2$ (g)
f. Al (s) + Fe$_2$O$_3$ (s) → Al$_2$O$_3$ (s) + Fe (s)　　　　　2 Al (s) + Fe$_2$O$_3$ (s) → Al$_2$O$_3$ (s) + 2 Fe (s)
g. aluminum + bromine → aluminum bromide

　　　　　　　　　　　　　　　　　　　　　　　　2 Al (s) + 3 Br$_2$ (l) → 2 AlBr$_3$ (s)

h. HgNO$_3$ (aq) + NaCl (aq) → Hg$_2$Cl$_2$ (s) + NaNO$_3$ (aq)

　　　　　　　　　　　　　　　　2 HgNO$_3$ (aq) + 2 NaCl (aq) → Hg$_2$Cl$_2$ (s) + 2 NaNO$_3$ (aq)

REDOX REACTIONS (SECTION 5.3)

☑5.10

		Calculations	O.N.
a.	H<u>Cl</u>O	H = +1　Cl = ?　O = –2　HClO = 0	+1
		(+1) + ? + (–2) = 0　⇒ ? = +1	
b.	Na<u>N</u>O$_2$	Na = +1　N = ?　O = –2　NaNO$_2$ = 0	+3
		(+1) + ? + 2(–2) = 0　⇒ ? = +3	
c.	<u>N</u>$_2$	N = ?　N$_2$ = 0	0
		2(?) = 0　⇒ ? = 0	

d. Ca(<u>Cl</u>O)₂   Ca = +2   Cl = ?   O = –2   Ca(ClO)₂ = 0   +1
                    (+2) + 2(?) + 2(–2) = 0   ⇒   ? = +1

e. <u>Al</u>³⁺   Al = ?   Al³⁺ = +3   +3
                    ? = +3

f. <u>N</u>₂O₅   N = ?   O = –2   N₂O₅ = 0   +5
                    2(?) + 5(–2) = 0   ⇒   ? = +5

5.12   **Oxidation Numbers**   **Highest O.N.**
  a. HNO₂   H = +1, N = +3, O = –2   nitrogen
  b. B₂O₃   B = +3, O = –2   boron
  c. CaCO₃   Ca = +2, C = +4, O = –2   carbon
  d. NaHSO₃   Na = +1, H = +1, S = +4, O = –2 sulfur
  e. Mg(ClO₄)₂   Mg = +2, Cl = +7, O = –2   chlorine
  f. KIO₃   K = +1, I = +5, O = –2   iodine

5.14   **O.N.**   **Change**   **Classification**
  a. 2 Al (s) + 6 <u>H</u>Cl (aq) → 2 AlCl₃ + 3 H₂ (g)   +1 → 0   gained e⁻   reduced
  b. <u>Na</u>₂CO₃ (s) + 2 HCl (aq) → 2 NaCl (aq) + CO₂ (g) +1 → +1 no change   neither
        + H₂O (l)
  c. <u>Zn</u> (s) + CuCl₂ (aq) → Cu (s) + ZnCl₂ (aq)   0 → +2   lost e⁻   oxidized
  d. <u>Cu</u>O (s) + H₂ (g) → Cu (s) + H₂O (g)   +2 → 0   gained e⁻   reduced
  e. 3 MnO₂ (s) + 4 <u>Al</u> (s) → 2 Al₂O₃ (s) + 3 Mn (s) 0 → +3   lost e⁻   oxidized

5.16 a. 4 Al (s) + 3 O₂ (g) → 2 Al₂O₃ (s)

Elements	Al	O
Reactant	0	0
Product	+3	-2

The oxidizing agent is O₂.
The reducing agent is Al.

b. 2 Na (s) + 2 H₂O (l) → 2 NaOH (aq) + H₂ (g)

Elements	Na	H	O
Reactant	0	+1	-2
Product	+1	+1, 0	-2

The oxidizing agent is H₂O.
The reducing agent is Na.

c. FeO (s) + CO (g) → Fe (s) + CO₂ (g)

Elements	Fe	O	C
Reactant	+2	-2	+2
Product	0	-2	+4

The oxidizing agent is FeO.
The reducing agent is CO.

d. 2 MnO₄⁻ (aq) + 10 Cl⁻ (aq) + 16 H⁺ (aq) → 5 Cl₂ (g) + 2 Mn²⁺ (aq) + 8 H₂O (l)

Elements	Mn	O	Cl	H
Reactant	+7	-2	-1	+1
Product	+2	-2	0	+1

The oxidizing agent is MnO₄⁻.
The reducing agent is Cl⁻.

e. 2 N₂O (g) + 3 O₂ (g) → 4 NO₂ (g)

Elements	N	O	O
Reactant	+1	-2	0
Product	+4	-2	-2

The oxidizing agent is O₂.
The reducing agent is N₂O.

f. $2\ TiCl_4\ (s) + Zn\ (s) \rightarrow 2\ TiCl_3\ (s) + ZnCl_2\ (s)$

Elements	Ti	Cl	Zn
Reactant	+4	-1	0
Product	+3	-1	+2

The oxidizing agent is $TiCl_4$.
The reducing agent is Zn.

☑5.18 $6\ NaOH\ (aq) + 2\ Al\ (s) \rightarrow 3\ H_2\ (g) + 2\ Na_3AlO_3\ (aq) + heat$

The oxidizing agent is NaOH because the oxidation number for the hydrogen changes from +1 to 0. The reducing agent is Al because the oxidation number for the aluminum changes from 0 to +3.

DECOMPOSITION, COMBINATION, AND REPACEMENT REACTIONS (SECTION 5.4-5.6)

☑5.20
a. $K_2CO_3\ (s) \rightarrow K_2O\ (s) + CO_2\ (g)$ nonredox: decomposition
b. $Ca\ (s) + 2\ H_2O\ (l) \rightarrow Ca(OH)_2\ (s) + H_2\ (g)$ redox: single-replacement
c. $BaCl_2\ (aq) + H_2SO_4\ (aq) \rightarrow BaSO_4\ (s) + 2\ HCl\ (aq)$ nonredox: double-replacement
d. $SO_2\ (g) + H_2O\ (l) \rightarrow H_2SO_3\ (aq)$ nonredox: combination
e. $2\ NO\ (g) + O_2\ (g) \rightarrow NO_2\ (g)$ redox: combination
f. $2\ Zn\ (s) + O_2\ (g) \rightarrow 2\ ZnO\ (s)$ redox: combination

5.22 $2\ NaHCO_3\ (s) \xrightarrow{Heat} Na_2CO_3\ (s) + H_2O\ (g) + CO_2\ (g)$

This reaction is a nonredox decomposition reaction.

5.24 $CH_4\ (g) + 2\ O_2\ (g) \rightarrow CO_2\ (g) + 2\ H_2O\ (g)$; This reaction is a redox reaction.

5.26 $Cl_2\ (aq) + H_2O\ (l) \rightarrow HOCl\ (aq) + HCl\ (aq)$; This is a redox reaction.

IONIC EQUATIONS (SECTION 5.7)

5.28
a. $CaCl_2$ $Ca^{2+}, 2\ Cl^-$ d. $LiOH$ Li^+, OH^-
b. $Mg(NO_3)_2$ $Mg^{2+}, 2\ NO_3^-$ e. K_2CrO_4 $2\ K^+, CrO_4^{2-}$
c. $(NH_4)_3PO_4$ $3\ NH_4^+, PO_4^{3-}$ f. $Ca(HCO_3)_2$ $Ca^{2+}, 2\ HCO_3^-$

5.30 ☑
a. $\underline{SO_2}\ (aq) + \underline{H_2O}\ (l) \rightarrow H_2SO_3\ (aq)$

    Total ionic equation: $SO_2\ (aq) + H_2O\ (l) \rightarrow 2\ H^+\ (aq) + SO_3^{2-}\ (aq)$
    Spectator ions: none
    Net ionic equation: $SO_2\ (aq) + H_2O\ (l) \rightarrow 2\ H^+\ (aq) + SO_3^{2-}\ (aq)$

☑ b. $CuSO_4\ (aq) + Zn\ (s) \rightarrow Cu\ (s) + ZnSO_4\ (aq)$

    Total ionic equation: $Cu^{2+}(aq) + SO_4^{2-}(aq) + Zn\ (s) \rightarrow Cu\ (s) + Zn^{2+}(aq) + SO_4^{2-}(aq)$
    Spectator ions: $SO_4^{2-}(aq)$
    Net ionic equation: $Cu^{2+}(aq) + Zn\ (s) \rightarrow Cu\ (s) + Zn^{2+}(aq)$

☑ c. $2\ KBr\ (aq) + 2\ H_2SO_4\ (aq) \rightarrow \underline{Br_2}\ (aq) + \underline{SO_2}\ (aq) + K_2SO_4\ (aq) + 2\ \underline{H_2O}\ (l)$

    Total ionic equation:

$2\ K^+(aq) + 2\ Br^-(aq) + 4\ H^+(aq) + 2\ SO_4^{2-}(aq) \rightarrow Br_2\ (aq) + SO_2\ (aq) + 2\ K^+\ (aq) + SO_4^{2-}\ (aq) + 2\ H_2O\ (l)$

Spectator ions: $2\ K^+$ (aq), SO_4^{2-} (aq)

Net ionic equation: $2\ Br^-$(aq) $+ 4\ H^+$(aq) $+ SO_4^{2-}$(aq) $\rightarrow Br_2$ (aq) $+ SO_2$ (aq) $+ 2\ H_2O$ (l)

d. $AgNO_3$ (aq) $+ NaOH$ (aq) $\rightarrow AgOH$ (s) $+ NaNO_3$ (aq)

Total ionic equation:

Ag^+(aq) $+ NO_3^-$(aq) $+ Na^+$(aq) $+ OH^-$(aq) $\rightarrow AgOH$ (s) $+ Na^+$(aq) $+ NO_3^-$(aq)

Spectator ions: Na^+(aq), NO_3^-(aq)

Net ionic equation: Ag^+(aq) $+ OH^-$(aq) $\rightarrow AgOH$ (s)

e. $BaCO_3$ (s) $+ 2\ HNO_3$ (aq) $\rightarrow Ba(NO_3)_2$ (aq) $+ \underline{CO_2}$ (g) $+ \underline{H_2O}$ (l)

Total ionic equation:

$BaCO_3$ (s) $+ 2\ H^+$(aq) $+ 2\ NO_3^-$(aq) $\rightarrow Ba^{2+}$(aq) $+ 2\ NO_3^-$(aq) $+ CO_2$ (g) $+ H_2O$ (l)

Spectator ions: $2\ NO_3^-$(aq)

Net ionic equation: $BaCO_3$ (s) $+ 2\ H^+$(aq) $\rightarrow Ba^{2+}$(aq) $+ CO_2$ (g) $+ H_2O$ (l)

f. $\underline{N_2O_5}$ (aq) $+ \underline{H_2O}$ (l) $\rightarrow 2\ HNO_3$ (aq)

Total ionic equation: N_2O_5 (aq) $+ H_2O$ (l) $\rightarrow 2\ H^+$(aq) $+ 2\ NO_3^-$(aq)

Spectator ions: none

Net ionic equation: N_2O_5 (aq) $+ H_2O$ (l) $\rightarrow 2\ H^+$(aq) $+ 2\ NO_3^-$(aq)

5.32 a. HI (aq) $+ KOH$ (aq) $\rightarrow KI$ (aq) $+ H_2O$ (l)

Total ionic equation: H^+ (aq) $+ I^-$(aq) $+ K^+$ (aq) $+ OH^-$(aq) $\rightarrow K^+$(aq) $+ I^-$(aq) $+ H_2O$ (l)

Spectator ions: K^+(aq), I^-(aq)

Net ionic equation: H^+ (aq) $+ OH^-$(aq) $\rightarrow H_2O$ (l)

b. H_2SO_4 (aq) $+ 2\ NH_4OH$ (aq) $\rightarrow (NH_4)_2SO_4$ (aq) $+ 2\ H_2O$ (l)

Total ionic equation:

$2\ H^+$(aq) $+ SO_4^{2-}$ (aq) $+ 2\ NH_4^+$ (aq) $+ 2\ OH^-$(aq) $\rightarrow 2\ NH_4^+$ (aq) $+ SO_4^{2-}$ (aq) $+ 2\ H_2O$ (l)

Spectator ions: $2\ NH_4^+$ (aq), SO_4^{2-} (aq)

Net ionic equation: $2\ H^+$ (aq) $+ 2\ OH^-$(aq) $\rightarrow 2\ H_2O$ (l); H^+ (aq) $+ OH^-$(aq) $\rightarrow H_2O$ (l)

c. HCl (aq) $+ LiOH$ (aq) $\rightarrow LiCl$ (aq) $+ H_2O$ (l)

Total ionic equation: H^+(aq) $+ Cl^-$(aq) $+ Li^+$(aq) $+ OH^-$(aq) $\rightarrow Li^+$(aq) $+ Cl^-$(aq) $+ H_2O$ (l)

Spectator ions: Li^+(aq), Cl^-(aq)

Net ionic equation: H^+ (aq) $+ OH^-$(aq) $\rightarrow H_2O$ (l)

The reactants and products are identical for all three net ionic equations.

ENERGY AND REACTIONS (SECTION 5.8)

☑5.34 The emergency hot pack becoming warm is an exothermic process because, as water is mixed with the solid, heat is released.

64   Chapter 5

5.36   By insulating the ice, the heat in the food will not be able to travel into the ice as easily. It would be better if the ice were not wrapped in a thick insulating blanket. Wrapping the blanket around the ice and the food would be a better arrangement for keeping the food cold.

THE MOLE AND CHEMICAL EQUATIONS (SECTION 5.9)

5.38   a. $S\ (s) + O_2\ (g) \rightarrow SO_2\ (g)$

Statement 1: 1 S atom + 1 O_2 molecule → 1 SO_2 molecule
Statement 4: 6.02×10^{23} S atoms + 6.02×10^{23} O_2 molecules → 6.02×10^{23} SO_2 molecules
Statement 5: 1 mole S + 1 mole O_2 → 1 mole SO_2
Statement 6: 32.07 g S + 32.00 g O_2 → 64.07 g SO_2

b. $Sr\ (s) + 2\ H_2O\ (l) \rightarrow Sr(OH)_2\ (s) + H_2\ (g)$

Statement 1: 1 Sr atom + 2 H_2O molecules → 1 $Sr(OH)_2$ formula unit + 1 H_2 molecule
Statement 4: 6.02×10^{23} Sr atoms + 12.04×10^{23} H_2O molecules →
$\qquad 6.02 \times 10^{23}$ $Sr(OH)_2$ formula units + 6.02×10^{23} H_2 molecules
Statement 5: 1 mole Sr + 2 moles H_2O → 1 mole $Sr(OH)_2$ + 1 mole H_2
Statement 6: 87.62 g Sr + 36.03 g H_2O → 121.64 g $Sr(OH)_2$ + 2.02 g H_2

c. $2\ H_2S\ (g) + 3\ O_2\ (g) \rightarrow 2\ H_2O\ (g) + 2\ SO_2\ (g)$

Statement 1: 2 H_2S molecules + 3 O_2 molecules → 2 H_2O molecules + 2 SO_2 molecules
Statement 4: 12.04×10^{23} H_2S molecules + 18.06×10^{23} O_2 molecules →
$\qquad 12.04 \times 10^{23}$ H_2O molecules + 12.04×10^{23} SO_2 molecules
Statement 5: 2 moles H_2S + 3 moles O_2 → 2 moles H_2O + 2 moles SO_2
Statement 6: 68.17 g H_2S + 96.00 g O_2 → 36.03 g H_2O + 128.14 g SO_2

d. $4\ NH_3\ (g) + 5\ O_2\ (g) \rightarrow 4\ NO\ (g) + 6\ H_2O\ (g)$

Statement 1: 4 NH_3 molecules + 5 O_2 molecules → 4 NO molecules + 6 H_2O molecules
Statement 4: 24.08×10^{23} NH_3 molecules + 30.10×10^{23} O_2 molecules →
$\qquad 24.08 \times 10^{23}$ NO molecules + 36.12×10^{23} H_2O molecules
Statement 5: 4 moles NH_3 + 5 moles O_2 → 4 moles NO + 6 moles H_2O
Statement 6: 68.13 g NH_3 + 160.00 g O_2 → 120.04 g NO + 108.09 g H_2O

e. $CaO\ (s) + 3\ C\ (s) \rightarrow CaC_2\ (s) + CO\ (g)$

Statement 1: 1 CaO formula unit + 3 C atoms → 1 CaC_2 formula unit + 1 CO molecule
Statement 4: 6.02×10^{23} CaO formula units + 18.06×10^{23} C atoms →
$\qquad 6.02 \times 10^{23}$ CaC_2 formula units + 6.02×10^{23} CO molecules
Statement 5: 1 mole CaO + 3 moles C → 1 mole CaC_2 + 1 mole CO
Statement 6: 56.08 g CaO + 36.03 g C → 64.10 g CaC_2 + 28.01 g CO

5.40 $2\ SO_2 + O_2 \rightarrow 2\ SO_3\ (g)$

Statement 1: 2 SO_2 molecules + 1 O_2 molecule → 2 SO_3 molecules
Statement 4: 12.04×10^{23} SO_2 molecules + 6.02×10^{23} O_2 molecules →
$\qquad 12.04 \times 10^{23}$ SO_3 molecules

Statement 5: 2 moles SO$_2$ + 1 mole O$_2$ → 2 moles SO$_3$
Statement 6 : 128.14 g SO$_2$ + 32.00 g O$_2$ → 160.14 g SO$_3$

Factors :

$$\frac{2 \text{ SO}_2 \text{ molecules}}{1 \text{ O}_2 \text{ molecule}}; \frac{1 \text{ O}_2 \text{ molecule}}{2 \text{ SO}_2 \text{ molecules}}; \frac{2 \text{ SO}_2 \text{ molecules}}{2 \text{ SO}_3 \text{ molecules}}; \frac{2 \text{ SO}_3 \text{ molecules}}{2 \text{ SO}_2 \text{ molecules}}; \frac{1 \text{ O}_2 \text{ molecule}}{2 \text{ SO}_3 \text{ molecules}};$$

$$\frac{2 \text{ SO}_3 \text{ molecules}}{1 \text{ O}_2 \text{ molecule}}; \frac{12.04 \times 10^{23} \text{ SO}_2 \text{ molecules}}{6.02 \times 10^{23} \text{ O}_2 \text{ molecules}}; \frac{6.02 \times 10^{23} \text{ O}_2 \text{ molecules}}{12.04 \times 10^{23} \text{ SO}_2 \text{ molecules}};$$

$$\frac{12.04 \times 10^{23} \text{ SO}_2 \text{ molecules}}{12.04 \times 10^{23} \text{ SO}_3 \text{ molecules}}; \frac{12.04 \times 10^{23} \text{ SO}_3 \text{ molecules}}{12.04 \times 10^{23} \text{ SO}_2 \text{ molecules}}; \frac{6.02 \times 10^{23} \text{ O}_2 \text{ molecules}}{12.04 \times 10^{23} \text{ SO}_3 \text{ molecules}};$$

$$\frac{12.04 \times 10^{23} \text{ SO}_3 \text{ molecules}}{6.02 \times 10^{23} \text{ O}_2 \text{ molecules}}; \frac{2 \text{ mole SO}_2}{1 \text{ mole O}_2}; \frac{1 \text{ mole O}_2}{2 \text{ mole SO}_2}; \frac{2 \text{ mole SO}_2}{2 \text{ mole SO}_3}; \frac{2 \text{ mole SO}_3}{2 \text{ mole SO}_2}; \frac{1 \text{ mole O}_2}{2 \text{ moles SO}_3};$$

$$\frac{2 \text{ moles SO}_3}{1 \text{ mole O}_2}; \frac{128.14 \text{ g SO}_2}{32.00 \text{ g O}_2}; \frac{32.00 \text{ g O}_2}{128.14 \text{ g SO}_2}; \frac{128.14 \text{ g SO}_2}{160.14 \text{ g SO}_3}; \frac{160.14 \text{ g SO}_3}{128.14 \text{ g SO}_2}; \frac{32.00 \text{ g O}_2}{160.14 \text{ g SO}_3}; \frac{160.14 \text{ g SO}_3}{32.00 \text{ g O}_2}$$

This list does not include all possible factors.

☑5.42 CaCO$_3$ (s) → CaO (s) + CO$_2$ (g)

$$500 \text{ g CaO} \left(\frac{1 \text{ mole CaO}}{56.08 \text{ g CaO}}\right)\left(\frac{1 \text{ mole CaCO}_3}{1 \text{ mole CaO}}\right)\left(\frac{100.09 \text{ g CaCO}_3}{1 \text{ mole CaCO}_3}\right) = 892.3858 \text{ g CaCO}_3$$

$$\approx 900 \text{ g CaCO}_3 \text{ (if 500 g has 1 SF)}$$

5.44 2 Al (s) + 3 Br$_2$ (l) → 2 AlBr$_3$ (s)

$$50.1 \text{ g Al}\left(\frac{1 \text{ mole Al}}{26.98 \text{ g Al}}\right)\left(\frac{3 \text{ moles Br}_2}{2 \text{ moles Al}}\right)\left(\frac{159.80 \text{ g Br}_2}{1 \text{ mole Br}_2}\right) = 445.106375093 \text{ g Br}_2$$

$$\approx 445 \text{ g Br}_2 \text{ with SF}$$

5.46
$$50.1 \text{ g Al}\left(\frac{1 \text{ mole Al}}{26.98 \text{ g Al}}\right)\left(\frac{1 \text{ mole AlBr}_3}{1 \text{ mole Al}}\right)\left(\frac{266.68 \text{ g AlBr}_3}{1 \text{ mole AlBr}_3}\right) = 495.206375093 \text{ g AlBr}_3$$

$$\approx 495 \text{ g AlBr}_3 \text{ with SF}$$
or
$$50.1 \text{ g Al} + 445 \text{ g Br}_2 = 495.1 \text{ g Al Br}_3 \approx 495 \text{ g AlBr}_3 \text{ (SF)}$$

5.48 TiCl$_4$ (s) + 2 Mg (s) → Ti (s) + 2 MgCl$_2$ (s)

$$1.00 \text{ kg Ti}\left(\frac{1000 \text{ g Ti}}{1 \text{ kg Ti}}\right)\left(\frac{1 \text{ mole Ti}}{47.87 \text{ g Ti}}\right)\left(\frac{2 \text{ moles Mg}}{1 \text{ mole Ti}}\right)\left(\frac{24.31 \text{ g Mg}}{1 \text{ mole Mg}}\right) = 1015.66743263 \text{ g Mg}$$

$$\approx 1.02 \times 10^3 \text{ g Mg with SF}$$

5.50 C$_6$H$_{12}$O$_2$ (aq) + 8 O$_2$ (aq) → 6 CO$_2$ (aq) + 6 H$_2$O (l)

$$1.00 \text{ mol caproic acid}\left(\frac{8 \text{ moles O}_2}{1 \text{ mol caproic acid}}\right)\left(\frac{32.00 \text{ g O}_2}{1 \text{ mole O}_2}\right) = 256 \text{ g O}_2$$

THE LIMITING REACTANT (SECTION 5.10)

☑5.52 a. N$_2$ (g) + 2 O$_2$ (g) → 2 NO$_2$ (g)
O$_2$ is the limiting reactant as shown by the calculations on the next page:

66   Chapter 5

$$1.25 \text{ moles N}_2 \div 1 \text{ (N}_2 \text{ coefficient in the balanced equation)} = 1.25$$

$$50.0 \text{ g O}_2 \left(\frac{1 \text{ mole O}_2}{32.00 \text{ g O}_2}\right) = 1.5625 \text{ moles O}_2 \div 2 \text{ (O}_2 \text{ coeff. balanced eqn)} = 0.78125$$

b. Only 71.9 g NO₂ can be produced because O₂ is the limiting reactant. Enough N₂ is present to make 115 g of NO₂, but not enough oxygen is present to react with the excess amount of N₂.

$$50.0 \text{ g O}_2 \left(\frac{1 \text{ mole O}_2}{32.00 \text{ g O}_2}\right)\left(\frac{2 \text{ moles NO}_2}{2 \text{ moles O}_2}\right)\left(\frac{46.01 \text{ g NO}_2}{1 \text{ mole NO}_2}\right) = 71.9 \text{ g NO}_2$$

$$1.25 \text{ moles N}_2 \left(\frac{2 \text{ moles NO}_2}{1 \text{ moles N}_2}\right)\left(\frac{46.01 \text{ g NO}_2}{1 \text{ mole NO}_2}\right) = 115 \text{ g NO}_2$$

5.54   NH₃ (aq) + CO₂ (aq) + H₂O (l) → NH₄HCO₃ (aq)
144 g NH₄HCO₃ is the maximum grams that can be produced by these reactants. CO₂ is the limiting reactant and the NH₃ and H₂O are both excess reactants.

$$50.0 \text{ g NH}_3 \left(\frac{1 \text{ mole NH}_3}{17.03 \text{ g NH}_3}\right)\left(\frac{1 \text{ mole NH}_4\text{HCO}_3}{1 \text{ mole NH}_3}\right)\left(\frac{79.06 \text{ g NH}_4\text{HCO}_3}{1 \text{ mole NH}_4\text{HCO}_3}\right) = 232 \text{ g NH}_4\text{HCO}_3$$

$$80.0 \text{ g CO}_2 \left(\frac{1 \text{ mole CO}_2}{44.01 \text{ g CO}_2}\right)\left(\frac{1 \text{ mole NH}_4\text{HCO}_3}{1 \text{ mole CO}_2}\right)\left(\frac{79.06 \text{ g NH}_4\text{HCO}_3}{1 \text{ mole NH}_4\text{HCO}_3}\right) = 144 \text{ g NH}_4\text{HCO}_3$$

$$2.00 \text{ moles H}_2\text{O} \left(\frac{1 \text{ mole NH}_4\text{HCO}_3}{1 \text{ moles H}_2\text{O}}\right)\left(\frac{79.06 \text{ g NH}_4\text{HCO}_3}{1 \text{ mole NH}_4\text{HCO}_3}\right) = 158 \text{ g NH}_4\text{HCO}_3$$

REACTION YIELDS (SECTION 5.11)

☑5.56 $\dfrac{12.18 \text{ g}}{15.93 \text{ g}} \times 100 = 76.46\%$ yield

5.58 $\dfrac{9.04 \text{ g}}{7.59 \text{ g} + 4.88 \text{ g}} \times 100 = 72.5\%$ yield

5.60   2 HgO (s) → 2 Hg (l) + O₂ (g)

$$\frac{5.95 \text{ g}}{7.22 \text{ g HgO}\left(\dfrac{1 \text{ mole HgO}}{216.59 \text{ g HgO}}\right)\left(\dfrac{1 \text{ mole Hg}}{1 \text{ mole HgO}}\right)\left(\dfrac{200.59 \text{ g Hg}}{1 \text{ mole Hg}}\right)} \times 100 = 88.9\% \text{ yield}$$

ADDITIONAL ACTIVITIES

Section 5.1 Review:
(1) Punctuate the following chemical reactions.
(2) Add the appropriate symbols for the states of matter.
(3) Do the following chemical reactions violate the law of conservation of matter? If so, correct the mistakes.

Reaction A: Solid white phosphorus reacts with oxygen gas to produce solid tetraphophorus decaoxide.

P₄ ( )   O₂ ( )   P₄O₁₀ ( )

Reaction B: Aqueous sodium hydroxide neutralizes aqueous sulfuric acid and forms aqueous sodium sulfate and water.

NaOH ( )   H$_2$SO$_4$ ( )   Na$_2$SO$_4$ ( )   H$_2$O ( )

Section 5.2 Review:
Complete the diagram by adding the reaction types (combination, decomposition, single replacement, and double replacement) to the appropriate locations. The circle on the left represents reaction types that can be classified only as nonredox reactions. The circle on the right represents reactions types that can be classified only as redox reactions. The area where they overlap represents reaction types that can be redox or nonredox reactions. The area outside of the circles represents reaction types that cannot be classified as redox or nonredox reactions.

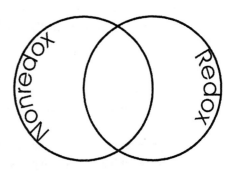

Section 5.3 Review:
(1) Calculate the oxidation numbers for all of the elements in the chemical formulas in the reactions from the Section 5.1 Review.
(2) Which of the reactions in Section 5.1 Review are redox reactions?
(3) Which element(s) from (2) were oxidized? Which element(s) from (2) were reduced?
(4) Identify the oxidizing agent and the reducing agent in the redox reaction(s) from (2).

Section 5.4 Review:
(1) What does it mean for something to "decompose"?
(2) Is the label "decomposition" an appropriate description for this type of reaction?
(3) Are the reactions in the Section 5.1 Review decomposition reactions?

Section 5.5 Review:
(1) What does it mean to "combine"?
(2) Is the label "combination" an appropriate description for this type of reaction?
(3) Do the terms "addition reaction" and "synthesis reaction" also have a descriptive meaning for this type of reaction?
(4) Are the reactions in the Section 5.1 Review combination reactions?

Section 5.6 Review:
(1) What does it mean to "replace"?
(2) What does it mean to "substitute"?
(3) Do the terms "single-replacement reaction" and "substitution reaction" describe the process that occurs during these reactions?
(4) Are the reactions in the Section 5.1 Review single-replacement reactions?
(5) What does "metathesis" mean?
(6) Do the terms "double-replacement reaction" and "metathesis reaction" describe the process that occurs during these reactions?
(7) Are the reactions in the Section 5.1 Review double-replacement reactions?

Section 5.7 Review:
(1) Write the total ionic equation for Reaction B from the Section 5.1 Review.
(2) What are the overall charges for the reactant and product sides of the reaction?
(3) Does the law of conservation of matter hold in the total ionic equation?
(4) Identify the spectator ions in the total ionic equation.
(5) Write the net ionic equation for Reaction B.

68   Chapter 5

(6) Can the net ionic equation for Reaction B be simplified? If so, simplify the equation.
(7) What are the overall charges for the reactant and product sides of the reaction?
(8) Does the law of conservation of matter hold in the net ionic equation?

Section 5.8 Review:
(1) What do the prefixes *exo-* and *endo-* mean?
(2) What does the root word *therm* mean?
(3) What common measuring device (found in a laboratory) uses the root word *therm*?
(4) Would the word energy be written as a reactant or a product in an endothermic reaction?

Section 5.9 Review:
(1) Draw a reaction map similar to Figure 5.11 to convert between grams of reactant and molecules of product.
(2) Use your reaction map from (1) to determine how many molecules of tetraphosphorus decaoxide can be formed if 5.9 g of white phosphorus react completely with oxygen as shown in Reaction A from the Section 5.1 Review.
(3) Use your reaction map from (1) to determine how many molecules of tetraphosphous decaoxide can be formed if 5.9 g of oxygen react completely with white phosphorus as shown in Reaction A from the Section 5.1 Review.

Section 5.10 Review:
(1) If 5.9 g of white phosphorus react with 5.9 g of oxygen, how many molecules of tetraphosphorus decaoxide can be formed?
(2) What are the identities of the limiting reactant and the excess reactant in (1)?

Section 5.11 Review:
(1) If only 10.0 g of P_4O_{10} are formed from the reaction described in (1) of the Section 5.10 Review, what is the percent yield?
(2) Other than poor lab technique, what could account for this apparent violation of the law of conservation of matter?

Tying It All Together with a Laboratory Application:
The "copper cycle" is a series of chemical reactions that begins with copper metal, forms several intermediate compounds, and finally returns copper metal.

In a fume hood, 0.48 g of copper metal reacts with 5.0×10^{-2} moles of nitric acid to form copper (II) nitrate, water, and nitrogen dioxide. The equation is:

(1) _____ Cu (s) + _____ HNO_3 (aq) → _____ $Cu(NO_3)_2$ (aq) + _____ H_2O (l) + _____ NO_2 (g)

The (2) _____ (percent or theoretical) yield of brown nitrogen dioxide gas is (3) _____ g. (4) _____ is the excess reactant. This is a (5) _____ reaction. Copper is the (6) _____ agent. (7) _____ is reduced.

Once the copper has completely reacted, a blue solution remains. Sodium hydroxide is added to the solution. The following reaction occurs:

(8) _____ $Cu(NO_3)_2$ (aq) + _____ NaOH (aq) → + _____ $Cu(OH)_2$ (s) + _____ $NaNO_3$ (aq)

This reaction can be classified both (9) _____ and _____. Write the (10) total ionic equation and (11) net ionic equation for this reaction. (12) Identify the spectator ions in the equation.

Chemical Reactions 69

The sodium nitrate solution is decanted from the copper (II) hydroxide solid which is then transferred to an evaporating dish. The copper (II) hydroxide is gently heated above a beaker of boiling water and the following reaction occurs:

(13) _____ Cu(OH)$_2$ (s) → _____ CuO (s) + _____ H$_2$O (l)

This reaction can be classified as both (14) _____ and _____. This reaction occurs because the copper hydroxide is heated. This is an (15) _____ (endothermic or exothermic) reaction.

Sulfuric acid (H$_2$SO$_4$) is then added to the solid copper (II) oxide and aqueous copper (II) sulfate and liquid water are formed. The reactants for this reaction are (16) _____ and the products for this reaction are (17) _____. (18) Write the balanced molecular equation for this reaction. This reaction can be classified as both (19) _____ and _____.

A coil of zinc wire is added to the blue copper sulfate solution. The following reaction occurs:

(20) _____ CuSO$_4$ (aq) + _____ Zn (s) → _____ Cu (s) + _____ ZnSO$_4$ (aq)

This reaction can be classified as both (21) _____ and _____. (22) Which type of reaction was not represented in the copper cycle?

The mass of copper that theoretically could have been recovered from the copper cycle is (23) _____ because the copper compounds were always the (24) _____ reactants in the reactions. The actual amount of copper recovered was 0.42 g. The percent yield for this reaction cycle is (25) _____.
(26) Why was the percent yield not 100%?

SOLUTIONS FOR THE ADDITIONAL ACTIVITIES

Section 5.1 Review:
Reaction A: P$_4$ (s) + 5 O$_2$ (g) → P$_4$O$_{10}$ (s); the coefficient of 5 before the O$_2$ balances the reaction
Reaction B: 2 NaOH (aq) + H$_2$SO$_4$ (aq) → Na$_2$SO$_4$ (aq) + 2 H$_2$O (l); the coefficients of 2 before the NaOH and the H$_2$O balance the reaction

Section 5.2 Review:

Section 5.3 Review:
(1) Reaction A: P$_4$ (s) + 5 O$_2$ (g) → P$_4$O$_{10}$ (s);
         0   0   +5 -2

Reaction B: 2 Na O H (aq) + H$_2$ S O$_4$ (aq) →
              +1 -2 +1   +1 +6 -2

                        Na$_2$ S O$_4$ (aq) + 2 H$_2$ O (l)
                         +1 +6 -2   +1 -2

(2) Reaction A is a redox reaction. Reaction B is a nonredox reaction.
(3) Phosphorus is oxidized. Oxygen is reduced.
(4) Oxygen is the oxidizing agent. Phosphorus is the reducing agent.

Section 5.4 Review:
(1) When something decomposes, it breaks into smaller pieces. (2) During a decomposition reaction, a single reactant breaks apart into two or more products which are smaller pieces than the original reactant. This label is appropriate. (3) Neither of the reactions from the Section 5.1 Review is a decomposition reaction.

Section 5.5 Review:
(1) When two or more materials are joined or brought into a union, they are said to combine.
(2) Combination reactions occur when two or more reactants join to form one product. The label is appropriate. (3) The term addition reaction implies that two or more reactants are joined together. The term synthesis reaction also implies elements are put together to form a whole (the product). Both of

these terms have descriptive meaning for this type of reaction. (4) Reaction A from the Section 5.1 Review is a combination reaction.

Section 5.6 Review:
(1) Replace means to put something in the position of a previous object. (2) Substitute means to let something use the place of another. (3) Both single-replacement reaction and substitution reaction imply that one element will take the place of another element during the reaction. These names reflect the process that occurs. (4) Neither of the reactions in the Section 5.1 Review is a single-replacement reaction. (5) Metathesis means transposition or interchange. (6) Double-replacement reaction and metathesis reaction both imply that two elements trade their positions as the reaction progresses. These names reflect the process that occurs. (7) Reaction B from the Section 5.1 Review is a double-replacement reaction.

Section 5.7 Review:
(1) $2\ Na^+(aq) + 2\ OH^-(aq) + 2\ H^+(aq) + SO_4^{2-}(aq) \rightarrow 2\ Na^+(aq) + SO_4^{2-}(aq) + 2\ H_2O\ (l)$
(2) Reactant side = 2(+1) + 2(-1) + 2(+1) + 1(-2) = 0; Product side = 2(+1) + 1(-2) + 2(0) = 0
(3) The law of conservation of matter holds for the total ionic equation.
(Reactant side: 2 Na, 4 H, 6 O, and 1 S; Product side: 2 Na, 1 S, 6 O, and 4 H)
(4) Spectator ions = $2\ Na^+(aq)$ and $SO_4^{2-}(aq)$
(5) Net ionic equation = $2\ OH^-(aq) + 2\ H^+(aq) \rightarrow 2\ H_2O\ (l)$
(6) Simplified net ionic equation = $OH^-(aq) + H^+(aq) \rightarrow H_2O\ (l)$
(7) Reactant side = (-1) + (+1) = 0, Product side = 0
(8) The law of conservation of matter holds for the net ionic equation. (Reactant side: 1 O and 2 H, Product side: 2 H and 1 O)

Section 5.8 Review:
(1) *exo-* means outside, outer, outer part; *endo-* means in, within, inner; (2) *therm-* means heat; (3) thermometer; (4) Energy is a reactant in an endothermic reaction.

Section 5.9 Review:

Section 5.10 Review:
(1) 2.2×10^{22} molecules P_4O_{10} can be formed
(2) limiting reactant = oxygen, excess reactant = white phosphorus

Section 5.11 Review:
(1) $\dfrac{10.0\ g}{5.9\ g\ O_2 \left(\dfrac{1\ mole\ O_2}{32.00\ g\ O_2}\right)\left(\dfrac{1\ mole\ P_4O_{10}}{5\ mole\ O_2}\right)\left(\dfrac{283.90\ g\ P_4O_{10}}{1\ mole\ P_4O_{10}}\right)} \times 100 = 96\%$ yield

(2) Side reactions may have occurred. The mass of the products from the side reactions added to the mass of the P_4O_{10} should equal the theoretical mass.

Chemical Reactions 71

Tying It All Together with a Laboratory Application:
(1) 1, 4, 1, 2, 2 (3) 0.70 g (5) redox (7) HNO₃ (9) nonredox, double replacement
(2) theoretical (4) HNO₃ (6) reducing (8) 1, 2, 1, 2

(10) Cu^{2+} (aq) + 2 NO_3^- (aq) + 2 Na^+ (aq) + 2 OH^- (aq) → $Cu(OH)_2$ (s) + 2 Na^+ (aq) + 2 NO_3^- (aq)

(11) Cu^{2+} (aq) + 2 OH^- (aq) → $Cu(OH)_2$ (s)

(12) 2 Na^+ (aq), 2 NO_3^- (aq)
(13) 1, 1, 1
(14) nonredox, decomposition
(15) endothermic
(16) sulfuric acid, copper (II) oxide
(17) copper (II) sulfate, water
(18) H_2SO_4 (aq) + CuO (s) → $CuSO_4$ (aq) + H_2O (l)

(19) nonredox, double replacement
(20) 1, 1, 1, 1
(21) redox, single replacement
(22) combination
(23) 0.48 g
(24) limiting
(25) 88%

(26) This cycle of reactions requires excellent lab technique. Some of the copper may not have completely reacted in the first reaction. Some of the copper hydroxide may not have been transferred to the evaporating dish. Some of the copper may not have been scraped off the zinc wire. Also, side reactions may have occurred.

SELF-TEST QUESTIONS
Multiple Choice

1. Which of the following is a reactant in the reaction: $2H^+ + CaCO_3 \rightarrow H_2O + Ca^{2+} + CO_2$
   a. H_2O b. H^+ c. Ca^{2+} d. CO_2

2. What is the coefficient to the left of H_2 when the following equation is balanced?
   $$Na + H_2O \rightarrow NaOH + H_2$$
   a. 1 b. 2 c. 3 d. 4

3. A decomposition reaction can also be classified as
   a. a combination reaction
   b. a single replacement reaction
   c. a double replacement reaction
   d. a redox or nonredox reaction

4. The oxidation number of a monatomic ion is always
   a. +1 b. 0 c. -1 d. equal to the charge on the ion

5. In a redox reaction, the reducing agent
   a. is reduced
   b. is oxidized
   c. gains electrons
   d. contains an element whose oxidation number decreases

6. Identify the spectator ion in the following reaction: $Cl_2 + 2 K^+ + 2 Br^- \rightarrow 2 K^+ + 2 Cl^- + Br_2$
   a. Br^- b. Cl_2 c. K^+ d. Cl^-

7. Which of the following statements is consistent with the balanced equation: $2 SO_2 + O_2 \rightarrow 2 SO_3$
   a. one mol SO_2 reacts with one mol O_2
   b. two mol SO_2 produces two mol SO_3
   c. 64.07 g SO_2 reacts with 32.0 g O_2
   d. 32.0 g O_2 produces 80.07 g SO_3

8. According to the reaction $N_2 + 3 H_2 \rightarrow 2 NH_3$, how many grams of H_2 are needed to produce 4.0 moles of NH_3?
   a. 3.0 b. 6.0 c. 9.0 d. 12.0

72 Chapter 5

9. For the reaction 2 H₂ + O₂ → 2 H₂O, how many moles of H₂O could be obtained by reacting 64.0 grams of O₂ and 8.0 grams of H₂?
    a. 2.0 b. 4.0 c. 8.0 d. 72

10. Nitrogen and hydrogen react as follows to form ammonia: N₂ + 3 H₂ → 2 NH₃. In a reaction mixture consisting of 1.50 g H₂ and 6.00 g N₂, it is true that
    a. H₂ is the limiting reactant
    b. N₂ is the limiting reactant
    c. N₂ is present in excess
    d. an exact reacting ratio is present

True-False
11. Oxidation numbers never change in combination reactions.
12. The sum of the oxidation numbers of elements in a compound always equals zero.
13. The loss of hydrogen corresponds to an oxidation process.
14. The number of reactants is greater than the number of products in all decomposition reactions.
15. A reaction that releases heat is classified as endothermic.
16. A balanced equation represents a statement of the law of conservation of matter.
17. The percentage yield of a reaction cannot exceed 100%, if the measured product is pure.

Matching
Match the oxidation number of the underlined element to the correct value given as a response.
18. H₂$\underline{P}$O₄   a. 0
19. $\underline{Br}$₂   b. +1
20. $\underline{C}$O₂   c. +2
21. $\underline{Cs}$₂O   d. +3
22. $\underline{Cr}$₂O₃   e. +4
23. $\underline{S}$O₃²⁻   f. none of the above
24. $\underline{Ba}$²⁺
25. $\underline{Na}$

Match each of the reactions below with the correct category from the responses.
26. 3 Fe + 2 O₂ → Fe₃O₄   a. decomposition
27. CuO + H₂ → Cu + H₂O   b. single replacement
28. CuCO₃ → CuO + CO₂   c. double replacement
29. KOH + HBr → H₂O + KBr   d. combination
30. 2 Ag₂O → 4 Ag + O₂

Match each of the reactions below with the correct category from the responses. For redox reactions, also identify the reducing agent.
31. 3 Fe + 2 O₂ → Fe₃O₄   a. redox reaction
32. CuO + H₂ → Cu + H₂O   b. nonredox reaction
33. CuCO₃ → CuO + CO₂
34. KOH + HBr → H₂O + KBr
35. 2 Ag₂O → 4 Ag + O₂

Chapter 6: The States of Matter

CHAPTER OUTLINE
6.1 Observed Properties of Matter
6.2 The Kinetic Molecular Theory of Matter
6.3 The Solid State
6.4 The Liquid State
6.5 The Gaseous State
6.6 The Gas Laws
6.7 Pressure, Temperature, and Volume Relationships
6.8 The Ideal Gas Law
6.9 Dalton's Law
6.10 Graham's Law
6.11 Changes in State
6.12 Evaporation and Vapor Pressure
6.13 Boiling and the Boiling Point
6.14 Sublimation and Melting
6.15 Energy and the States of Matter

LEARNING OBJECTIVES/ASSESSMENT
When you have completed your study of this chapter, you should be able to:
1. Do calculations based on the property of density. (Section 6.1; Exercise 6.2)
2. Use the kinetic molecular theory to explain the properties of matter in different states. (Sections 6.3, 6.4, and 6.5; Exercises 6.12 and 6.16)
3. Do calculations based on Boyle's law, Charles's law, the combined gas law, Dalton's law, and Graham's law. (Sections 6.6, 6.7, 6.9, and 6.10; Exercises 6.20, 6.28, 6.58, and 6.60)
4. Do calculations based on the ideal gas law. (Section 6.8; Exercise 6.48)
5. Do calculations based on the energy changes that accompany the heating, cooling, or changing of state of a substance. (Section 6.11, 6.12, 6.13, and 6.15; Exercise 6.64, 6.68, 6.70, and 6.74)

SOLUTIONS FOR THE END OF CHAPTER EXERCISES
OBSERVED PROPERTIES OF MATTER (SECTION 6.1)

☑6.2 a. Sea water (d = 1.03 g/mL) $85.0 \text{ g} \left(\dfrac{1 \text{ mL}}{1.03 \text{ g}} \right) = 82.5 \text{ mL}$

b. Methyl alcohol (d = 0.792 g/mL) $85.0 \text{ g} \left(\dfrac{1 \text{ mL}}{0.792 \text{ g}} \right) = 107 \text{ mL}$

c. Concentrated sulfuric acid (d = 1.84 g/mL) $85.0 \text{ g} \left(\dfrac{1 \text{ mL}}{1.84 \text{ g}} \right) = 46.2 \text{ mL}$

6.4 The volume of water increases by 2 mL on heating as shown by the calculation below:
$500 \text{ mL} \left(\dfrac{1.00 \text{ g}}{1 \text{ mL}} \right) = 500 \text{ g}; \; 500 \text{ g} \left(\dfrac{1 \text{ mL}}{0.996 \text{ g}} \right) = 502 \text{ mL}; \; 502 \text{ mL} - 500 \text{ mL} = 2 \text{ mL}$

6.6 a. The density of the gas will decrease as the gas is heated. The mass of the gas will not change, but the volume of the gas will increase. Volume and density have an inverse relationship; therefore, the density will decrease.

b.
in 1 step: $\text{density}_2 = \dfrac{\text{volume}_1 \times \text{density}_1}{\text{volume}_2} \Rightarrow \dfrac{1.50 \text{ L} \left(\dfrac{1.98 \text{ g}}{1 \text{ L}} \right)}{1.78 \text{ L}} = \dfrac{2.97 \text{ g}}{1.78 \text{ L}} = 1.67 \tfrac{\text{g}}{\text{L}}$

or in 2 steps: $\text{mass} = \text{volume}_1 \times \text{density}_1 \Rightarrow 1.50 \text{ L} \left(\dfrac{1.98 \text{ g}}{1 \text{ L}} \right) = 2.97 \text{ g}$

$\text{density}_2 = \dfrac{\text{mass}}{\text{volume}_2} \Rightarrow \dfrac{2.97 \text{ g}}{1.78 \text{ L}} = 1.67 \tfrac{\text{g}}{\text{L}}$ The mass of the sample is the same before and after heating.

74   Chapter 6

THE KINETIC MOLECULAR THEORY OF MATTER (SECTION 6.2)

6.8   As the ball is moving upward, it has kinetic energy that is transferred into potential energy as the ball slows and reaches the highest point in its path. At its highest point, the ball stops and all of the energy is potential energy. As the ball falls from its highest point, the potential energy is transferred into kinetic energy. When the ball is caught, the kinetic energy is transferred into the hand of the catcher.

6.10 The kinetic energy of helium and methane molecules is the same as shown in the calculations below. The kinetic energy of all gases is the same at a given temperature.

$$KE = \tfrac{1}{2}mv^2$$

$$KE_{He} = \tfrac{1}{2}(4.00 \text{ u})(1.26 \times 10^5 \tfrac{cm}{s})^2 = 3.1752 \times 10^{10} \tfrac{u \cdot cm^2}{s^2} = 3.18 \times 10^{10} \tfrac{u \cdot cm^2}{s^2}$$

$$KE_{CH_4} = \tfrac{1}{2}(16.04 \text{ u})(6.30 \times 10^4 \tfrac{cm}{s})^2 = 3.183138 \times 10^{10} \tfrac{u \cdot cm^2}{s^2} = 3.18 \times 10^{10} \tfrac{u \cdot cm^2}{s^2}$$

THE SOLID, LIQUID, AND GASEOUS STATES (SECTION 6.3-6.5)

☑6.12   a. The molecules of a liquid possess kinetic and potential energy. The kinetic energy is not great enough to overcome the attractive forces between the molecules; therefore, the molecules are able to flow together into the shape of the container, but the volume of the liquid remains constant.
   b. Solid and liquids are composed of molecules with considerable attractive forces between the molecules. These attractive forces bring the molecules close together and cause them to be difficult to further compress.
   c. Gases are composed of molecules with high kinetic energy and low potential energy. The molecules are small compared to the amount of space occupied by the gas. The molecules strike each other and the walls of the container; however, all of the collisions are elastic and no net energy is lost from the system. Because the molecules are in constant motion, the same average number of gas molecules will strike the walls of the container at any given time and the elastic collisions result in uniform pressure on the walls of its container.

6.14 The particles in a solid have the lowest kinetic energy and the highest potential energy.
The particles in a liquid have higher kinetic energy and lower potential energy than a solid, as well as lower kinetic energy and higher potential energy than a gas.
The particles in a gas have the highest kinetic energy and the lowest potential energy.

☑6.16   a. Temperature changes influence the volume of this state substantially.   gas
   b. In this state, constituent particles are less free to move about than in other states.   solid
   c. Pressure changes influence the volume of this state more than that of the other two states.   gas
   d. This state is characterized by an indefinite shape and a low density.   gas

THE GAS LAWS (SECTION 6.6)

6.18   a. atm

$$27.7 \text{ in. Hg}\left(\frac{1 \text{ atm}}{29.9 \text{ in. Hg}}\right) = 0.926 \text{ atm}$$

   b. torr

$$27.7 \text{ in. Hg}\left(\frac{1 \text{ atm}}{29.9 \text{ in. Hg}}\right)\left(\frac{760 \text{ torr}}{1 \text{ atm}}\right) = 704 \text{ torr}$$

The States of Matter 75

c. psi $\quad 27.7 \text{ in. Hg}\left(\dfrac{1 \text{ atm}}{29.9 \text{ in. Hg}}\right)\left(\dfrac{14.7 \text{ psi}}{1 \text{ atm}}\right) = 13.6 \text{ psi}$

d. bars $\quad 27.7 \text{ in. Hg}\left(\dfrac{1.01 \text{ bars}}{29.9 \text{ in. Hg}}\right) = 0.936 \text{ bars}$

☑6.20 a. atm $\quad 190 \text{ psi}\left(\dfrac{1 \text{ atm}}{14.7 \text{ psi}}\right) = 12.9 \text{ atm}$ (if 0 in 190 is significant)

$\approx 13 \text{ atm}$ (if 0 in 190 is not significant)

b. bars $\quad 190 \text{ psi}\left(\dfrac{1 \text{ atm}}{14.7 \text{ psi}}\right)\left(\dfrac{1.01 \text{ bars}}{1 \text{ atm}}\right) = 13.1 \text{ bars}$ (if 0 in 190 is significant)

$\approx 13 \text{ bars}$ (if 0 in 190 is not significant)

c. mm Hg $\quad 190 \text{ psi}\left(\dfrac{1 \text{ atm}}{14.7 \text{ psi}}\right)\left(\dfrac{760 \text{ mm Hg}}{1 \text{ atm}}\right) = 9.82 \times 10^3 \text{ mm Hg}$ (if 0 in 190 is significant)

$\approx 9.8 \times 10^3 \text{ mm Hg}$ (if 0 in 190 is not significant)

d. in. Hg $\quad 190 \text{ psi}\left(\dfrac{1 \text{ atm}}{14.7 \text{ psi}}\right)\left(\dfrac{29.9 \text{ in. Hg}}{1 \text{ atm}}\right) = 386 \text{ mm Hg}$ (if 0 in 190 is significant)

$\approx 3.9 \times 10^2 \text{ mm Hg}$ (if 0 in 190 is not significant)

6.22 a. The melting point of lithium metal, 180.5°C to kelvins. $\quad 180.5°C + 273 = 454 \text{ K}$
b. The freezing point of liquid helium, 0.9 kelvins to degrees Celsius. $\quad 0.9 \text{ K} - 273 = -272°C$
c. The boiling point of liquid hydrogen, -252.8°C to kelvins. $\quad -252.8°C + 273 = 20 \text{ K}$

PRESSURE, TEMPERATURE, AND VOLUME RELATIONSHIPS (SECTION 6.7)

6.24 Gas A: $\dfrac{(1.50 \text{ atm})(2.00 \text{ L})}{300 \text{ K}} = \dfrac{(P_f)(3.00 \text{ L})}{450 \text{ K}}$; Gas B: $\dfrac{(2.35 \text{ atm})(1.97 \text{ L})}{293 \text{ K}} = \dfrac{(1.09 \text{ atm})(V_f)}{310 \text{ K}}$;

$P_f = 1.50 \text{ atm}$ $\quad V_f = 4.49 \text{ L}$

Gas C: $\dfrac{(9.86 \text{ atm})(11.7 \text{ L})}{500 \text{ K}} = \dfrac{(5.14 \text{ atm})(9.90 \text{ L})}{T_f}$

$T_f = 221 \text{ K}$

6.26 $\dfrac{(650 \text{ torr})(250 \text{ mL})}{35.0°C + 273} = \dfrac{(760 \text{ torr})(V_f)}{0.0°C + 273} \Rightarrow V_f = 190 \text{ mL}$

☑6.28 $\dfrac{(1.25 \text{ atm})(3.00 \text{ L})}{0.00°C + 273} = \dfrac{(P_f)(1.00 \text{ L})}{25.0°C + 273} \Rightarrow P_f = 4.09 \text{ atm}$

6.30 $(1.00 \text{ atm})\left(\dfrac{14.7 \text{ psi}}{1 \text{ atm}}\right)(V_i) = (30.0 \text{ psi})(12.9 \text{ L}) \Rightarrow V_i = 26.3 \text{ L}$

6.32 $(650 \text{ torr})(800 \text{ mL}) = (760 \text{ torr})(V_f) \Rightarrow V_f = 684 \text{ mL}$

76   Chapter 6

6.34 $\dfrac{4.2 \text{ L}}{25°C+273} = \dfrac{V_f}{85°C+273} \Rightarrow V_f = 5.0 \text{ L}$

6.36 $\dfrac{V_i}{150°C+273} = \dfrac{2.0 \text{ L}}{40°C+273} \Rightarrow V_i = 2.7 \text{ L}$

6.38 $(1.00 \text{ atm})(2000 \text{ L}) = (P_f)(25.0 \text{ L}) \Rightarrow P_f = 80.0 \text{ atm}$

6.40 $\dfrac{(0.98 \text{ atm})(8000 \text{ ft}^3)}{23°C+273} = \dfrac{(400 \text{ torr})\left(\frac{1 \text{ atm}}{760 \text{ torr}}\right)(V_f)}{5.3°C+273} \Rightarrow V_f = 1.4 \times 10^4 \text{ ft}^3$

6.42 $(2.50 \text{ atm})(200 \text{ mL}) = (60.0 \text{ kPa})\left(\dfrac{1 \text{ atm}}{101 \text{ kPa}}\right)(V_f) \Rightarrow V_f = 842 \text{ mL}$

6.44 $(760 \text{ torr})(2.00 \text{ L}) = (4.00 \text{ atm})\left(\dfrac{760 \text{ torr}}{1 \text{ atm}}\right)(V_f) \Rightarrow V_f = 0.500 \text{ L}$

$\text{density} = \dfrac{\text{mass}}{\text{volume}} \Rightarrow \dfrac{2.50 \text{ g}}{0.500 \text{ L}} = 5.00 \tfrac{g}{L}$

THE IDEAL GAS LAW (SECTION 6.8)

6.46 a. $PV = nRT \Rightarrow (800 \text{ torr})\left(\tfrac{1 \text{ atm}}{760 \text{ torr}}\right)(500 \text{ mL})\left(\tfrac{1 \text{ L}}{1000 \text{ mL}}\right) = n\left(0.0821 \tfrac{L \cdot atm}{mol \cdot K}\right)((85.5+273) \text{ K})$

$n = \dfrac{(800 \text{ torr})\left(\tfrac{1 \text{ atm}}{760 \text{ torr}}\right)(500 \text{ mL})\left(\tfrac{1 \text{ L}}{1000 \text{ mL}}\right)}{\left(0.0821 \tfrac{L \cdot atm}{mol \cdot K}\right)((85.5+273) \text{ K})} = 0.0179 \text{ moles}$

b. $PV = nRT \Rightarrow P(2.00 \text{ L}) = (0.855 \text{ mol})\left(0.0821 \tfrac{L \cdot atm}{mol \cdot K}\right)((50.0+273) \text{ K})$

$P = \dfrac{(0.855 \text{ mol})\left(0.0821 \tfrac{L \cdot atm}{mol \cdot K}\right)((50.0+273) \text{ K})}{(2.00 \text{ L})} = 11.3 \text{ atm}$

c. $PV = nRT \Rightarrow (3.85 \text{ atm})V = (2.15 \text{ mol})\left(0.0821 \tfrac{L \cdot atm}{mol \cdot K}\right)((30.0+273) \text{ K})$

$V = \dfrac{(2.15 \text{ mol})\left(0.0821 \tfrac{L \cdot atm}{mol \cdot K}\right)((30.0+273) \text{ K})}{(3.85 \text{ atm})} = 13.9 \text{ L}$

☑6.48 $PV = nRT \Rightarrow P(1.25 \text{ L}) = (12.5 \text{ g})\left(\tfrac{1 \text{ mole}}{64.07 \text{ g SO}_2}\right)\left(0.0821 \tfrac{L \cdot atm}{mol \cdot K}\right)((38.0+273) \text{ K})$

$P = \dfrac{(12.5 \text{ g})\left(\tfrac{1 \text{ mole}}{64.07 \text{ g SO}_2}\right)\left(0.0821 \tfrac{L \cdot atm}{mol \cdot K}\right)((38.0+273) \text{ K})}{(1.25 \text{ L})} = 3.99 \text{ atm}$

6.50 $PV = nRT \Rightarrow (400 \text{ psi})\left(\tfrac{1 \text{ atm}}{14.7 \text{ psi}}\right)(1.50 \text{ L}) = n\left(0.0821 \tfrac{L \cdot atm}{mol \cdot K}\right)((30.0+273) \text{ K})$

$n = \dfrac{(400 \text{ psi})\left(\tfrac{1 \text{ atm}}{14.7 \text{ psi}}\right)(1.50 \text{ L})}{\left(0.0821 \tfrac{L \cdot atm}{mol \cdot K}\right)((30.0+273) \text{ K})} = 1.6407715991 \text{ moles}$

$\text{mass} = 1.6407715991 \text{ moles}\left(\dfrac{16.04 \text{ g CH}_4}{1 \text{ mole CH}_4}\right) = 26.3 \text{ g}$

6.52 $PV = nRT \Rightarrow (1 \text{ atm})(85 \text{ L}) = n(0.0821 \frac{\text{L·atm}}{\text{mol·K}})(273 \text{ K})$

$n = \dfrac{(1 \text{ atm})(85 \text{ L})}{(0.0821 \frac{\text{L·atm}}{\text{mol·K}})(273 \text{ K})} = 3.8 \text{ moles}$

6.54 $PV = nRT \Rightarrow (1 \text{ atm})(3.96 \text{ L}) = n(0.0821 \frac{\text{L·atm}}{\text{mol·K}})(273 \text{ K})$

$n = \dfrac{(1 \text{ atm})(3.96 \text{ L})}{(0.0821 \frac{\text{L·atm}}{\text{mol·K}})(273 \text{ K})} = 0.1766808 \text{ moles} \Rightarrow MW = \dfrac{8.12 \text{ g}}{0.1766808 \text{ moles}} = 46.0 \frac{\text{g}}{\text{mol}}$

6.56 The gas is CO_2 (MW = 44.01 g/mole) according to the calculation below:

$PV = nRT$

$(640 \text{ torr})(\frac{1 \text{ atm}}{760 \text{ torr}})(114.0 \text{ mL})(\frac{1 \text{ L}}{1000 \text{ mL}}) = n(0.0821 \frac{\text{L·atm}}{\text{mol·K}})((20 + 273) \text{ K})$

$n = \dfrac{(640 \text{ torr})(\frac{1 \text{ atm}}{760 \text{ torr}})(114.0 \text{ mL})(\frac{1 \text{ L}}{1000 \text{ mL}})}{(0.0821 \frac{\text{L·atm}}{\text{mol·K}})((20+273) \text{ K})} = 0.00399 \text{ moles}$

$MW = \dfrac{0.176 \text{ g}}{0.00399 \text{ moles}} = 44.1 \frac{\text{g}}{\text{mol}}$

DALTON'S LAW (SECTION 6.9)

☑6.58 $P_{N_2} + P_{O_2} + P_{CO_2} = P_T \Rightarrow 810 \text{ torr} + 920 \text{ torr} + P_{CO_2} = 2100 \text{ torr} \Rightarrow P_{CO_2} = 370 \text{ torr}$

GRAHAM'S LAW (SECTION 6.10)

☑6.60 The molecular mass of oxygen is 16 times the molecular mass of hydrogen according to the calculation below:

$\dfrac{\text{effusion rate A}}{\text{effusion rate B}} = \sqrt{\dfrac{\text{molecular mass of B}}{\text{molecular mass of A}}}$

$4 = \sqrt{\dfrac{MM_{O_2}}{MM_{H_2}}} \Rightarrow (4)^2 = \left(\sqrt{\dfrac{MM_{O_2}}{MM_{H_2}}}\right)^2 \Rightarrow 16 = \dfrac{MM_{O_2}}{MM_{H_2}}$

The molecular mass of oxygen according to the periodic table is 32.00 u and the molecular mass of hydrogen according to the periodic table is 2.01 u. The molecular mass of oxygen is approximately 16 times greater than the molecular mass of hydrogen.

6.62 The helium balloon is the one that appeared to be going "flat." The helium gas was able to escape more easily than the nitrogen gas through the porous rubber surface of the balloon because helium has a smaller molecular mass than the nitrogen. Lighter molecules move more rapidly than heavier molecules at the same temperature.

CHANGES IN STATE (SECTION 6.11)

☑6.64
a. Condensation   (g) → (l); remove heat   exothermic
b. Liquefaction   (g) → (l); remove heat   exothermic
c. Boiling   (l) → (g); add heat   endothermic

6.66 A change in state is the process by which a material gains enough energy or loses enough energy to change between two of the three phases of matter (solid, liquid, or gas).

78 Chapter 6

EVAPORATION AND VAPOR PRESSURE (SECTION 6.12)

☑6.68   Methylene chloride is a volatile liquid. When it was sprayed in the mouth, the methylene chloride absorbed the heat from the tissue and evaporated. The tissue became cold and was anesthetized.

BOILING AND THE BOILING POINT (SECTION 6.13)

☑6.70   Water and ethylene glycol have unique boiling points. Water has a boiling point of 100°C, while ethylene glycol has a boiling point that is higher than 100°C. To determine the identity of the two boiling liquids, use a thermometer to measure the boiling point.

6.72   I would throw the potato into the campfire. The temperature of the burning wood will be higher than the temperature of the boiling water. Water boils at a lower temperature at higher elevations because the barometric pressure is lower at higher elevations. The temperature of combusting wood is not affected by elevation.

ENERGY AND THE STATES OF MATTER (SECTION 6.15)

☑6.74   a. 210 g of copper from 40°C to 95°C

$$\text{heat} = (210 \text{ g})\left(0.093 \tfrac{\text{cal}}{\text{g·°C}}\right)(95°\text{C} - 40°\text{C})$$
$$= 1.1 \times 10^3 \text{ cal}$$

b. 150 g of mercury from 120°C to 300°C

$$\text{heat} = (150 \text{ g})\left(0.033 \tfrac{\text{cal}}{\text{g·°C}}\right)(300°\text{C} - 120°\text{C})$$
$$= 8.91 \times 10^2 \text{ cal}$$

c. 2500 g of helium gas from 250°C to 900°C

$$\text{heat} = (2500 \text{ g})\left(1.25 \tfrac{\text{cal}}{\text{g·°C}}\right)(900°\text{C} - 250°\text{C})$$
$$= 2.0 \times 10^6 \text{ cal}$$

6.76   The solar heat storage system requires the materials are able to be melted and will release heat on resolidification. The melting point of potassium sulfate is too high for this to be a quality material for this application.

6.78   Step 1. $500 \text{ g}\left(0.51 \tfrac{\text{cal}}{\text{g·°C}}\right)(0°\text{C} - (-10°\text{C})) = 2550 \text{ cal} = 2.55 \times 10^3 \text{ cal}$

Step 2. $500 \text{ g}\left(80 \tfrac{\text{cal}}{\text{g}}\right) = 40000 \text{ cal} = 40.0 \times 10^3 \text{ cal}$

Step 3. $500 \text{ g}\left(1.00 \tfrac{\text{cal}}{\text{g·°C}}\right)(100°\text{C} - 0°\text{C}) = 50000 \text{ cal} = 50.0 \times 10^3 \text{ cal}$

Step 4. $500 \text{ g}\left(540 \tfrac{\text{cal}}{\text{g}}\right) = 270000 \text{ cal} = 270. \times 10^3 \text{ cal}$

Step 5. $500 \text{ g}\left(0.48 \tfrac{\text{cal}}{\text{g·°C}}\right)(120°\text{C} - 100°\text{C}) = 4800 \text{ cal} = 4.80 \times 10^3 \text{ cal}$

2550 cal + 40000 cal + 50000 cal + 270000 cal + 4800 cal = 367350 cal = 3.67×10^5 cal

ADDITIONAL ACTIVITIES

Section 6.1 Review:
(1) Can the density of 3 g of gold be greater than the density of 150 g of water? Explain.
(2) Compare and contrast the shape and compressibility of gold and water at room temperature.
(3) Compare and contrast the thermal expansion of gold and water.
(4) Describe the density, shape, compressibility, and thermal expansion of helium gas relative to water and gold.

Section 6.2 Review:
(1) Are the ions that occupy lattice sites in a sodium chloride crystal lattice in motion? If so, how does the movement compare with the movement of water molecules in a glass of water and the movement of helium atoms inside a balloon?
(2) Relate the movement within the three states of matter to forms of dancing.
(3) Which phase of matter has the strongest cohesive forces?
(4) Which phase of matter has the strongest disruptive forces?

Section 6.3 Review:
(1) Refer to Figure 6.3 or draw your own representation of the solid state. Make sure that the picture reflects the molecular nature of matter.
(2) Using the picture from (1), determine why solids have a high density, definite shape, small compressibility, and very small thermal expansion.

Section 6.4 Review:
(1) Refer to Figure 6.3 or draw your own representation of the liquid state. Make sure that the picture reflects the molecular nature of matter.
(2) Using the picture from (1), determine why liquids have a high density, indefinite shape, small compressibility, and small thermal expansion.

Section 6.5 Review:
(1) Refer to Figure 6.3 or draw your own representation of the gaseous state. Make sure that the picture reflects the molecular nature of matter.
(2) Using the picture from (1), determine why gases have a low density, indefinite shape, large compressibility, and moderate thermal expansion.

Section 6.6 Review:
(1) If a barometer were made with water instead of mercury, would the glass tube need to be taller or shorter than for a mercury barometer? Mercury is more dense than water.
(2) Identify which unit of the following pairs is larger: atm or torr, torr or mmHg, mmHg or atm.
(3) If the temperature in degrees Celsius doubles from 20°C to 40°C, does the Kelvin temperature also double?
(4) If the pressure in atmospheres doubles from 1 atm to 2 atm, does the pressure in mmHg also double?
(5) If the volume in liters doubles from 0.500 L to 1.00 L, does the volume in milliliters also double?
(6) What are the differences between the conversions in (3), (4), and (5)?

Section 6.7 Review:
(1) If the temperature is held constant and the pressure on a helium balloon is increased, what will happen to the volume of the balloon?
(2) If the temperature is held constant as a valve is opened on tank of nitrogen, what will happen to the pressure of the gas as it leaves the tank and enters the room?
(3) What will happen to the volume of a helium balloon that is filled inside an air conditioned building and then placed in a hot delivery car? Assume the pressure in the building and the car is the same.
(4) Write an equation like 6.8 in the textbook for Boyle's law and for Charles's law.
(5) Do any conversions need to be made as long as the same units are used for both the "before" and "after" sides of the combined gas law? Explain.

Section 6.8 Review:
(1) Rearrange the ideal gas law to isolate each of the variables (P, V, n, T). You will need to write four separate equations, one for each variable.

80 Chapter 6

(2) Could you calculate the molecular weight of a gas if you knew the mass, pressure, volume, and temperature of the sample? Explain using Equations 6.9 and 6.10.
(3) Why would interparticle forces cause gases to deviate from ideal behavior?

Section 6.9 Review:
(1) Rearrange the ideal gas law to isolate the ratio of pressure to number of moles.
(2) What happens to the pressure as the number of moles increases? Assume the volume and temperature of the container remain constant.
(3) If the all of the gases in a mixture are assumed to be ideal gases, what will have more effect on the pressure: the identities of the gases or the total number of moles of gas?

Section 6.10 Review:
(1) Label the following pictures as examples of effusion or diffusion.

 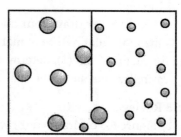

(2) When two gases are at the same temperature, they have the same kinetic energy. Do they also have the same velocity? Explain.

Section 6.11 Review:
(1) Redraw Figure 6.10 to show the relative kinetic energy levels associated with the different states of matter.
(2) Which direction do the arrows point in the endothermic picture from (1)?
(3) Which direction do the arrows point in the exothermic picture from (1)?

Section 6.12 Review:
(1) Draw a graph to indicate the rates of evaporation and condensation in a closed container over time.
(2) What is the relationship between interparticle forces and vapor pressure?

Section 6.13 Review:
(1) What is the relationship between interparticle forces and boiling point?
(2) Water boils at 91.4°C in La Paz, Bolivia, and at 100.0°C in San Francisco, CA. Approximately how much longer will it take for food to cook in La Paz, Bolivia?

Section 6.14 Review:
(1) What are the differences between melting, sublimation, and decomposition?
(2) How are vapor pressures of solids and interparticle forces related?
(3) Which would you expect to have a higher vapor pressure: an ionic solid or a molecular solid? Explain.

Section 6.15 Review:
(1) Figure 6.14 contains two plateaus and three angled sections. Briefly describe the process that occurs in each section of the curve.
(2) Why does the unit specific heat include a temperature term, but the units for the heat of fusion and heat of vaporization do not?

The States of Matter 81

Tying It All Together with a Laboratory Application:

A deflated balloon is placed over the mouth of an Erlenmeyer flask that contains a 1.578 g ice cube at -10°C. The density of ice at -10°C is 0.99815 g/cm³; therefore the volume of the ice is (1) _____. The shape of the ice (2) _____ affected by the Erlenmeyer flask. The specific heat of ice is $2.1 \frac{J}{g \cdot °C}$, so as the ice cube is gently warmed to 0°C, the ice absorbs (3) _____ J. The density of ice at 0°C is 0.99987 g/cm³. Solids have a (4) _____ thermal expansion.

The heat of (5) _____ for ice is 330 J/g. The ice will (6) _____ into water if (7) _____ J are added at 0°C. During this transition (called a (8) _____ (state of matter or phase change)), the temperature of the water (9) _____. The vapor pressure of the water (10) _____ the vapor pressure of the ice. The water is slowly heated to 100°C. Water has a specific heat of $4.184 \frac{J}{g \cdot °C}$. The water absorbs (11) _____ J as the temperature increases. The heat of (12) _____ for water is 2.3 kJ/g. In order to create steam, (13) _____ kJ is added to the water. This transition is called (14) _____ or _____ .

The steam has a(n) (15) _____ shape and a(n) (16) _____ volume. Consequently, the balloon starts to fill as the water becomes steam. The pressure inside the flask and balloon is (17) _____ the pressure in the laboratory. The pressure in the laboratory is 0.9845 atm or (18) _____ mmHg. Once all of the water is converted to steam, the volume of the steam will (19) _____ if the temperature is increased because the pressure will remain (20) _____.

If the experimental apparatus is cooled, the steam will (21) _____ or _____ into water. Further cooling would return the water to ice by (22) _____ or _____ it. These phase transitions are (23) _____ because heat is removed.

At temperatures below STP, H₂O is in the (24) _____ state of matter.

The state of matter with the highest kinetic energy is (25) _____, regardless of the substance. (26) If this experiment was repeated with the same number of moles of dry ice (solid carbon dioxide), would the same phase changes occur? (27) Will the values for density and other quantitative values given for water be the same for carbon dioxide? (Hint: Carbon dioxide is a gas at 1 atm and 195 K or (28) _____ °C.)

If the balloons filled with steam and carbon dioxide were both punctured with the same size pin hole, the (29) _____ would effuse more quickly because it has a (30) _____ molecular mass. (31) Is all of the pressure inside the balloons from the steam and carbon dioxide?

SOLUTIONS FOR THE ADDITIONAL ACTIVITIES

Section 6.1 Review:

(1) Yes, the density of gold is greater than the density of water because the densities of gold and water are constant regardless of their sample size. Density is an intensive physical property.

(2) Gold is a solid at room temperature. It will maintain its shape, regardless of the container it occupies. Water, on the other hand, is a liquid at room temperature and it will take the shape of the container it occupies. Neither gold nor water is very compressible. Water should be more compressible than gold because water is a liquid and liquids are generally more compressible than solids.

(3) The thermal expansion of solids and liquids is small. Neither gold nor water will expand much on heating.

(4) Helium gas has a much lower density than water or gold. Like water, helium gas will take the shape of its container. Helium gas can be highly compressed unlike gold or water. Helium gas will noticeably expand on heating, unlike gold or water.

82 Chapter 6

Section 6.2 Review:

(1) Yes, the ions in a sodium chloride crystal lattice are in motion. Their motion is restricted, but they are still in motion. The movement of water molecules in a glass of water is less restricted than the movement of the ions in the crystal lattice. The movement of helium atoms inside a balloon is much less restricted than the movement of the water molecules or the ions in the crystal.

(2) The movement of the ions in the sodium chloride resembles any type of dancing that involves people being close together and moving very little. People at a concert who are swaying to the music, but keep their feet planted on the floor are an example of the type of restricted movement observed in a crystal. The movement of water molecules in a glass of water resembles any type of dancing that involves people moving relatively quickly and in a group of dancers. Waltzing is similar to the type of movement associated with the water molecules in a glass of water because the molecules revolve as they slip past other molecules. The movement of helium atoms in a balloon resembles any type of dancing that involves people moving extremely fast and dancing alone. Breakdancing is an example of this type of movement.

(3) Solids have the greatest cohesive forces. The attractive forces associated with potential energy are greater than the disruptive forces of kinetic energy.

(4) Gases have the strongest disruptive forces. Gases experience few attractive forces associated with potential energy because of their high kinetic energy and constant motion.

Section 6.3 Review:

(1)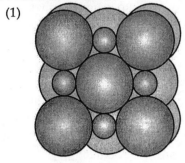

(2) Solids have a high density because the atoms, ions, or molecules they contain are close together. They have a definite shape because the particles they contain occupy specific lattice sites. The particles do not leave the lattice sites because of the interparticle forces discussed in Chapter 4. Solids have small compressibility because the particles are already extremely close together. Solids have very small thermal expansion because the particles cannot deviate very far from their original lattice sites without the solid breaking apart. The interparticles forces keep the particles close to their original lattice sites.

Section 6.4 Review:

(1)

(2) Liquids have a high density because the atoms or molecules they contain are close together. They have an indefinite shape because the particles they contain are not connected in an extended structure. The particles do experience weak interparticle forces. Liquids have small compressibility because the particles are already close together. Liquids have small thermal expansion because the interparticle forces hold the particles together. If these cohesive forces are overcome, the particles would escape into the gas phase.

Section 6.5 Review:
(1)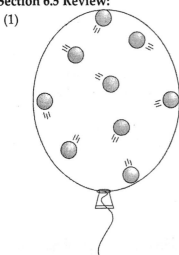

(2) Gases have a low density because the atoms or molecules they contain are far apart. They have an indefinite shape because the particles they contain are not connected in an extended structure and the particles experience only very weak interparticle forces. The disruptive forces dominate the gaseous state. Gases have large compressibility because the particles are far apart. Gases have moderate thermal expansion because particles are already far apart and are not held together by strong interparticle forces. Increasing the temperature of a gas provides additional kinetic energy which in turn increases the disruptive forces within the gas.

Section 6.6 Review:
(1) A water based barometer would need to be taller than a mercury based barometer. The weight of the water in the tube will be balanced by the weight of air pressing on the water pool. The volume of water needed to equal the mass of air pushing on the pool will be larger than the volume of mercury needed because water is less dense than mercury. Consequently, a taller tube is needed for the water than the mercury.

(2) atm or torr: atm is larger; torr or mmHg: they are equal; mmHg or atm: atm is larger; 1 atm = 760 torr = 760 mmHg

(3) No, the Kelvin temperature does not double. The Kelvin temperature only increases by 1.07 times.

$$K = 20°C + 273 = 293 \text{ K} \quad K = 40°C + 273 = 313 \text{ K} \quad \frac{313 \text{ K}}{293 \text{ K}} = \frac{1.07}{1}$$

(4) Yes, the pressure in mmHg also doubles.

$$1 \text{ atm}\left(\frac{760 \text{ mmHg}}{1 \text{ atm}}\right) = 760 \text{ mmHg} \quad 2 \text{ atm}\left(\frac{760 \text{ mmHg}}{1 \text{ atm}}\right) = 1520 \text{ mmHg} \quad \frac{1520 \text{ mmHg}}{760 \text{ mmHg}} = \frac{2}{1}$$

(5) Yes, the volume in milliliters also doubles.

$$0.500 \text{ L}\left(\frac{1000 \text{ mL}}{1 \text{ L}}\right) = 500 \text{ mL} \quad 1.00 \text{ L}\left(\frac{1000 \text{ mL}}{1 \text{ L}}\right) = 1000 \text{ mL} \quad \frac{1000 \text{ mL}}{500 \text{ mL}} = \frac{2}{1}$$

(6) The temperature conversion involves addition, but the pressure and the milliliter conversions involve multiplication. The ratio of two temperatures in degrees Celsius and Kelvin are not necessarily the same, but the ratio of two pressures or two volumes are the same regardless of which units are used.

Section 6.7 Review:
(1) The volume of the balloon will decrease as the pressure increases.
(2) The pressure will decrease as the volume increases.
(3) The volume of the balloon will increase as the temperature increases.
(4) Boyle's Law: $P_1 V_1 = P_2 V_2$, Charles's Law: $\frac{V_1}{T_1} = \frac{V_2}{T_2}$
(5) The temperature must be in Kelvins on both the "before" and "after" sides of the combined gas law. For pressure and volume, the units are not important as long as both sides have the same units. The conversion for temperature is additive, but the conversion for pressure and volume is multiplicative as shown in the Section 6.6 Review.

84 Chapter 6

Section 6.8 Review:

(1) $P = \dfrac{nRT}{V}$, $V = \dfrac{nRT}{P}$, $n = \dfrac{PV}{RT}$, $T = \dfrac{PV}{nR}$

(2) Yes, the molecular weight of a gas can be calculated using Equation 6.10 which requires the mass of the gas as well as the number of moles of the gas. While, the number of moles of the gas are not known, Equation 6.9 can be used to calculate the moles of gas because the pressure, volume, and temperature of the sample are known.

(3) Interparticle forces between gas molecules will cause them to have cohesive type forces. These cohesive forces are not accounted for in the ideal gas law. The ideal gas law is modeled after a gas that is governed by disruptive forces.

Section 6.9 Review:

(1) $\dfrac{P}{n} = \dfrac{RT}{V}$

(2) The pressure will increase as the number of moles increases, if the temperature and volume remain constant.

(3) The total number of moles of gas will have a greater effect on the pressure because all of the gases are governed by the ideal gas law, if they are ideal gases. Their actual identity is not important.

Section 6.10 Review:

(1) = effusion;   = diffusion

(2) Two gases that have the same kinetic energy will only have the same velocity if the mass of the gas molecules is the same. If the masses of the molecules are different, then the lighter gas will have a faster velocity than the heavier gas.

Section 6.11 Review:

(1)   Endothermic   Exothermic

(2) The arrows in the endothermic picture all point upwards because additional energy is required.

(3) The arrows in the exothermic picture all point downwards because energy is released.

Section 6.12 Review:

(1)

(2) Vapor pressure decreases as interparticle forces increase. Fewer molecules can escape from the liquid phase into the gas phase when they experience attraction for other liquid molecules.

The States of Matter 85

Section 6.13 Review:

(1) As interparticle forces increase, the boiling point increases. Boiling occurs when the vapor pressure of the liquid is equal to atmospheric pressure. Vapor pressure decreases as interparticle forces increase; therefore, more energy is needed to counteract the cohesive nature of the interparticle forces and raise the vapor pressure to the atmospheric pressure.

(2) $(100.0°C - 91.4°C)\left(\dfrac{2 \times \text{time}}{10°C}\right) = 1.72 \times \text{time}$; It takes approximately 1.72 times longer for food to cook in La Paz, Bolivia, than it does in San Francisco, California.

Section 6.14 Review:

(1) When heating a solid, it might turn into a liquid. This process is called melting and is a physical change.
When heating a solid, it might turn directly into a gas. This process is called sublimation and is a physical change.
When heating a solid, it might go through a chemical change to produce other compounds. This process is called decomposition.

(2) The vapor pressure of a solid is higher for solids with weak interparticle forces.

(3) A molecular solid will have a higher vapor pressure than an ionic solid because dispersion forces, dipolar forces, and hydrogen bonding are all weaker interparticle forces than ionic bonding.

Section 6.15 Review:

(1) Section AB = heating solid, Section BC = melting, Section CD = heating liquid, Section DE = boiling, Section EF = heating gas

(2) The specific heat is the amount of heat required to heat a gram of material by 1°C. The heat of fusion is the heat required per gram of material to melt a solid. The heat of vaporization is the heat required per gram of material to boil a liquid. Melting and boiling occur at one temperature for a specific material. The material stays at the melting point or boiling point until the phase change has finished. Heating of a solid to its melting point, a liquid to its boiling point, or a gas occurs across a temperature range. Consequently, specific heat requires a temperature term in its unit.

Tying It All Together with a Laboratory Application:

(1) 1.581 cm³   (6) melt   (11) 660.2 J   (15) indefinite   (20) constant
(2) is not   (7) 520 J   (12) vaporization   (16) indefinite   (21) liquefy, condense
(3) 33 J   (8) phase change   (13) 3.6 kJ   (17) equal to   (22) freezing, crystallizing
(4) small   (9) stays the same   (14) evaporation,   (18) 748.2 mmHg   (23) exothermic
(5) fusion   (10) is larger than   vaporization   (19) increase   (24) solid
                                                                                               (25) gaseous

(26) No, carbon dioxide does not melt at atmospheric pressure, it sublimes.

(27) No, density, specific heat, heat of fusion, heat of vaporization, melting point, boiling point, and sublimation point are unique for each substance.

(28) -78°C   (29) water   (30) smaller

(31) No, the flasks were not evacuated before the balloons were added, so air was in the flasks. The majority of the pressure in both flasks is from the water and carbon dioxide, but part of the pressure in both flasks is from air which is a mixture of several different gases (oxygen, nitrogen, argon, etc.).

SELF-TEST QUESTIONS

Multiple Choice

1. The density of ether is 0.736 g/mL. How much would 20.0 mL of ether weigh?
   a. 14.7 g   b. 1.47 g   c. 2.72 g   d. 27.2 g

86   Chapter 6

2. Calculate the density of a swimmer who weighs 40.0 kg and occupies a volume of 45.0 liters.
    a. 0.889 g/mL   b. 0.0088 g/mL   c. 1.12 g/mL   d. 0.0112 g/mL

3. As a pure liquid is heated, its temperature increases and it becomes less dense. Therefore, which of the following is true?
    a. its potential energy increases
    b. its kinetic energy increases
    c. both its potential and kinetic energy increase
    d. its potential energy increases while kinetic energy decreases

4. Which of the following is an endothermic process?
    a. freezing of water
    b. melting of tin
    c. condensation of steam
    d. solidification of liquid sulfur

5. Which of the following is a property of the gaseous state?
    a. it has a lower density than the solid phase
    b. it has large compressibility
    c. it has moderate thermal expansion
    d. more than one response is correct.

6. When the temperature is 0°C, a balloon has a volume of 5 liters. If the temperature is changed to 50°C and if we assume that the pressure inside the balloon equals atmospheric pressure at all times, which of the following will be true?
    a. the new volume will be larger
    b. the new volume will be smaller
    c. there will be no change in volume
    d. the new volume could be calculated from Boyle's law

7. A steel cylinder of CO_2 gas has a pressure of 20 atmospheres (atm) at 20°C. If the cylinder is heated to 70°C, what will the pressure be?
    a. it will be greater than 20 atm
    b. it will be exactly 20 atm
    c. it will be less than 20 atm
    d. it cannot be calculated

8. A balloon is filled with exactly 5.0 liters of helium gas in a room where the temperature is 20°C. What volume will the balloon have when it is taken outside and cooled to 10°C?
    a. 2.5 L   b. 4.8 L   c. 5.2 L   d. 10.0 L

9. If the specific heat of water is 1.00 calorie per gram °C $\left(\frac{cal}{g \cdot °C}\right)$ and the heat of vaporization is 540 cal/g, how much heat would it take to raise the temperature of 10.0 g of water 10°C?
    a. 10 cal   b. 100 cal   c. 5400 cal   d. 54,000 cal

10. The heat of fusion of water is 80.0 cal/g. Therefore, what happens when 1.00 g of ice is melted?
    a. 80 calories of heat would be absorbed
    b. no heat change would take place
    c. 80 calories of heat would be released
    d. what would happen is not predictable

11. Calculate the amount of heat necessary to convert 5.00 g of water at 100°C into steam at 100°C. The heat of vaporization of water is 540 cal/g. The specific heat of steam is 0.480 calories per gram °C $\left(\frac{cal}{g \cdot °C}\right)$.
    a. 24.0 cal   b. 545 cal   c. 2700 cal   d. 2724 cal

12. What volume would 1 mole of gaseous CH_4 occupy at standard temperature and 0.82 atmospheres of pressure? Remember that R = 0.0821 $\frac{L \cdot atm}{mol \cdot K}$.
    a. 0.100 L
    b. 1.00 L
    c. 2.70 L
    d. 27.3 L

13. A sample of gas is found to occupy a volume of 6.80 L at a temperature of 30.0°C and a pressure of 640 torr. How many moles of gas are in the sample? Remember that R = 0.0821 $\frac{L \cdot atm}{mol \cdot K}$.
    a. 0.230 mol
    b. 4.33 mol
    c. 1769 mol
    d. 2.33 mol

14. A 1.20 L sample of gas is under a pressure of 15.0 atm. What volume would the sample occupy if the pressure were lowered to 5.20 atm and the temperature was kept constant?
    a. 0.289 L
    b. 0.416 L
    c. 3.46 L
    d. 34.6 L

True-False
15. According to Graham's law, a gas with molecules four times as heavy as a second gas will diffuse two time faster than the second gas.
16. Solids have a definite volume and a definite shape.
17. The density of a material depends on the amount of material.
18. The average particle speed decreases as temperature decreases.
19. A pressure of 1.50 atm can also be expressed as 1.14×10^3 torr.
20. If the volume of a gas is held constant, increasing the temperature would result in no increase in pressure.
21. The total pressure of a sample of oxygen saturated with water vapor is equal to the partial pressure of the oxygen plus the partial pressure of the water vapor.
22. The state of matter in which disruptive forces predominate is the gaseous state.
23. When the combined gas law is used in a calculation, the gas pressures must be expressed in atm.
24. When the vapor pressure of a liquid is equal to the atmospheric pressure, the liquid will freeze.
25. Boiling points of liquids decrease as atmospheric pressure decreases.
26. When a liquid is placed in a closed container, evaporation continues for a time and then stops.
27. The specific heat of a substance is the same for all three phases of matter.

Matching
Match the states of matter given as responses to the following descriptions.
28. state has a high density
29. state has a large compressibility
30. cohesive forces dominate slightly
31. cohesive forces predominate over disruptive forces
32. sample takes shape of its container

a. solid
b. liquid
c. gas
d. two or more states

Match the terms given as responses to the following descriptions.
33. molten steel changes to a solid
34. water in an open container disappears after a time
35. ice on a car windshield disappears without melting

a. sublimation
b. evaporation
c. freezing

Chapter 7: Solutions and Colloids

CHAPTER OUTLINE
7.1 Physical States of Solutions
7.2 Solubility
7.3 The Solution Process
7.4 Solution Concentrations
7.5 Solution Preparation
7.6 Solution Stoichiometry
7.7 Solution Properties
7.8 Colloids
7.9 Colloid Formation and Destruction
7.10 Dialysis

LEARNING OBJECTIVES/ASSESSMENT
When you have completed your study of this chapter, you should be able to:
1. Classify mixtures as solutions or non-solutions based on their appearance. (Section 7.1; Exercise 7.4)
2. Predict in a general way the solubility of solutes in solvents on the basis of molecular polarity. (Section 7.3; Exercise 7.16)
3. Calculate solution concentrations in units of molarity, weight/weight percent, weight/volume percent, and volume/volume percent. (Section 7.4; Exercises 7.22 b, 7.30 c, 7.34 a, and 7.38 c)
4. Describe how to prepare solutions of specific concentration using pure solutes and solvent, or solutions of greater concentration than the one desired. (Section 7.5; Exercises 7.46 and 7.48 b)
5. Do stoichiometric calculations based on solution concentrations. (Section 7.6; Exercise 7.56)
6. Do calculations based on the colligative solution properties of boiling point, freezing point, and osmotic pressure. (Section 7.7; Exercises 7.64 a & c and 7.74)
7. Describe the characteristics of colloids. (Section 7.8 and 7.9; Exercise 7.80)
8. Describe the process of dialysis and compare it to the process of osmosis. (Section 7.10; Exercise 7.82)

SOLUTIONS FOR THE END OF CHAPTER EXERCISES
PHYSICAL STATES OF SOLUTIONS (SECTION 7.1)

7.2

		Solvent	Solutes
a.	Liquid laundry bleach: sodium hypochlorite 5.25%, inert ingredients 94.75%	water	sodium hypochlorite
b.	Rubbing alcohol: isopropyl alcohol 70%	isopropyl alcohol	water
c.	Hydrogen peroxide: 3% hydrogen peroxide	water	hydrogen peroxide
d.	After-shave: SD alcohol, water, glycerin, fragrance, menthol, benzophenone-1, coloring	SD alcohol	water, glycerin, fragrance, menthol, benzophenone-1, coloring

☑7.4

		Classification	Solvent	Solutes
a.	Foggy air	not a solution (cannot see through it)		
b.	Tears	solution	water	sodium chloride, other ions
c.	Freshly squeezed orange juice	not a solution (cannot see through it)		
d.	Strained tea	solution	water	extractions from tea leaves (color, caffeine, flavor)
e.	Creamy hand lotion	not a solution (cannot see through it)		

SOLUBILITY (SECTION 7.2)
7.6   a.   25 mL of cooking oil and 25 mL of vinegar – the resulting mixture is cloudy and gradually separates into two layers.   immiscible

90 Chapter 7

    b. 25 mL of water and 10 mL of rubbing alcohol – the resulting mixture is clear and colorless   soluble

    c. 25 mL of chloroform and 1 g of roofing tar – the resulting mixture is clear but dark brown in color   soluble

7.8   a. A solution to which a small piece of solute is added, and it dissolves.   unsaturated
     b. A solution to which a small piece of solute is added, and much more solute comes out of solution.   supersaturated
     c. The final solution resulting from the process in part (b).   saturated

7.10   This solution could become supersaturated by slowly lowering the temperature of the solution or by allowing some of the solvent to evaporate.

7.12   a. barium nitrate, $Ba(NO_3)_2$ (8.7 g at 20°C)   soluble
     b. aluminum oxide, Al_2O_3 (9.8 x 10^{-5} g at 29°C)   insoluble
     c. calcium sulfate, $CaSO_4$ (0.21 g at 30°C)   slightly soluble
     d. manganese chloride, $MnCl_2$ (72.3 g at 25°C)   very soluble
     e. lead bromide, $PbBr_2$ (0.46 g at 0°C)   slightly soluble

THE SOLUTION PROCESS (SECTION 7.3)

7.14   I would add water to the sample, stir the mixture, and filter out the solid from the heterogeneous mixture. The calcium carbonate is insoluble in water and would be the solid (residue) from the filtration of the heterogeneous mixture. Calcium chloride is very soluble and would dissolve in the water.

☑7.16   a. (tetrahedral)   soluble in benzene

     b. Ne   soluble in benzene

     c. (triangular-based pyramid)   soluble in water

     d. (flat triangle)   soluble in benzene

7.18 This molecule is mostly nonpolar and grease is nonpolar. Like dissolves like; therefore, Freon-114 will dissolve grease.

SOLUTION CONCENTRATIONS (SECTION 7.4)

7.20   a. 1.25 L of solution contains 0.455 mol of solute. $\dfrac{0.455 \text{ mol}}{1.25 \text{ L}} = 0.364 \text{ M}$

Solutions and Colloids 91

b. 250 mL of solution contains 0.215 mol of solute

$$250 \text{ mL}\left(\frac{1\text{ L}}{1000\text{ mL}}\right) = 0.250 \text{ L}$$

$$M = \frac{\text{moles}}{\text{L}} \Rightarrow \frac{0.215 \text{ mol}}{0.250 \text{ L}} = 0.860 \text{ M}$$

or

$$\frac{0.215 \text{ mol}}{250 \text{ mL}\left(\frac{1\text{ L}}{1000\text{ mL}}\right)} = 0.860 \text{ M}$$

c. 0.175 mol of solute is put into a container and enough distilled water is added to give 100 mL of solution

$$100 \text{ mL}\left(\frac{1\text{ L}}{1000\text{ mL}}\right) = 0.100 \text{ L}$$

$$M = \frac{\text{moles}}{\text{L}} \Rightarrow \frac{0.175 \text{ mol}}{0.100 \text{ L}} = 1.75 \text{ M}$$

or

$$\frac{0.175 \text{ mol}}{100 \text{ mL}\left(\frac{1\text{ L}}{1000\text{ mL}}\right)} = 1.75 \text{ M}$$

7.22 a. A sample of solid KBr weighing 11.9 g is put in enough distilled water to give 200 mL of solution.

$$11.9 \text{ g}\left(\frac{1\text{ mole}}{119.00 \text{ g}}\right) = 0.100 \text{ mol}$$

$$200 \text{ mL}\left(\frac{1\text{ L}}{1000\text{ mL}}\right) = 0.200 \text{ L}$$

$$M = \frac{\text{mole}}{\text{L}} \Rightarrow \frac{0.100 \text{ mol}}{0.200 \text{ L}} = 0.500 \text{ M}$$

or

$$\frac{11.9 \text{ g}\left(\frac{1\text{ mole}}{119.00 \text{ g}}\right)}{200 \text{ mL}\left(\frac{1\text{ L}}{1000\text{ mL}}\right)} = 0.500 \text{ M}$$

☑ b. A 14.2 g sample of solid Na₂SO₄ is dissolved in enough water to give 500 mL of solution.

$$14.2 \text{ g}\left(\frac{1\text{ mole}}{142.05 \text{ g}}\right) = 0.0100 \text{ moles}$$

$$500 \text{ mL}\left(\frac{1\text{ L}}{1000\text{ mL}}\right) = 0.500 \text{ L}$$

$$M = \frac{\text{moles}}{\text{L}} \Rightarrow \frac{0.0100 \text{ mol}}{0.500 \text{ L}} = 0.0200 \text{ M}$$

or

$$\frac{14.2 \text{ g}\left(\frac{1\text{ mole}}{142.05 \text{ g}}\right)}{500 \text{ mL}\left(\frac{1\text{ L}}{1000\text{ mL}}\right)} = 0.200 \text{ M}$$

c. A 10.0-mL sample of solution is evaporated to dryness and leaves 0.29 g of solid residue that is identified as Li₂SO₄.

$$0.29 \text{ g}\left(\frac{1\text{ mole}}{109.95 \text{ g}}\right) = 0.0026 \text{ moles}$$

$$10.0 \text{ mL}\left(\frac{1\text{ L}}{1000\text{ mL}}\right) = 0.0100 \text{ L}$$

$$M = \frac{\text{moles}}{\text{L}} \Rightarrow \frac{0.0026 \text{ mol}}{0.0100 \text{ L}} = 0.26 \text{ M}$$

or

$$\frac{0.29 \text{ g}\left(\frac{1\text{ mole}}{109.95 \text{ g}}\right)}{10.0 \text{ mL}\left(\frac{1\text{ L}}{1000\text{ mL}}\right)} = 0.26 \text{ M}$$

92 Chapter 7

7.24

a. How many moles of solute are contained in 1.75 L of 0.215 M solution?

$$1.75 \text{ L} \left(\frac{0.215 \text{ moles}}{1 \text{ L}} \right) = 0.376 \text{ moles}$$

b. How many moles of solute are contained in 250 mL of 0.300 M solution?

$$250 \text{ mL} \left(\frac{1 \text{ L}}{1000 \text{ mL}} \right) \left(\frac{0.300 \text{ moles}}{1 \text{ L}} \right) = 0.0750 \text{ moles}$$

c. How many mL of 0.350 M solution contains 0.200 mol of solute?

$$0.200 \text{ mol} \left(\frac{1 \text{ L}}{0.350 \text{ moles}} \right) \left(\frac{1000 \text{ mL}}{1 \text{ L}} \right) = 571 \text{ mL}$$

7.26

a. How many grams of solid would be left behind if 25.0 mL of 0.400 M NaCl solution was evaporated to dryness?

$$25.0 \text{ mL} \left(\frac{1 \text{ L}}{1000 \text{ mL}} \right) \left(\frac{0.400 \text{ moles}}{1 \text{ L}} \right) \left(\frac{58.44 \text{ g}}{1 \text{ mole}} \right) = 0.584 \text{ g}$$

b. How many liters of 0.255 M HCl solution is needed to provide 0.0400 mol of HCl?

$$0.0400 \text{ mol} \left(\frac{1 \text{ L}}{0.255 \text{ moles}} \right) = 0.157 \text{ L}$$

c. How many mL of 0.500 M NaNO₃ solution contains 50.0 g of solute?

$$50.0 \text{ g} \left(\frac{1 \text{ mole}}{85.00 \text{ g}} \right) \left(\frac{1 \text{ L}}{0.500 \text{ moles}} \right) \left(\frac{1000 \text{ mL}}{1 \text{ L}} \right) = 1.18 \times 10^3 \text{ mL}$$

7.28

a. 5.3 g of sugar and 100 mL of water

$$100 \text{ mL water} \left(\frac{1.00 \text{ g}}{1 \text{ mL}} \right) = 100 \text{ g water}$$

solution mass = 5.3 g sugar + 100 g water = 105.3 g (without SF)

$$\%(w/w) = \frac{\text{solute mass}}{\text{solution mass}} \times 100 \Rightarrow \frac{5.3 \text{ g}}{105.3 \text{ g}} \times 100 = 5.0\%(w/w)$$

or

$$\frac{5.3 \text{ g sugar}}{5.3 \text{ g sugar} + 100 \text{ mL water} \left(\frac{1.00 \text{ g}}{1 \text{ mL}} \right)} \times 100 = 5.0\%(w/w)$$

b. 5.3 g of any solute and 100 mL of water

$$100 \text{ mL water} \left(\frac{1.00 \text{ g}}{1 \text{ mL}} \right) = 100 \text{ g water}$$

solution mass = 5.3 g solute + 100 g water = 105.3 g (without SF)

$$\%(w/w) = \frac{\text{solute mass}}{\text{solution mass}} \times 100 \Rightarrow \frac{5.3 \text{ g}}{105.3 \text{ g}} \times 100 = 5.0\%(w/w)$$

or

$$\frac{5.3 \text{ g solute}}{5.3 \text{ g solute} + 100 \text{ mL water} \left(\frac{1.00 \text{ g}}{1 \text{ mL}} \right)} \times 100 = 5.0\%(w/w)$$

c. 5.3 g of any solute and 100 g of any solvent

$$\frac{5.3 \text{ g solute}}{5.3 \text{ g solute} + 100 \text{ g solvent}} \times 100 = 5.0\%(w/w)$$

7.30 a. 15.0 g of salt is dissolved in 200 mL of water.

$$200 \text{ mL water}\left(\frac{1.00 \text{ g}}{1 \text{ mL}}\right) = 200 \text{ g water}$$

solution mass = 15.0 g salt + 200 g water = 215.0 g (without SF)

$$\%(w/w) = \frac{\text{solute mass}}{\text{solution mass}} \times 100 \Rightarrow \frac{15.0 \text{ g}}{215.0 \text{ g}} \times 100 = 6.98\%(w/w)$$

or

$$\frac{15.0 \text{ g salt}}{15.0 \text{ g salt} + 200 \text{ mL water}\left(\frac{1.00 \text{ g}}{1 \text{ mL}}\right)} \times 100 = 6.98\%(w/w)$$

b. 0.150 mol of solid NH₄Cl is dissolved in 150 mL of water.

$$\text{mass solute} = 0.150 \text{ mol}\left(\frac{53.49 \text{ g}}{1 \text{ mole}}\right) = 8.0235 \text{ g (without SF)}$$

$$\text{mass of water} = 150 \text{ mL}\left(\frac{1.00 \text{ g}}{1 \text{ mL}}\right) = 150 \text{ g}$$

mass of solution = 8.0235 g + 150 g = 158.0235 g

$$\%(w/w) = \frac{\text{mass of solute}}{\text{mass of solution}} \times 100 \Rightarrow \frac{8.0235 \text{ g}}{158.0235 \text{ g}} \times 100 = 5.08\%(w/w)$$

or

$$\frac{0.150 \text{ mol}\left(\frac{53.49 \text{ g}}{1 \text{ mole}}\right)}{0.150 \text{ mol}\left(\frac{53.49 \text{ g}}{1 \text{ mole}}\right) + 150 \text{ mL water}\left(\frac{1.00 \text{ g}}{1 \text{ mL}}\right)} \times 100 = 5.08\%(w/w)$$

☑ c. 75.0 g of solid is dissolved in 500 mL of water

$$\text{mass of water} = 500 \text{ mL water}\left(\frac{1.00 \text{ g}}{1 \text{ mL}}\right) = 500 \text{ g}$$

mass of solution = 75.0 g solute + 500 g water = 575.0 g (without SF)

$$\%(w/w) = \frac{\text{mass solute}}{\text{mass solution}} \times 100 \Rightarrow \frac{75.0 \text{ g}}{575.0 \text{ g}} \times 100 = 13.0\%(w/w)$$

or

$$\frac{75.0 \text{ g solute}}{75.0 \text{ g solute} + 500 \text{ mL water}\left(\frac{1.00 \text{ g}}{1 \text{ mL}}\right)} \times 100 = 13.0\%(w/w)$$

d. 20.0 mL of ethyl alcohol (density = 0.789 g/mL) is mixed with 15.0 mL of water.

$$\text{mass of ethyl alcohol} = 20.0 \text{ mL ethyl alcohol}\left(\frac{0.789 \text{ g}}{1 \text{ mL}}\right) = 15.78 \text{ g}$$

$$\text{mass of water} = 15.0 \text{ mL water}\left(\frac{1.00 \text{ g}}{1 \text{ mL}}\right) = 15.0 \text{ g}$$

mass of solution = 15.78 g ethyl alcohol + 15.0 g water = 30.78 g

$$\%(w/w) = \frac{\text{mass of solute}}{\text{mass of solution}} \times 100 \Rightarrow \frac{15.78 \text{ g}}{30.78 \text{ g}} \times 100 = 51.3\%(w/w)$$

or

$$\frac{20.0 \text{ mL ethyl alcohol}\left(\frac{0.789 \text{ g}}{1 \text{ mL}}\right)}{20.0 \text{ mL ethyl alcohol}\left(\frac{0.789 \text{ g}}{1 \text{ mL}}\right) + 15.0 \text{ mL water}\left(\frac{1.00 \text{ g}}{1 \text{ mL}}\right)} \times 100 = 51.3\%(w/w)$$

94   Chapter 7

7.32 a. 15.0 g of solute is dissolved in enough water to give 100 mL of solution. The density of the resulting solution is 1.20 g/mL.

mass of solution = 100 mL solution $\left(\frac{1.20 \text{ g}}{1 \text{ mL}}\right)$ = 120 g

$\%(w/w) = \dfrac{\text{mass of solute}}{\text{mass of solution}} \times 100 \Rightarrow \dfrac{15.0 \text{ g}}{120 \text{ g}} \times 100 = 12.5\%(w/w)$

or

$\dfrac{15.0 \text{ g solute}}{100 \text{ mL solution}\left(\frac{1.20 \text{ g}}{1 \text{ mL}}\right)} \times 100 = 12.5\%(w/w)$

b. A 10.0-mL solution sample with a density of 1.15 g/mL leaves 1.54 g of solid residue when evaporated.

mass of solution = 10.0 mL solution $\left(\frac{1.15 \text{ g}}{1 \text{ mL}}\right)$ = 11.5 g

$\%(w/w) = \dfrac{\text{mass of solute}}{\text{mass of solution}} \times 100 \Rightarrow \dfrac{1.54 \text{ g}}{11.5 \text{ g}} \times 100 = 13.4\%(w/w)$

or

$\dfrac{1.54 \text{ g solute}}{10.0 \text{ mL solution}\left(\frac{1.15 \text{ g}}{1 \text{ mL}}\right)} \times 100 = 13.4\%(w/w)$

c. A 50.0-g sample of solution on evaporation leaves a 3.12 g residue of MgCl₂.

$\dfrac{3.12 \text{ g MgCl}_2}{50.0 \text{ g solution}} \times 100 = 6.24\%(w/w)$

7.34 ☑a. 150 mL of solution contains 12.0 mL of alcohol.

$\dfrac{12.0 \text{ mL alcohol}}{150 \text{ mL solution}} \times 100 = 8.00\%(v/v)$

b. 150 mL of solution contains 12.0 mL of any soluble liquid solute.

$\dfrac{12.0 \text{ mL solute}}{150 \text{ mL solution}} \times 100 = 8.00\%(v/v)$

c. 6.0 fluid ounces of oil is added to 2.0 gallons (256 fluid ounces) of gasoline.

$\dfrac{6.0 \text{ fluid ounces oil}}{6.0 \text{ fluid ounces oil} + 256 \text{ fluid ounces gasoline}} \times 100$

$= 2.3\%(v/v)$

d. A solution of alcohol and water is separated by distillation. A 150-mL solution sample gives 51.3 mL of alcohol.

$\dfrac{51.3 \text{ mL alcohol}}{150 \text{ mL solution}} \times 100 = 34.2\%(v/v)$

7.36 $(5.0 \text{ L})(0.50\%(v/v)) = (5.0 \text{ L})(0.0050) = 0.025 \text{ L}\left(\dfrac{1000 \text{ mL}}{1 \text{ L}}\right) = 25 \text{ mL}$

Note: This number may seem small when compared with the volume of a shot of alcohol. (Typically, 1 shot is 1.5 fl. oz. or 44 mL.) This question assumed that the drink was pure ethanol and that it was completely absorbed into the bloodstream. These assumptions do not take into account that most alcoholic beverages are not pure ethanol (many spirits are approximately 40% alcohol by volume) and that the alcohol is distributed throughout the body (not just the bloodstream).

7.38 a. 150 mL of solution contains 7.50 g of dissolved solid Na₂SO₄.

$\dfrac{7.50 \text{ g Na}_2\text{SO}_4}{150 \text{ mL solution}} \times 100 = 5.00\%(w/v)$

Solutions and Colloids 95

b. 150 mL of solution contains 7.50 g of any dissolved solid solute.

$$\frac{7.50 \text{ g solute}}{150 \text{ mL solution}} \times 100 = 5.00\%(w/v)$$

☑ c. 350 mL of solution contains 30.7 g of dissolved solid solute.

$$\frac{30.7 \text{ g solute}}{350 \text{ mL solution}} \times 100 = 8.77\%(w/v)$$

7.40 $\dfrac{65 \text{ g KBr}}{65 \text{ g KBr} + 100 \text{ g H}_2\text{O}} \times 100 = 39\%(w/w) \approx 40\%(w/w)$

SOLUTION PREPARATION (SECTION 7.5)

7.42 a. $200 \text{ mL}\left(\dfrac{1 \text{ L}}{1000 \text{ mL}}\right)\left(\dfrac{0.150 \text{ moles Na}_2\text{SO}_4}{1 \text{ L solution}}\right)\left(\dfrac{142.05 \text{ g Na}_2\text{SO}_4}{1 \text{ mole Na}_2\text{SO}_4}\right) = 4.26 \text{ g Na}_2\text{SO}_4$

I would mass 4.26 g Na₂SO₄ and add it to a 200 mL volumetric flask. I would add enough water to dissolve the Na₂SO₄, then add water up to the mark on the volumetric flask, cap, and shake to ensure the solution is homogeneous.

b. $250 \text{ mL}\left(\dfrac{1 \text{ L}}{1000 \text{ mL}}\right)\left(\dfrac{0.250 \text{ moles Zn(NO}_3)_2}{1 \text{ L solution}}\right)\left(\dfrac{189.41 \text{ g Zn(NO}_3)_2}{1 \text{ mole Zn(NO}_3)_2}\right) = 11.8 \text{ g Zn(NO}_3)_2$

I would mass 11.8 g Zn(NO₃)₂ and add it to a 250 mL volumetric flask. I would add enough water to dissolve the Zn(NO₃)₂, then add water up to the mark on the volumetric flask, cap, and shake to ensure the solution is homogeneous.

c. $150 \text{ g} \times 2.25\%(w/w) = 150 \text{ g} \times 0.0225 = 3.38 \text{ g NaCl}$

$150 \text{ g} - 3.38 \text{ g} = 146.62 \text{ g H}_2\text{O}\left(\dfrac{1 \text{ mL}}{1.00 \text{ g H}_2\text{O}}\right) = 146.62 \text{ mL H}_2\text{O} \approx 147 \text{ mL H}_2\text{O}$

I would mass 3.38 g NaCl and add it to a 250 mL Erlenmeyer flask. I would measure 147 mL of water and add enough of it to the Erlenmeyer flask to dissolve the salt, then I would add the rest of the water and swirl the flask to ensure that the solution is homogeneous.

d. $125 \text{ mL} \times 0.75\%(w/v) = 0.938 \text{ g KCl} \approx 0.94 \text{ g KCl}$

I would add 0.94 g KCl to a 125 mL volumetric flask. I would add enough water to dissolve the KCl, then dilute the mixture to the mark on the volumetric flask. I would then cap the flask and shake it to ensure the solution is homogeneous.

7.44 mass of solution = 15.0 g ethyl alcohol + 45.0 g water = 60.0 g

$\%(w/w) = \dfrac{\text{mass of solute}}{\text{mass of solution}} \times 100 \Rightarrow \dfrac{15.0 \text{ g}}{60.0 \text{ g}} \times 100 = 25.0\%(w/w)$

or

$\dfrac{15.0 \text{ g ethyl alcohol}}{15.0 \text{ g ethyl alcohol} + 45.0 \text{ g water}} \times 100 = 25.0\%(w/w)$

96   Chapter 7

$$\text{volume of solution} = \frac{\text{mass of solution}}{\text{density of solution}} \Rightarrow \frac{60.0 \text{ g}}{0.952 \frac{\text{g}}{\text{mL}}} = 63.025 \text{ mL (without SF)}$$

$$\%(w/v) = \frac{\text{mass of solute}}{\text{volume of solution}} \times 100 \Rightarrow \frac{15.0 \text{ g}}{63.025 \text{ mL}} \times 100 = 23.8\%(w/v)$$

or

$$\frac{15.0 \text{ g ethyl alcohol}}{(15.0 \text{ g ethyl alcohol} + 45.0 \text{ g water})\left(\frac{1 \text{ mL}}{0.952 \text{ g}}\right)} \times 100 = 23.8\%(w/v)$$

☑7.46 a. $250 \text{ mL}\left(\frac{1 \text{ L}}{1000 \text{ mL}}\right)\left(\frac{1.75 \text{ moles Li}_2\text{CO}_3}{1 \text{ L solution}}\right)\left(\frac{73.89 \text{ g Li}_2\text{CO}_3}{1 \text{ mole Li}_2\text{CO}_3}\right) = 32.3 \text{ g Li}_2\text{CO}_3$

b. $200 \text{ mL}\left(\frac{1 \text{ L}}{1000 \text{ mL}}\right)\left(\frac{3.50 \text{ moles NH}_3}{1 \text{ L solution}}\right) = 0.700 \text{ moles NH}_3$

c. $250 \text{ mL} \times 12.5\%(v/v) = 250 \text{ mL} \times 0.125 = 31.3 \text{ mL}$

d. $50.0 \text{ mL} \times 4.20\%(w/v) = 50.0 \text{ mL} \times 0.0420 = 2.10 \text{ g CaCl}_2$

7.48 a. $(6.00 \text{ M})(5.00 \text{ L}) = (18.0 \text{ M})(V_c)$
$V_c = 1.67 \text{ L}$

I would add 3.00 L of water to a 5.00 L volumetric flask, then add 1.67 L of 18.0 M H₂SO₄ to the flask, let it cool, and dilute to the mark with water. (Always add acid to water, not the reverse!)

☑b. $(0.500 \text{ M})(250 \text{ mL}) = (3.00 \text{ M})(V_c)$
$V_c = 41.7 \text{ mL}$

I would add 41.7 mL of 3.00 M CaCl₂ to a 250 mL volumetric flask, then dilute to the mark with water. I would be sure to shake well.

c. $(1.50\%(w/v))(200 \text{ mL}) = (10.0\%(w/v))(V_c)$
$V_c = 30.0 \text{ mL}$

I would add 30.0 mL of 10.0%(w/v) KBr to a 200 mL volumetric flask, then dilute to the mark with water. I would be sure to shake well.

d. $(10.0\%(w/v))(500 \text{ mL}) = (50.0\%(w/v))(V_c)$
$V_c = 100 \text{ mL}$

I would add 100 mL of 50.0%(w/v) alcohol to a 500 mL volumetric flask, then dilute to the mark with water. I would be sure to shake well.

7.50 a. 2.00 L $(100 \text{ mL})(0.225 \text{ M}) = (2.00 \text{ L})\left(\frac{1000 \text{ mL}}{1 \text{ L}}\right)(C_d)$
$C_d = 0.0113 \text{ M}$

b. 225 mL $(100 \text{ mL})(0.225 \text{ M}) = (225 \text{ mL})(C_d)$
$C_d = 0.100 \text{ M}$

c. 600 mL $\quad (100 \text{ mL})(0.225 \text{ M}) = (600 \text{ mL})(C_d)$
$$C_d = 0.0375 \text{ M}$$

d. 850 mL $\quad\quad\quad\quad\quad\quad\quad\quad\quad\quad\quad (100 \text{ mL})(0.225 \text{ M}) = (850 \text{ mL})(C_d)$
$$C_d = 0.0265 \text{ M}$$

SOLUTION STOICHIOMETRY (SECTION 7.6)

7.52 $\quad Na_2CO_3 \text{ (aq)} + 2 \text{ HCl (aq)} \rightarrow 2 \text{ NaCl (aq)} + H_2O \text{ (l)}$

$$250 \text{ mL HCl} \left(\frac{1 \text{ L}}{1000 \text{ mL}}\right)\left(\frac{1.25 \text{ moles}}{1 \text{ L HCl}}\right)\left(\frac{1 \text{ mole } Na_2CO_3}{2 \text{ moles HCl}}\right)\left(\frac{105.99 \text{ g } Na_2CO_3}{1 \text{ moles } Na_2CO_3}\right)$$
$$= 16.6 \text{ g } Na_2CO_3$$

7.54 $\quad NaCl \text{ (aq)} + AgNO_3 \text{ (aq)} \rightarrow NaNO_3 \text{ (aq)} + AgCl \text{ (s)}$

$$25.0 \text{ mL NaCl} \left(\frac{1 \text{ L}}{1000 \text{ mL}}\right)\left(\frac{0.200 \text{ moles}}{1 \text{ L NaCl}}\right)\left(\frac{1 \text{ mole } AgNO_3}{1 \text{ mole NaCl}}\right)\left(\frac{1 \text{ L } AgNO_3}{0.250 \text{ moles}}\right)\left(\frac{1000 \text{ mL}}{1 \text{ L } AgNO_3}\right)$$
$$= 20.0 \text{ mL } AgNO_3$$

☑7.56 $\quad 2 NH_3 \text{ (aq)} + H_2SO_4 \text{ (aq)} \rightarrow (NH_4)_2SO_4 \text{ (aq)}$

$$30.0 \text{ mL } H_2SO_4 \left(\frac{1 \text{ L}}{1000 \text{ mL}}\right)\left(\frac{0.190 \text{ moles } H_2SO_4}{1 \text{ L } H_2SO_4}\right)\left(\frac{2 \text{ moles } NH_3}{1 \text{ mole } H_2SO_4}\right)\left(\frac{1 \text{ L } NH_3}{0.225 \text{ moles } NH_3}\right)\left(\frac{1000 \text{ mL } NH_3}{1 \text{ L } NH_3}\right)$$
$$= 50.7 \text{ mL } NH_3$$

7.58 $\quad NaOH \text{ (aq)} + HCl \text{ (aq)} \rightarrow NaCl \text{ (q)} + H_2O \text{ (l)}$

$$30.0 \text{ mL HCl} \left(\frac{1 \text{ L}}{1000 \text{ mL}}\right)\left(\frac{0.125 \text{ moles HCl}}{1 \text{ L HCl}}\right)\left(\frac{1 \text{ mole NaOH}}{1 \text{ mole HCl}}\right)\left(\frac{1 \text{ L NaOH}}{0.135 \text{ moles NaOH}}\right)\left(\frac{1000 \text{ mL NaOH}}{1 \text{ L NaOH}}\right)$$
$$= 27.8 \text{ mL NaOH}$$

7.60 $\quad CaCO_3 \text{ (s)} + 2 \text{ HCl (aq)} \rightarrow CO_2 \text{ (g)} + CaCl_2 \text{ (aq)} + H_2O \text{ (l)}$

$$250 \text{ mL HCl} \left(\frac{1 \text{ L}}{1000 \text{ mL}}\right)\left(\frac{0.10 \text{ moles HCl}}{1 \text{ L HCl}}\right)\left(\frac{1 \text{ mole } CaCO_3}{2 \text{ mole HCl}}\right)\left(\frac{100.09 \text{ g } CaCO_3}{1 \text{ moles } CaCO_3}\right)$$
$$= 1.25 \text{ g } CaCO_3 \approx 1.3 \text{ g } CaCO_3$$

SOLUTION PROPERTIES (SECTION 7.7)

7.62 As solutes are added to a solvent, the freezing point of the solution is depressed when compared to the freezing point of the pure solvent. The ice cream mixture contains several solutes dissolved in water and the freezing point is less than the freezing point of water. By adding salt to the ice and water, the freezing point of this solution is also decreased. This

allows the ice cream mixture to transfer heat to the salt and ice water mixture, and eventually, freeze.

7.64 ☑a. KCl, a strong electrolyte $\quad\quad\quad\quad BP_{solution} = BP_{solvent} + \Delta t_b \quad\quad \Delta t_b = nK_bM$
$$= 100.0°C + (2)(0.52 \tfrac{°C}{M})(1.50 \text{ M}) = 101.56°C$$
$$FP_{solution} = FP_{solvent} - \Delta t_f \quad\quad \Delta t_f = nK_fM$$
$$= 0.0°C - (2)(1.86 \tfrac{°C}{M})(1.50 \text{ M}) = -5.58°C$$

b. glycerol, a nonelectrolyte

$$BP_{solution} = BP_{solvent} + \Delta t_b \qquad \Delta t_b = nK_bM$$
$$= 100.0°C + (1)(0.52 \tfrac{°C}{M})(1.50 \text{ M}) = 100.78°C$$
$$FP_{solution} = FP_{solvent} - \Delta t_f \qquad \Delta t_f = nK_fM$$
$$= 0.0°C - (1)(1.86 \tfrac{°C}{M})(1.50 \text{ M}) = -2.79°C$$

☑c. $(NH_4)_2SO_4$, a strong electrolyte

$$BP_{solution} = BP_{solvent} + \Delta t_b \qquad \Delta t_b = nK_bM$$
$$= 100.0°C + (3)(0.52 \tfrac{°C}{M})(1.50 \text{ M}) = 102.34°C$$
$$FP_{solution} = FP_{solvent} - \Delta t_f \qquad \Delta t_f = nK_fM$$
$$= 0.0°C - (3)(1.86 \tfrac{°C}{M})(1.50 \text{ M}) = -8.37°C$$

d. $Al(NO_3)_3$, a strong electrolyte

$$BP_{solution} = BP_{solvent} + \Delta t_b \qquad \Delta t_b = nK_bM$$
$$= 100.0°C + (4)(0.52 \tfrac{°C}{M})(1.50 \text{ M}) = 103.12°C$$
$$FP_{solution} = FP_{solvent} - \Delta t_f \qquad \Delta t_f = nK_fM$$
$$= 0.0°C - (4)(1.86 \tfrac{°C}{M})(1.50 \text{ M}) = -11.2°C$$

7.66 a. A 0.750 M solution of urea, a nonelectrolyte

$$BP_{solution} = BP_{solvent} + \Delta t_b \qquad \Delta t_b = nK_bM$$
$$= 100.0°C + (1)(0.52 \tfrac{°C}{M})(0.750 \text{ M}) = 100.39°C$$
$$FP_{solution} = FP_{solvent} - \Delta t_f \qquad \Delta t_f = nK_fM$$
$$= 0.0°C - (1)(1.86 \tfrac{°C}{M})(0.750 \text{ M}) = -1.40°C$$

b. A 0.150 M solution of $CaCl_2$, a strong electrolyte

$$BP_{solution} = BP_{solvent} + \Delta t_b \qquad \Delta t_b = nK_bM$$
$$= 100.0°C + (3)(0.52 \tfrac{°C}{M})(0.150 \text{ M}) = 100.23°C$$
$$FP_{solution} = FP_{solvent} - \Delta t_f \qquad \Delta t_f = nK_fM$$
$$= 0.0°C - (3)(1.86 \tfrac{°C}{M})(0.150 \text{ M}) = -0.84°C$$

c. A solution containing 120 g of ethylene glycol ($C_2H_6O_2$) per 400 mL of solution

$$\frac{120 \text{ g } C_2H_6O_2 \left(\dfrac{1 \text{ mole } C_2H_6O_2}{62.08 \text{ g } C_2H_6O_2}\right)}{400 \text{ mL}\left(\dfrac{1 \text{ L}}{1000 \text{ mL}}\right)} = 4.83247 \text{ M}$$

$$BP_{solution} = BP_{solvent} + \Delta t_b \qquad \Delta t_b = nK_bM$$
$$= 100.0°C + (1)(0.52 \tfrac{°C}{M})(4.83247 \text{ M}) = 102.51°C$$
$$FP_{solution} = FP_{solvent} - \Delta t_f \qquad \Delta t_f = nK_fM$$
$$= 0.0°C - (1)(1.86 \tfrac{°C}{M})(4.83247 \text{ M}) = -8.99°C$$

7.68 a. A 0.20 M solution of glycerol, a nonelectrolyte

$$\text{osmolarity} = nM = (1)(0.20 \text{ M}) = 0.20 \tfrac{\text{osmol}}{\text{L}}$$

b. A 0.20 M solution of $(NH_4)_2SO_4$, a strong electrolyte

$$\text{osmolarity} = nM = (3)(0.20 \text{ M}) = 0.60 \tfrac{\text{osmol}}{\text{L}}$$

c. A solution containing 33.7 g of LiCl (a strong electrolyte) per 500 mL

$$\text{osmolarity} = nM = (2)\frac{33.7 \text{ g LiCl}\left(\frac{1 \text{ mole LiCl}}{42.39 \text{ g LiCl}}\right)}{500 \text{ mL solution}\left(\frac{1 \text{ L}}{1000 \text{ mL}}\right)} = 3.18 \frac{\text{osmol}}{\text{L}}$$

7.70 $\Pi = nMRT = (0.300 \text{ M})(0.0821 \tfrac{\text{L·atm}}{\text{mol·K}})((25.0+273)\text{K}) = 7.34 \text{ atm}$

$= (0.300 \tfrac{\text{moles}}{\text{L}})(0.0821 \tfrac{\text{L·atm}}{\text{mol·K}})((25.0+273)\text{K})\left(\frac{760 \text{ torr}}{1 \text{ atm}}\right) = 5.58 \times 10^3 \text{ torr}$

$= (0.300 \tfrac{\text{moles}}{\text{L}})(0.0821 \tfrac{\text{L·atm}}{\text{mol·K}})((25.0+273)\text{K})\left(\frac{760 \text{ mmHg}}{1 \text{ atm}}\right) = 5.58 \times 10^3 \text{ mmHg}$

7.72 $\Pi = nMRT = 3(0.200 \text{ M})(0.0821 \tfrac{\text{L·atm}}{\text{mol·K}})((25.0+273)\text{K}) = 14.7 \text{ atm}$

$= 3(0.200 \tfrac{\text{moles}}{\text{L}})(0.0821 \tfrac{\text{L·atm}}{\text{mol·K}})((25.0+273)\text{K})\left(\frac{760 \text{ torr}}{1 \text{ atm}}\right) = 1.12 \times 10^4 \text{ torr}$

$= 3(0.200 \tfrac{\text{moles}}{\text{L}})(0.0821 \tfrac{\text{L·atm}}{\text{mol·K}})((25.0+273)\text{K})\left(\frac{760 \text{ mmHg}}{1 \text{ atm}}\right) = 1.12 \times 10^4 \text{ mmHg}$

☑7.74 $\dfrac{95.0 \text{g}\left(\frac{1 \text{ mole}}{60.06 \text{ g}}\right)}{500 \text{ mL}\left(\frac{1 \text{ L}}{1000 \text{ mL}}\right)} = 3.1635 \text{ M}$

$\Pi = nMRT = 1(3.1635 \text{ M})(0.0821 \tfrac{\text{L·atm}}{\text{mol·K}})((25.0+273)\text{K}) = 77.4 \text{ atm}$

$= 1(3.1635 \text{ M})(0.0821 \tfrac{\text{L·atm}}{\text{mol·K}})((25.0+273)\text{K})\left(\frac{760 \text{ torr}}{1 \text{ atm}}\right) = 5.88 \times 10^4 \text{ torr}$

$= 1(3.1635 \text{ M})(0.0821 \tfrac{\text{L·atm}}{\text{mol·K}})((25.0+273)\text{K})\left(\frac{760 \text{ mmHg}}{1 \text{ atm}}\right) = 5.88 \times 10^4 \text{ mmHg}$

7.76 $FP_{\text{solution}} = FP_{\text{solvent}} - \Delta t_f \qquad \Delta t_f = nK_f M$

$-0.35°\text{C} = 0.0°\text{C} - nM(1.86 \tfrac{°\text{C}}{\text{M}})$

$nM = 0.188172 \text{ M}$

$\Pi = nMRT = (0.188172 \text{ M})(0.0821 \tfrac{\text{L·atm}}{\text{mol·K}})((25.0+273)\text{K}) = 4.6 \text{ atm}$

$= (0.188172 \tfrac{\text{moles}}{\text{L}})(0.0821 \tfrac{\text{L·atm}}{\text{mol·K}})((25.0+273)\text{K})\left(\frac{760 \text{ torr}}{1 \text{ atm}}\right) = 3.5 \times 10^3 \text{ torr}$

$= (0.188172 \tfrac{\text{moles}}{\text{L}})(0.0821 \tfrac{\text{L·atm}}{\text{mol·K}})((25.0+273)\text{K})\left(\frac{760 \text{ mmHg}}{1 \text{ atm}}\right) = 3.5 \times 10^3 \text{ mmHg}$

7.78 $nM = 2\left(\dfrac{5.30 \text{ g NaCl}\left(\frac{1 \text{ mole NaCl}}{58.44 \text{ g NaCl}}\right)}{750 \text{ mL}\left(\frac{1 \text{ L}}{1000 \text{ mL}}\right)}\right) + 2\left(\dfrac{8.20 \text{ g KCl}\left(\frac{1 \text{ mole KCl}}{74.55 \text{ g KCl}}\right)}{750 \text{ mL}\left(\frac{1 \text{ L}}{1000 \text{ mL}}\right)}\right) = 0.5351589 \tfrac{\text{osmol}}{\text{L}}$

$$\Pi = nMRT = \left(0.5351589 \tfrac{osmol}{L}\right)\left(0.0821 \tfrac{L \cdot atm}{mol \cdot K}\right)\left((25.0+273)K\right) = 13.1 \text{ atm}$$

$$= \left(0.5351589 \tfrac{osmol}{L}\right)\left(0.0821 \tfrac{L \cdot atm}{mol \cdot K}\right)\left((25.0+273)K\right)\left(\frac{760 \text{ torr}}{1 \text{ atm}}\right) = 9.95 \times 10^3 \text{ torr}$$

$$= \left(0.5351589 \tfrac{osmol}{L}\right)\left(0.0821 \tfrac{L \cdot atm}{mol \cdot K}\right)\left((25.0+273)K\right)\left(\frac{760 \text{ mmHg}}{1 \text{ atm}}\right) = 9.95 \times 10^3 \text{ mmHg}$$

COLLOIDS AND DIALYSIS (SECTIONS 7.8-7.10)

☑7.80   When the bag functions as an osmotic membrane, the solvent will travel from the solution with a lower solute concentration to the solution with a higher solute concentration. When the bag functions as a dialysis membrane, the solvent and the small molecules can flow through the membrane.

☑7.82   The dispersing medium in a colloidal suspension is the material present in the largest amount (like the solvent in a solution). The dispersed phase in a colloidal suspension is the material present in smaller amount (like the solute in a solution). The colloid emulsifying agent acts to keep the dispersing medium and the dispersed phase relatively homogeneous and stable.

ADDITIONAL ACTIVITIES

Section 7.1 Review:
(1) Draw a before and after picture of the process of making a salt water solution.
(2) What is the solute in (1)? What is the solvent in (1)?
(3) What process occurs in (1)?
(4) What ways can you think of to speed up the process in (1)?

Section 7.2 Review:
(1) What do the following prefixes and suffix mean: *in-, im-, super-, un-, -ility*?
(2) What is the difference between soluble and miscible?
(3) What is the relationship between solubility and temperature?
(4) What is the relationship between solubility and pressure?

Section 7.3 Review:
(1) Both solutes and solvents have interparticle forces. What has to occur in order for a solution to form?
(2) What does "like" mean in "like dissolves like"?
(3) Why does "like dissolves like"?
(4) What patterns are evident in Table 7.4?
(5) Describe the fastest way to make a glass of ice tea sweetened with solid sugar.
(6) Have the definitions of *endothermic* and *exothermic* changed since Chapter 5?

Section 7.4 Review:
(1) What is the difference between the molarity of a solution and the density of a solution?
(2) What do all of the formulas for molarity, %(w/w), %(w/v), and %(v/v) have in common?
(3) How many grams of solvent are in 200 g of a 12.0% (w/w) sugar solution?

Section 7.5 Review:
(1) What is the advantage of selling orange juice concentrate?
(2) What process must occur in order for the orange juice concentrate to become a beverage ready to drink?

(3) If the volume of orange juice concentrate is 16 fluid ounces and the volume of ready to drink orange juice is 64 fluid ounces, what the ratio of the concentration of the concentrate to the ready to drink beverage?

Section 7.6 Review:
(1) Describe the process of converting liters of solution to moles of solute using the molarity.
(2) Describe the process of converting moles of solute to liters of solution using the molarity.
(3) A solution has a concentration of 5.00 M, write two factors that could be used in dimensional analysis for this solution.
(4) Identify which factor from (3) could be used in the factor-unit method to convert from liters of solution to moles of solute.
(5) Identify which factor from (3) could be used in the factor-unit method to convert form moles of solute to liters of solution.

Section 7.7 Review:
(1) What is the difference between *dissolving* and *dissociating*?
(2) Do both *dissolving* and *dissociating* always occur at the same time?
(3) Describe the level of dissolving and dissociating for strong electrolytes, weak electrolytes, and nonelectrolytes.
(4) Compare the vapor pressure, boiling points, and freezing points of solutions and pure solvents.
(5) During osmosis, what flows through the semipermeable membrane?
(6) During osmosis, what is the direction of flow through the semipermeable membrane?
(7) I once left a bunch of grapes to soak in a container of water. I accidentally left the grapes overnight. When I returned, the grapes had split open. Explain what happened using your knowledge of colligative properties.

Section 7.8 Review:
(1) When making a gelatin dessert, a packet of gelatin flavored with sugar is added to 2 cups of boiling water. What is the dispersing medium in a gelatin dessert? What is the dispersed phase in a gelatin dessert?
(2) The gelatin dessert from (1) is placed in the refrigerator and allowed to cool. It becomes semisolid. What are two other names from this section that can be used to describe this dessert?
(3) What type of process (endothermic or exothermic) is described in (2)?
(4) What should happen if a flashlight is shown through the gelatin dessert after it cools? What is the name of this phenomenon?

Section 7.9 Review:
(1) Mix half a cup of water and half a cup of vegetable oil in a clear container. Stir rapidly for 30 seconds. Stop stirring and observe the oil and water mixture. What happens?
(2) Add the yolk from an egg to the mixture from (1). Stir rapidly for 30 seconds. Stop stirring and observe the oil, water, and egg-yolk mixture. What happens?
(3) What is the difference between the mixtures in (1) and (2)?

Section 7.10 Review:
(1) Why is it important to use a dialyzing membrane rather than a semipermeable membrane when treating people suffering from a kidney malfunction?
(2) How could you test to see if grape skin acts as a semipermeable membrane or a dialyzing membrane?

Chapter 7

Tying It All Together with a Laboratory Application:
A solution is made with iodine and benzene. Iodine is a (1) _____ molecule that dissolves well in benzene because benzene is (2) _____. Iodine is less (3) _____ in water than benzene because water is a (4) _____ molecule.

The actual (5) _____ of iodine in benzene is 16 g iodine per 100 g benzene which is a concentration of (6) _____ %(w/w). (7) _____ is the solute and (8) _____ is the solvent in this solution. If 0.81 g of iodine are added to 4.0 g of benzene, a(n) (9) _____ solution will form. (10) _____ of the iodine will dissolve. The (11) _____ can be increased by crushing or grinding the iodine and stirring the solution. The solubility will (12) _____ if the solution is heated. The solution cools and at room temperature, the solution is (13) _____. (14) What will happen if another crystal of iodine is added to the solution after it cools?

Iodine is a (15) _____ in benzene because the solution does not conduct electricity. This means that iodine (16) _____ (dissociates or does not dissociate) in benzene. The boiling point of pure benzene is 80.1°C and the K_b for benzene is 2.53°C/M. The boiling point of a 0.10 M iodine in benzene solution is (17) _____ °C. The freezing point of the solution is (18) _____ (greater than, less than, or equal to) the freezing point of pure benzene. The vapor pressure of the solution is (19) _____ (greater than, less than, or equal to) the vapor pressure of pure benzene. At 22°C, the osmotic pressure of the 0.10 M iodine in benzene solution is (20) _____ torr.

In order to make 250 mL of a 0.250 M iodine in benzene solution, (21) _____ g of iodine should be added to a 250 mL volumetric flask and dissolve in a small amount of benzene before benzene is added to the mark on the flask. This solution could be used to prepare a 500 mL of a 0.10 M, if (22) _____ mL of the 0.25 M solution are (23) _____ to 500 mL.

Iodine reacts with aluminum to form aluminum iodide as shown below:

(24) _____ Al (s) + _____ I_2 (benzene) → _____ AlI_3 (s)

If a coil of 0.35 g of thin aluminum wire is added to 25 mL of the 0.10 M iodine in benzene solution, (25) _____ g of AlI_3 can be formed because (26) _____ is the limiting reactant.

SOLUTIONS FOR THE ADDITIONAL ACTIVITIES
Section 7.1 Review:

(1) water + salt → salt water solution
(2) Salt is the solute. Water is the solvent. (Note: The solute is the smaller part of a solution and has six letters. The solvent is the larger part of a solution and has 8 letters.)
(3) Dissolving is the process that occurs when making a solution.
(4) Heating the water, increasing the surface area of the salt, and stirring the components are a few methods for speeding up the solution process.

Section 7.2 Review:
(1) *in-* = no, not, without, non-; *im-* = same as *in-*, but used before *b*, *m*, or *p*; *super-* = to a degree greater than normal; *un-* = not, lack of, the opposite of; *-ility* = the quality of being (as specified)

(2) Soluble means that two substances can form a solution. Miscible means that two liquids can form a solution.
(3) For solid and liquid solutes, solubility usually increases as temperature increases. For gaseous solutes, solubility usually decreases as temperature increases.
(4) For solid and liquid solutes, solubility is not influenced by pressure. For gaseous solutes, solubility usually increases as pressure increases.

Section 7.3 Review:
(1) In order for a solution to form, the interparticle forces between solute particles must be overcome by the solvent particles and the interparticle forces between the solvent particles must be overcome by the solute particles. Overcoming the existing interparticle forces is easiest when the solute and solvent particles will establish strong interparticle forces as they interact.
(2) Similar interparticle forces in the solute and the solvent are the "like" in "like dissolves like."
(3) "Like dissolves like" because the new interparticle forces between the solute and the solvent will be strong enough to overcome the interparticle forces within the solute and within the solvent. Polar solvents are attracted to polar or ionic solutes because both the solvent and the solute have either partial or full charges. Nonpolar solvents are attracted to nonpolar solutes because neither the solvent nor the solute has any partial or full charges.
(4) The two ions that primarily form water insoluble compounds are the carbonates and the phosphates. Both of these ions have large charges. Compounds that contain these ions would have strong ionic bonds because of the differences in the charges. The ions that are associated with water soluble compounds have small charges. The chlorides and the sulfates are mainly water soluble, but when they combine with heavy ions, they can form insoluble compounds. Water is unable to break the strong bonds these compounds form.
(5) While the tea is still hot, stir in the desired amount of solid sugar. For the fastest results, use extremely fine sugar. Once the sugar has dissolved, add ice. The sugar will dissolve much more rapidly in a warm solution than a cold solution.
(6) The definitions of *endothermic* and *exothermic* have not changed since Chapter 5. Endothermic solution processes require energy just like endothermic chemical reactions. Exothermic solution processes release energy just like exothermic chemical reactions.

Section 7.4 Review:
(1) The molarity of a solution is the ratio of moles of solute per liter of solution. The density of a solution is the ratio of mass of solution per milliliter of solution. Density is calculated based on the mass of the entire solution. Molarity is calculated based on the moles of solute.
(2) All of the formulas for molarity, %(w/w), %(w/v), and %(v/v) contain a fraction with solute in the numerator and solution in the denominator.

$$\frac{\text{solute}}{\text{solution}}; \quad M = \frac{\text{moles solute}}{\text{L solution}}; \quad \%(w/w) = \frac{\text{mass solute}}{\text{mass solution}} \times 100;$$

$$\%(w/v) = \frac{\text{mass solute in grams}}{\text{volume solution in mL}} \times 100; \quad \%(v/v) = \frac{\text{volume solute}}{\text{volume solution}} \times 100$$

(3) $200 \text{ g solution}(12.0\% \text{ sugar solution}) = 24.0 \text{ g sugar}; \quad 200 \text{ g solution} - 24.0 \text{ g sugar} = 176 \text{ g solvent}$

Section 7.5 Review:
(1) Orange juice concentrate does not take up as much space as ready to drink orange juice. Refrigeration space is costly for a grocer, so it is advantageous to sell orange juice concentrate because it takes less refrigeration space.
(2) The concentrate must be diluted before it is ready to drink.

(3) $(C_c)(V_c) = (C_d)(V_d) \Rightarrow \dfrac{(C_c)}{(C_d)} = \dfrac{(V_d)}{(V_c)} \Rightarrow \dfrac{(C_c)}{(C_d)} = \dfrac{64 \text{ fl. oz.}}{16 \text{ fl. oz.}} = \dfrac{4}{1}$

The orange juice concentrate is four times more concentrated than the ready to drink beverage because it has ¼ the volume of ready to drink orange juice.

Section 7.6 Review:
(1) To convert liters of solution to moles of solute using the molarity, multiply the volume in liters by the molarity.
(2) To convert moles of solute to liters of solution using the molarity, divide the moles of solute by the molarity (or multiply by the reciprocal of the molarity).
(3) $\dfrac{5.00 \text{ moles solute}}{1 \text{ L solution}}$ and $\dfrac{1 \text{ L solution}}{5.00 \text{ moles solute}}$
(4) $\text{liters}\left(\dfrac{5.00 \text{ moles solute}}{1 \text{ L solution}}\right) = \text{moles}$
(5) $\text{moles}\left(\dfrac{1 \text{ L solution}}{5.00 \text{ moles solute}}\right) = \text{liters}$

Section 7.7 Review:
(1) Dissolving is the process by which a solute and solvent form a solution. Dissociating is the process by which a solute breaks apart into ions in solution.
(2) Dissolving does not necessarily involve a dissociation process; however, when dissociation occurs, it accompanies the dissolving process. Ionic compounds and highly polar covalent compounds dissociate in polar solvents like water. Nonpolar covalent compounds or weakly polar covalent compounds do not dissociate in polar solvents like water; however, some of these compounds dissolve in water.
(3) Strong electrolytes both dissolve and dissociate completely. Weak electrolytes dissolve completely, but they do not dissociate completely. Nonelectrolytes dissolve completely, but they do not dissociate.
(4) Solutions have lower vapor pressures, higher boiling points, and lower freezing points than pure solvents.
(5) Solvent flows through the semipermeable membrane during osmosis.
(6) Solvent flows from the solution of lower solute concentration to the solution of higher solute concentration during osmosis.
(7) The "skin" on the grapes functions as a semipermeable membrane. The concentration of solutes was higher inside the grapes than in the container of water. The solvent (water) flowed through the semipermeable membranes into the grapes. Osmosis continued until the skin of the grapes could not hold any more water and the skin split.

Section 7.8 Review:
(1) The dispersing medium is water. The dispersed phase is the gelatin flavored with sugar. (Notice medium is the larger component in a colloid and it has 6 letters. Phase is the smaller component in a colloid and it has 5 letters.)
(2) The dessert is a sol and a gel. It is a sol because the dispersing medium is a liquid and the dispersed phase is a solid. It is a gel because it is a sol that becomes semisolid.
(3) The process of forming this gel is exothermic. The heat in the dessert is released as the gel forms.
(4) The path of the beam of light should be visible. This is called the Tyndall effect.

Section 7.9 Review:
(1) The oil and water begin to separate. An unstable emulsion exists briefly as a middle layer between the oil on top and the water on the bottom. Eventually, only two layers (oil and water) remain.
(2) The mixture becomes cloudy when the egg yolk is added. The mixture does not separate into layers. The consistency of the mixture is relatively homogenous.
(3) The egg yolk acts as an emulsifying agent (also known as a stabilizing agent) in the second test. The first mixture formed an unstable emulsion that coalesced into the original materials. Because the second mixture had the egg yolk to serve as an emulsifying agent it was more stable and the emulsion lasted longer.

Section 7.10 Review:
(1) Properly functioning kidneys filter out waste products from the blood which include small molecules and hydrated ions. If a semipermeable membrane were used to treat people suffering from a kidney malfunction, water could flow into the membrane and dilute the blood, but the waste products could not leave the blood. A dialyzing membrane allows for small molecules and hydrated ions to flow through the membrane as well as for water to flow into and out of the membrane. The dialyzing membrane functions more like a kidney than a semipermeable membrane does.
(2) After cleaning a grape, it could be placed into a container of distilled water. A "control" sample of distilled water should also be kept. The grape should be left in the container of distilled water for a length of time, but not long enough to split the grape skin. Once the grape is removed from the container, the water should be tested for small molecules. If the water that interacted with the grape contains small molecules or ions that are not present in the control distilled water, then the grape cell membranes are functioning as a dialyzing membrane. If the water that interacted with the grape is identical to the control water, then the grape skin is functioning as a semipermeable membrane.

Tying It All Together with a Laboratory Application:
(1) nonpolar
(2) nonpolar
(3) soluble
(4) polar
(5) solubility
(6) 14%(w/w)
(7) Iodine
(8) benzene
(9) saturated
(10) Most (0.64 g, 79%)
(11) rate of dissolving
(12) increase
(13) supersaturated

(14) The additional crystal of iodine will act as a seed crystal. All of the extra dissolved iodine will precipitate out of solution, leaving a saturated solution.

(15) nonelectrolyte
(16) does not dissociate
(17) 80.4°C
(18) less than
(19) less than
(20) 1.8×10^3 torr
(21) 15.9 g
(22) 200 mL
(23) diluted
(24) 2, 3, 2
(25) 0.68 g
(26) I_2 in benzene

SELF-TEST QUESTIONS
Multiple Choice
1. A solution is prepared by dissolving a small amount of sugar in a large amount of water. In this case sugar would be the
   a. filtrate
   b. solute
   c. precipitate
   d. solvent

2. The freezing point of a solution
   a. is lower than that of pure solvent
   b. is higher than that of pure solvent
   c. is the same as that of pure solvent
   d. cannot be measured

3. Which of the following would show the Tyndall effect?
   a. a solution of salt in water
   b. a solution of sugar in water
   c. a solution of CO_2 gas in water
   d. a colloidal suspension

4. A crystal of solid magnesium sulfate is placed into a solution of magnesium sulfate in water. It is observed that the crystal dissolves slightly. The original solution was
   a. saturated
   b. unsaturated
   c. supersaturated
   d. cannot determine from the data given

5. The weight/weight percent of sugar in a solution containing 25 grams of sugar and 75 grams of water would be:
   a. 25%
   b. 33%
   c. 75%
   d. 80%

6. How many moles of $Mg(NO_3)_2$ are contained in 500 mL of 0.400 M solution?
   a. 0.400
   b. 0.200
   c. 0.800
   d. 2.00

7. How many grams of $Mg(NO_3)_2$ must be dissolved in water to give 500 mL of 0.400 M solution?
   a. 29.7
   b. 17.3
   c. 59.3
   d. 172.6

8. Which of the following aqueous solutions would be expected to have the highest boiling point at 1 atm pressure?
   a. 1 M NaCl
   b. 1 M $C_{12}H_{22}O_{11}$ (sucrose)
   c. pure water
   d. 1 M AlF_3

9. Calculate the freezing point of a water solution that contains 1.60 grams of methyl alcohol (a nonelectrolyte) CH_3OH, in each 100.0 mL. The K_f for water is 1.86°C/M.
   a. 0.93°C
   b. -0.93°C
   c. -1.86°C
   d. -3.72°C

10. How many moles of solute would be needed to prepare 250 mL of 0.150 M solution?
    a. 37.5
    b. 0.150
    c. 0.0375
    d. 0.600

11. What volume of 0.200 M silver nitrate solution ($AgNO_3$) would have to be diluted to form 500 mL of 0.050 M solution?
    a. 200 mL
    b. 2.0×10^{-5}
    c. 80.0 mL
    d. 125 mL

12. Calculate the osmolarity of a 0.400 M NaCl solution.
    a. 0.400
    b. 0.800
    c. 0.200
    d. 0.100

13. What is the osmotic pressure (in torr) of a 0.0015 M NaCl solution at 25°C? $R = 62.4 \frac{L \cdot torr}{mol \cdot K}$.
    a. 4.68
    b. 2.34
    c. 27.9
    d. 55.8

True-False
14. Paint is a colloid.
15. Gases dissolve better in liquids when the liquids are warm.
16. A solution can contain only one solute.
17. A solution is a homogeneous mixture.
18. A colligative property is used to prevent winter freeze-up of cars.
19. Water is a good solvent for both ionic and polar covalent materials.
20. A nonpolar solute would be relatively insoluble in water.
21. On the basis of their solubility in each other, water and gasoline molecules have similar polarities.
22. Potassium nitrate is soluble in water.
23. Soluble solutes always dissolve rapidly.
24. The larger a solute particle, the faster it will dissolve in a solvent.

25. An exothermic solution process releases heat.
26. A solution has a higher boiling point than the pure solvent.
27. The number of moles of sodium chloride in 1.00 L of 0.300 M NaCl (aq) is greater than the number of moles of sodium chloride in 50.0 mL of 6.00 M NaCl (aq).
28. The concentration of a solution increases when it is diluted.
29. The colligative properties depend only on the concentration of solute particles in a solution, not the identity of the solute particles.
30. Semipermeable membranes allow solute molecules to pass from a solution of lower concentration to a solution of higher concentration.

Matching
31. a liquid dispersed in a gas
32. a liquid dispersed in a liquid
33. mayonnaise is an example
34. a gas dispersed in a liquid
35. gelatin dessert is an example

a. foam
b. emulsion
c. aerosol
d. sol

Chapter 8: Reaction Rates and Equilibrium

CHAPTER OUTLINE
8.1 Spontaneous and Nonspontaneous Processes
8.2 Reaction Rates
8.3 Molecular Collisions
8.4 Energy Diagrams
8.5 Factors that Influence Reaction Rates
8.6 Chemical Equilibrium
8.7 The Position of Equilibrium
8.8 Factors That Influence Equilibrium Position

LEARNING OBJECTIVES/ASSESSMENT
When you have completed your study of this chapter, you should be able to:
1. Use the concepts of energy and entropy to predict the spontaneity of processes and reaction. (Section 8.1; Exercise 8.6)
2. Calculate reaction rates from experimental data. (Section 8.2; Exercise 8.14)
3. Use the concept of molecular collisions to explain reaction characteristics. (Section 8.3; Exercise 8.20)
4. Represent and interpret the energy relationships for reactions by using energy diagrams. (Section 8.4; Exercise 8.26)
5. Explain how factors such as reactant concentrations, temperature, and catalysts influence reaction rates. (Section 8.5; Exercise 8.30)
6. Write equilibrium expressions based on reaction equations. (Section 8.7; Exercise 8.38)
7. Do calculations based on equilibrium expressions. (Section 8.7; Exercise 8.44)
8. Use Le Châtelier's principle to predict the influence of changes in concentration, and reaction temperature on the position of equilibrium for a reaction. (Section 8.8; Exercise 8.50)

SOLUTIONS FOR THE END OF CHAPTER EXERCISES
SPONTANEOUS AND NONSPONTANEOUS PROCESSES (SECTION 8.1)

8.2
a.	The space shuttle leaves its pad and goes into orbit.	nonspontaneous	Rocket engines must continually operate
b.	The fuel in a booster rocket of the space shuttle burns.	spontaneous	Once the fuel is ignited, it will continue to burn. No additional energy has to be provided.
c.	Water boils at 100°C and 1 atm pressure.	nonspontaneous	Heat must be continually supplied to maintain boiling.
d.	Water temperature increases to 100°C at 1 atm pressure.	nonspontaneous	Increasing the temperature of water requires a continual supply of energy.
e.	Your bedroom becomes orderly.	nonspontaneous	A room will not become orderly on its own. Cleaning requires energy.

8.4
a.	An automobile being pushed up a slight hill (from point of view of the one pushing)	exergonic	The person pushing the car gives energy to the car.
b.	Ice melting (from point of view of the ice)	endergonic	Melting ice requires energy.
c.	Ice melting (from point of view of surroundings of the ice)	exergonic	The surroundings release heat into the ice.
d.	Steam condensing to liquid water (from point of view of the steam)	exergonic	Heat must be released from the steam.
e.	Steam condensing to liquid water (from point of view of surroundings of the steam)	endergonic	Heat must be absorbed by the surroundings.

8.6
a. On a cold day, water freezes. | energy decreases / entropy decreases | spontaneous (if the temperature is less than 0°C)
b. A container of water at 40°C cools to room temperature. | energy decreases / entropy decreases slightly | spontaneous
c. The odor from an open bottle of perfume spreads throughout a room. | energy increases slightly / entropy increases | spontaneous

8.8
a. The highest entropy example from this set is the two opposing football teams when the whistle is blown, ending the play. This is the highest entropy because the football players from the two teams are most interspersed.
b. From the gold's perspective, the highest entropy example is a 10% copper/gold alloy because the gold atoms are more separated in this material.

From the copper's perspective, the highest entropy example is a 2% copper/gold alloy because the copper atoms are more separated in this material.
c. The purse on the ground with the contents scattered is the highest entropy example from this set. This is the highest entropy example because the disorder is highest in this example.
d. The coins in a piggy bank are the example from this set with the highest entropy. The random arrangement of coins in the piggy bank represents greater disorder than the other two examples.
e. The dozen loose pearls in a box are the example from this set with the highest entropy. The random order of the pearl represents greater disorder than the other two examples.

REACTION RATES (SECTION 8.2)

8.10
a. The cooking of a pizza in a hot oven — slow*
b. The change in leaf color from green to read or orange in the autumn — very slow*
c. The drying of a wet swim suit hung out in the sun — slow*
d. The burning of a match after it has been started — fast*
e. The growing of grass during a warm summer — very slow*

*The definitions of very slow, slow, and fast are subjective. Your answers to this question may differ.

8.12
a. The melting of a block of ice | The changing height of the block or the changing mass of the block could be measured.
b. The setting (hardening) of concrete | The ability of an object to penetrate or mark the surface of the concrete could be measured.
c. The burning of a candle | The changing height of the candle or the changing mass of the candle could be measured.

8.14
a. Pure A and B are mixed, and after 12.0 minutes the measured concentration of C is 0.396 mol/L. | $\dfrac{0.396\text{ M} - 0.000\text{ M}}{12.0 \text{ min}} = 0.0330 \dfrac{\text{M}}{\text{min}}$
b. Pure A, B, and C are mixed together at equal concentrations of 0.300 M. After 8.00 minutes, the concentration of C is found to be 0.455 M. | $\dfrac{0.455\text{ M} - 0.300\text{ M}}{8.00 \text{ min}} = 0.0194 \dfrac{\text{M}}{\text{min}}$

8.16 $\dfrac{\left(\dfrac{2.97 \times 10^{-2}\text{ mol}}{250 \text{ mL}\left(\frac{1\text{ L}}{1000\text{ mL}}\right)}\right) - 0.000\text{ M}}{30.0 \text{ min}} = 0.00396 \dfrac{\text{M}}{\text{min}}$

8.18 $\left(2.77\times10^{-2}\text{ atm}\right)\left(500\text{ mL}\right)\left(\frac{1\text{ L}}{1000\text{ mL}}\right) = n\left(0.0821\frac{\text{L·atm}}{\text{mol·K}}\right)\left((25.0+273)\text{K}\right)$

$$n = 5.66\times10^{-4}\text{ moles}$$

$$\frac{\frac{5.66\times10^{-4}\text{ moles}}{500\text{ mL}\left(\frac{1\text{ L}}{1000\text{ mL}}\right)} - 0\text{ M}}{750\text{ sec}} = 1.51\times10^{-6}\frac{\text{M}}{\text{sec}}$$

MOLECULAR COLLISIONS (SECTION 8.3)

☑8.20   a. In order for a reaction to occur, the reactant molecules must collide. Fewer collisions will occur if the concentration is decreased.

b. In order for a reaction to occur, the reactant molecules must collide with the right amount of energy. If the temperature is lowered, the reactants will lose energy.

c. In order for a reaction to occur, the reactant molecules must collide. More collisions will occur if the concentration of either of the reactants is increased.

8.22   As temperature increases, the molecules move faster. The faster the molecules move, the more likely they are to collide.

As the temperature increases, the molecules gain energy. The more energy the molecules have, the more they are likely to react once they collide.

ENERGY DIAGRAMS (SECTION 8.4)

8.24   a. Endothermic (endergonic) reaction with activation energy.

b. Enothermic (endergonic) reaction without activation energy.

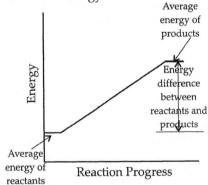

Both of these energy diagrams have the same average energy of the reactants, average energy of products, and energy difference between reactants and products. The main difference between these two energy diagrams is that the first diagram has an activation energy and the second diagram does not.

☑8.26

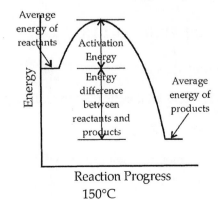

Reaction Progress
room temperature

Reaction Progress
150°C

112 Chapter 8

Both of these energy diagrams have the same average energy of the reactants, average energy of products, and energy difference between reactants and products (500 kJ/mol reactant). The reaction that occurs at room temperature does not require activation energy (or has a small activation energy). The reaction that occurs at 150°C has a significant activation energy.

FACTORS THAT INFLUENCE REACTION RATES (SECTION 8.5)

8.28 a. $2\,I^-\,(aq) + Pb^{2+}(aq) \rightarrow PbI_2\,(s)$
Rapid; The reactants are ions and the product is insoluble in water.

b. $Br^-\,(aq) + I^-\,(aq) \rightarrow BrI^{2-}\,(aq)$
Won't react; These ions are stable and do not have affinity for each other.

c. $NH_3\,(g) + HCl\,(g) \rightarrow NH_4Cl\,(s)$
Rapid; The reactant molecules are in the gas phase and will collide frequently.

d. $CaO\,(s) + CO_2\,(g) \rightarrow CaCO_3\,(s)$
The reaction rate depends on the surface area of the CaO. The larger the surface area of the solid, the faster the reaction will occur.

☑8.30 To speed up a reaction, I might (1) heat the reactants in order to increase the energy of the reactants and the frequency of collision, (2) stir the reactants in order to increase the frequency of collision, (3) increase the surface area of the reactants to increase the number of collisions, and (4) add a catalyst to lower the activation energy for the reaction. Only three of these steps are needed for a complete answer.

8.32 The reaction rate doubles for every 10°C increase in temperature. This reaction would only take 7.5 minutes at 40°C.

$$\text{change in temperature} = 40°C - 20°C = 20°C$$

The reaction rate doubles for every 10°C temperature increase.

$$20°C \text{ change in temperature} \left(\frac{1 \text{ double of the reaction rate}}{10°C} \right) = 2 \text{ doubles of the reaction rate}$$

or 1 quadruple of the reaction rate

The reaction rate will be 4 times faster; therefore, the time required will be one–quarter as long.

$$30 \text{ min} \left(\frac{1}{4} \right) = 7.5 \text{ min}$$

8.34 Catalysts speed up reactions by lowering the activation energy required for a reaction. The activation energy may be lowered because the catalyst provides a surface on which the reaction can occur or is part of a lower-energy reaction intermediate that is released after the reaction is complete.

CHEMICAL EQUILIBRIUM (SECTION 8.6)

8.36 a. H_2 + I_2 ⇌ $2\,HI$
colorless gas   violet gas   colorless gas

The color of the gas mixture will stop changing once equilibrium is reached.

b. solid sugar + water ⇌ sugar solution

The amount of solid sugar will become constant, once the mixture has reached equilibrium.

Reaction Rates and Equilibrium 113

c. N_2 + $2\,O_2$ ⇌ $2\,NO_2$
colorless gas   colorless gas   red-brown gas

Both the color and the pressure of the gas mixture will stop changing once equilibrium is reached.

THE POSITION OF EQUILIBRIUM (SECTION 8.7)

☑8.38 a. $2\,CO + O_2 \rightleftarrows 2\,CO_2$ $\qquad K_{eq} = \dfrac{[CO_2]^2}{[CO]^2[O_2]}$

b. $N_2O_4 \rightleftarrows 2\,NO_2$ $\qquad K_{eq} = \dfrac{[NO_2]^2}{[N_2O_4]}$

c. $2\,C_2H_6 + 7\,O_2 \rightleftarrows 4\,CO_2 + 6\,H_2O$ $\qquad K_{eq} = \dfrac{[CO_2]^4[H_2O]^6}{[C_2H_6]^2[O_2]^7}$

d. $2\,NOCl \rightleftarrows 2\,NO + Cl_2$ $\qquad K_{eq} = \dfrac{[NO]^2[Cl_2]}{[NOCl]^2}$

e. $2\,Cl_2O_5 \rightleftarrows O_2 + 4\,ClO_2$ $\qquad K_{eq} = \dfrac{[O_2][ClO_2]^4}{[Cl_2O_5]^2}$

8.40 a. $Fe^{3+} + 6\,CN^- \rightleftarrows Fe(CN)_6^{3-}$ $\qquad K_{eq} = \dfrac{\left[Fe(CN)_6^{3-}\right]}{[Fe^{3+}][CN^-]^6}$

b. $Ag^+ + 2\,NH_3 \rightleftarrows Ag(NH_3)_2^+$ $\qquad K_{eq} = \dfrac{\left[Ag(NH_3)_2^+\right]}{[Ag^+][NH_3]^2}$

c. $Au^{3+} + 4\,Cl^- \rightleftarrows AuCl_4^-$ $\qquad K_{eq} = \dfrac{[AuCl_4^-]}{[Au^{3+}][Cl^-]^4}$

8.42 a. $K_{eq} = \dfrac{[CO_2][H_2O]^2}{[CH_4][O_2]^2}$ $\qquad CH_4 + 2\,O_2 \rightleftarrows CO_2 + 2\,H_2O$

b. $K_{eq} = \dfrac{[CH_4][H_2O]}{[H_2]^3[CO]}$ $\qquad 3\,H_2 + CO \rightleftarrows CH_4 + H_2O$

c. $K_{eq} = \dfrac{[O_2]^3}{[O_3]^2}$ $\qquad 2\,O_3 \rightleftarrows 3\,O_2$

d. $K_{eq} = \dfrac{[NH_3]^4[O_2]^7}{[NO_2]^4[H_2O]^6}$ $\qquad 4\,NO_2 + 6\,H_2O \rightleftarrows 4\,NH_3 + 7\,O_2$

☑8.44 $K_{eq} = \dfrac{[Br_2][Cl_2]}{[BrCl]^2} = \dfrac{[0.26\,M][0.26\,M]}{[0.38\,M]^2} = 0.47$

8.46 $K_{eq} = \dfrac{[NO]^2[Cl_2]}{[NOCl]^2} = \dfrac{[0.92\,M]^2[0.20\,M]}{[1.31\,M]^2} = 0.099$

114 Chapter 8

8.48 a. K = 5.9   [reactants] smaller than [products]
    b. K = 3.3 x 10^6 [reactants] smaller than [products]
    c. K = 2.7 x 10^{-4} [reactants] larger than [products]
    d. K = 0.0000558   [reactants] larger than [products]

FACTORS THAT INFLUENCE EQUILIBRIUM POSITION (SECTION 8.8)

☑ 8.50 a. Ag^+ (aq) + Cl^- (aq) ⇌ AgCl (s) ; some Ag^+ is removed.   shift to the left
    b. 2 HI (g) + heat ⇌ H_2 (g) + I_2 (g) ; the system is heated.   shift to the right
    c. 6 Cu (s) + N_2 (g) + heat ⇌ 2 Cu_3N (s) ; the system is cooled and some
     N_2 is removed.   shift to the left

8.52 a. Cu^{2+} (aq)   +   4 NH_3 (aq)   ⇌ $Cu(NH_3)_4^{2+}$ (aq);
      blue   colorless   dark purple
    some NH_3 is added to the equilibrium mixture.

    The equilibrium will shift to the right and the mixture will become less blue and more purple.

   b. Pb^{2+} (aq)   +   2 Cl^- (aq)   ⇌ $PbCl_2$ (s)   +   heat;
      colorless   colorless   white solid
    the equilibrium mixture is cooled.

    The equilibrium will shift to the right and more precipitate will form.

   c. C_2H_4 + I_2 ⇌ $C_2H_4I_2$ +   heat;
      colorless gas   violet gas   colorless gas
    some $C_2H_4I_2$ is removed from the equilibrium mixture.

    The equilibrium will shift to the right and the mixture will become less violet and more heat will be produced.

   d. C_2H_4 + I_2 ⇌ $C_2H_4I_2$ +   heat;
      colorless gas   violet gas   colorless gas
    the equilibrium mixture is cooled.

    The equilibrium will shift to the right and the mixture will become less violet and more heat will be produced.

   e.   heat   +   4 NO_2 +   6 H_2O ⇌   7 O_2 +   4 NH_3;
          brown gas   colorless gas   colorless gas   colorless gas
    a catalyst is added, and NH_3 is added to the equilibrium mixture.

    The catalyst does not have any effect on the equilibrium; however, adding NH_3 shifts the equilibrium to the left and this produces heat, lowers the pressure, and increases the brown color of the equilibrium mixture.

8.54 a. 2 N_2O (g) + 3 O_2 (g) ⇌ 4 NO_2 (g) + heat; O_2 is added

    The equilibrium will shift to the right. The concentrations of N_2O and O_2 will decrease and the concentration of NO_2 will increase.

b. $2\ SO_3\ (g) + heat \rightleftarrows 2\ SO_2\ (g) + O_2\ (g)$; the system is heated

The equilibrium will shift to the right. The concentration of SO_3 will decrease and the concentrations of SO_2 and O_2 will increase.

c. $2\ CO\ (g) + O_2\ (g) \rightleftarrows 2\ CO_2\ (g) + heat$; the system is cooled

The equilibrium will shift to the right. The concentrations of CO and O_2 will decrease and the concentration of CO_2 will increase.

8.56 $N_2\ (g) + 3\ H_2\ (g) \rightleftarrows 2\ NH_3\ (g) + heat$

a.	Some N_2 is added.	to the right
b.	The temperature is increased.	to the left
c.	Some NH_3 is removed.	to the right
d.	Some H_2 is removed.	to the left
e.	A catalyst is added.	no shift
f.	The temperature is increased, and some H_2 is removed.	to the left

ADDITIONAL ACTIVITIES

Section 8.1 Review:
(1) Place the following pictures in the order of increasing entropy.
(2) The three pictures below can be put into 6 different sequences. Identify the sequence that depicts the most nonspontaneous transitions from first picture to second picture and from second picture to third picture. Explain you order using the terms endergonic or exergonic and entropy.
(3) Does the reverse order of (2) show a spontaneous process? Explain.

A

B

C

Section 8.2 Review:
(1) What is meant by the "rate of travel" in a car? What units are used?
(2) What is meant by a "long distance rate" in regard to telephones? What units are used?
(3) What do (1) and (2) have in common with a reaction rate?
(4) What is different about (1), (2), and reaction rates?

Section 8.3 Review:
(1) What are the three assumptions that serve as a basis for most reaction mechanisms?
(2) Why does orientation not matter for spherical reactants with uniform charges?
(3) What is the difference between *internal energy* and *activation energy*?

Section 8.4 Review:
Assume activation energy is required for both processes and draw/label energy diagrams for the following:

(1) $CH_4 (g) + 2 O_2 (g) \rightarrow CO_2 (g) + 2 H_2O (l) + 890$ kJ
(2) $NH_4NO_3 (s) + 26.5$ kJ $\rightarrow NH_4NO_3 (aq)$

Section 8.5 Review:
(1) What is meant by the "nature of the reactants"?
(2) What is the relationship between concentration, collisions, and reaction rates?
(3) What is the relationship between temperature and reaction rates?
(4) How does a catalyst affect reaction rates?
(5) How does an inhibitor affect reaction rates?

Section 8.6 Review:
(1) What is equal at equilibrium?
(2) What is not equal at equilibrium?
(3) Why is a double arrow used for an equilibrium reaction?

Section 8.7 Review:
(1) Write an equilibrium expression using a double arrow and the words *reactants* and *products*.
(2) Where are the reactants found in (1) (relative to the arrow)?
(3) Where are the products found in (1) (relative to the arrow)?
(4) Write an equilibrium expression for (1).
(5) What are the exponents in (4)?
(6) If K were 5×10^7 for (4), would there be more reactants or products at equilibrium?
(7) Would the equilibrium in (6) lie to the left or to the right?
(8) If K were 7×10^{-5} for (4), would there be more reactants or products at equilibrium?
(9) Would the equilibrium in (8) lie to the left or to the right?

Section 8.8 Review:
Equilibrium can be thought of as a seesaw. The pivot point (fulcrum) of the seesaw can be adjusted so that the forward and reverse reactions balance regardless of whether more reactants or products are present at equilibrium.

For convenience, let's use the middle picture to represent equilibrium in general. Remember, that the concentrations of reactants and products are not equal for most equilibrium systems.

(1) How will the seesaw tip if more reactants are added to the system at equilibrium? What change must be made to reestablish equilibrium?
(2) How will the seesaw tip if more products are added to the system at equilibrium? What change must be made to reestablish equilibrium?
(3) How will the seesaw tip if some reactants are removed from the system at equilibrium? What change must be made to reestablish equilibrium?
(4) How will the seesaw tip if some products are removed from the system at equilibrium? What change must be made to reestablish equilibrium?
(5) If the forward reaction is endothermic and heat is added, which way will the seesaw tip? What change must be made to reestablish equilibrium?
(6) If the forward reaction is exothermic and heat is added, which way will the seesaw tip? What change must be made to reestablish equilibrium?
(7) If the reverse reaction is endothermic and heat is removed, which way will the seesaw tip? What change must be made to reestablish equilibrium?

(8) If the reverse reaction is exothermic and heat is removed, which way will the seesaw tip? What change must be made to reestablish equilibrium?

(9) If a catalyst is added, which way will the seesaw tip? What change must be made to reestablish equilibrium?

(10) If an inhibitor is added, which way will the seesaw tip? What change must be made to reestablish equilibrium?

(11) Which of the scenarios above will change the equilibrium constant, K?

Tying It All Together with a Laboratory Application:

A beaker is placed on a small puddle of water on a piece of wood. A solution is prepared in the beaker by mixing 3.8 g of solid NH_4Cl with 50 mL of water. When the laboratory student attempts to lift the beaker from the piece of wood, she discovers that the beaker is stuck to the wood because the puddle of water has frozen. (1) Sketch an energy diagram of the freezing process. The formation of this solution is a(n) (2) _____ (endothermic or exothermic) process. It is also a spontaneous process. The level of entropy (3) _____ as the solution is prepared.

The ammonium ion establishes an equilibrium in the solution:

$$NH_4^+(aq) \rightleftharpoons H^+(aq) + NH_3(aq)$$

The equilibrium constant expression for this reaction is (4) _____. The equilibrium constant has a value of 5.6×10^{-10}. Equilibrium lies to the (5) _____ because the concentration of reactants is (6) _____ (greater than, less than, or equal to) the concentration of products at equilibrium. The reaction rate of the forward reaction is (7) _____ (greater than, less than, or equal to) the reaction rate of the reverse reaction at equilibrium.

A solution of silver nitrate is added to the reaction mixture, solid silver chloride forms, and precipitates to the bottom of the beaker.

$$AgNO_3(aq) + Cl^-(aq) \rightarrow AgCl(s) + NO_3^-(aq)$$

This reaction occurs (8) _____ than if solid silver nitrate were mixed with solid ammonium chloride because the reactants are both aqueous ions. The ions are spheres of uniform charge, thus the orientation of the ions (9) _____ important for an effective collision. The addition of silver nitrate (10) _____ the equilibrium established by the ammonium ion.

A solution of hydrochloric acid (HCl) is added to the reaction mixture. Hydrochloric acid dissociates in water to produce (11) _____ and _____ ions. This affects the equilibrium established by the ammonium ion because (12) _____ have been added. In order to reestablish equilibrium, the equilibrium will shift (13) _____. The concentration of NH_3 will (14) _____.

SOLUTIONS FOR THE ADDITIONAL ACTIVITIES

Section 8.1 Review:

(1) increasing entropy: B < A < C

(2) nonspontaneous process: C → A → B; Picture C is like a disordered stack of wood. Picture C has the highest entropy. To stack the wood as in picture A requires energy and is an endergonic process. The entropy is decreased. Turning a stack of wood (A) into a specific design (B) requires energy (is endergonic) and decreases entropy.

(3) The progression B → A → C is not spontaneous because the transition between picture B and picture A requires a significant activation energy. Both picture A and picture B are highly ordered. Picture B does have higher potential energy than picture A. The processes A → C and B → C are spontaneous processes because the entropy increases and the energy decreases.

Section 8.2 Review:
(1) The speed (or velocity) of the car is the "rate of travel." Speed is usually measured in miles per hour or kilometers per hour.
(2) The cost of the call per minute is a "long distance rate." The units are cents per minute in the USA.
(3) Speeds, long distance rates, and reaction rates all represent the change in a unit divided by time.
(4) For car speeds, the unit that changes is miles or distance per unit time. For long distance rates, the unit that changes is money per unit time. For reaction rates, the unit that changes is the concentration per unit time.

Section 8.3 Review:
(1) Molecules must collide in order to react. Molecules must collide with enough energy to react. Some molecules must collide at the correct orientation to react.
(2) No matter how the two spherical reactants collide, the orientation is the same.
(3) Internal energy is the energy of molecular vibrations. Chemical bonds can stretch, bend, and wag. Activation energy is the minimum energy required for molecules to react.

Section 8.4 Review:

Section 8.5 Review:
(1) The "nature of the reactants" refers to the state of matter as well as the type of bonding in the reactants. Ions in solution and gases usually react very quickly. Solids and liquids react more slowly unless the surface area of the solid is very large and the liquids are well mixed.
(2) As concentration increases, the number of collisions increases. The more collisions that occur, the higher the reaction rate is. As concentration increases, reaction rate increases.
(3) As temperature increases, the number of collisions increases. Also, the number of collisions with enough energy to be effective collisions increases. Consequently, the reaction rate increases.
(4) A catalyst increases reaction rates by lowering the activation energy.
(5) An inhibitor increases reaction rates by increasing the activation energy or by decreasing the number of effective collisions.

Section 8.6 Review:
(1) The rates of the forward and the reverse reactions are equal at equilibrium.
(2) The concentrations of reactants and products may not be equal at equilibrium.
(3) In equilibrium, the reaction proceeds from left to right and from right to left. The double arrow indicates both the forward and reverse reactions occur simultaneously.

Section 8.7 Review:
(1) reactants $\rightleftharpoons$ products
(2) Reactants are to the left of the arrow.
(3) Products are to the right of the arrow.
(4) $K = \dfrac{[\text{products}]}{[\text{reactants}]}$

(5) The exponents are both equal to 1 because neither the reactants nor the products in (1) had a coefficient. The exponents in the equilibrium constant expression are always equal to the coefficients from the chemical equation.
(6) More products would be present at equilibrium if K were 5×10^7 for (4).
(7) The equilibrium would lie to the right.
(8) More reactants would be present at equilibrium if K were 7×10^{-5} for (4).
(9) The equilibrium would lie to the left.

Section 8.8 Review:

(1) ; In order to rebalance the seesaw, some of the reactants will need to convert to products. The equilibrium must shift to the right.

(2) ; In order to rebalance the seesaw, some of the products will need to convert to reactants. The equilibrium must shift to the left.

(3) ; In order to rebalance the seesaw, some of the products will need to convert to reactants. The equilibrium must shift to the left.

(4) ; In order to rebalance the seesaw, some of the reactants will need to convert to products. The equilibrium must shift to the right.

(5) R + heat $\rightleftharpoons$ P; ; In order to rebalance the seesaw, the heat must be used and some of the reactants will need to convert to products. The equilibrium must shift to the right.

(6) R $\rightleftharpoons$ P + heat; ; In order to rebalance the seesaw, the heat must be used and some of the products will need to convert to reactants. The equilibrium must shift to the left.

(7) R $\rightleftharpoons$ P + heat; ; In order to rebalance the seesaw, heat must be produced and some of the reactants will need to convert to products. The equilibrium must shift to the right.

(8) R + heat $\rightleftharpoons$ P; ; In order to rebalance the seesaw, heat must be produced and some of the products will need to convert to reactants. The equilibrium must shift to the left.

(9) ; A catalyst will not tip the seesaw. A catalyst will speed up the forward and reverse reaction rates, but it will not affect the concentration of reactants or products.

(10) ; An inhibitor will not tip the seesaw. An inhibitor will slow down the forward and reverse reaction rates, but it will not affect the concentration of reactants or products.

(11) All of the scenarios involving heat (5, 6, 7, and 8) will change the value of K.

Tying It All Together with a Laboratory Application:

(1)
(2) endothermic
(3) increases
(4) $K = \dfrac{[H^+][NH_3]}{[NH_4^+]}$
(5) left
(6) greater than
(7) equal to
(8) faster
(9) is not
(10) does not affect
(11) H^+, Cl^-
(12) products
(13) left
(14) decrease

SELF-TEST QUESTIONS
Multiple Choice

1. A catalyst
   a. is not used up in a reaction
   b. changes the rate of reaction
   c. affects the forward reaction the same as it affects the reverse reaction
   d. all of the above

2. Which of the following responses correctly arranges the states of matter for a pure substance in the order of decreasing entropy?
   a. gas, liquid, solid
   b. liquid, solid, gas
   c. solid, liquid, gas
   d. solid, gas, liquid

3. Four processes occur as the following changes take place in energy and entropy. Which process is definitely nonspontaneous?
   a. energy decrease and entropy increase
   b. energy decrease and entropy decrease
   c. energy increase and entropy increase
   d. energy increase and entropy decrease

4. A carrot cooks in 15 minutes in boiling water (100°C). How long will it take to cook a carrot inside a pressure cooker where the temperature is 10°C greater (110°C)?
   a. 3.7 minutes
   b. 5.0 minutes
   c. 7.5 minutes
   d. 30 minutes

5. The gases A and B react as follows: A + B → C
   What is the average rate of reaction if pure A and B are mixed and after 30.0 seconds the concentration of C is found to be 4.75×10^{-3} M?
   a. 1.58×10^{-4} M/sec
   b. 2.38×10^{-3} M/sec
   c. 9.50×10^{-3} M/sec
   d. 1.43×10^{-1} M/sec

Questions 6 and 7 refer to the following reaction, which is assumed to be at equilibrium:
$$\text{heat} + 2\,NO + O_2 \rightleftarrows 2\,NO_2$$
In each case choose the response which best indicates the effects resulting from the described change in conditions.

6. The reaction mixture is heated.
   a. equilibrium shifts to the left
   b. equilibrium shifts to the right
   c. equilibrium does not shift
   d. the effect cannot be predicted

7. A catalyst is added to the reaction mixture.
   a. equilibrium shifts to the left
   b. equilibrium shifts to the right
   c. equilibrium does not shift
   d. the effect cannot be predicted

8. In the equilibrium constant expression for the reaction: $2 N_2O_5 \rightleftarrows 4 NO_2 + O_2$, the exponent on the concentration of N_2O_5 is
   a. 1
   b. 2
   c. 0
   d. can't be determined

9. A sample of ICl is placed in a container and equilibrium is established according to the reaction:
   $$2 ICl \rightleftarrows I_2 + Cl_2$$
   where all the materials are gases. Analysis of the equilibrium mixture gas the following molar concentrations: [ICl] = 0.26, [I_2] = [Cl_2]=0.09. What is the value of K, the equilibrium constant for the reaction?
   a. 0.031
   b. 0.12
   c. 0.35
   d. 14.8

10. The equilibrium shifts to the left when an equilibrium mixture is cooled. The forward reaction must be
    a. exothermic
    b. endothermic
    c. high entropy
    d. low entropy

True-False
11. The entropy of a cluttered room is higher than that of an orderly room.
12. A reaction rate can be thought of as the speed of a reaction.
13. Effective molecular collisions are those that allow molecules to collide but not react.
14. Increasing the concentration of reactants usually increases the reaction rate.
15. Activation energy is the energy difference between reactants and products.
16. If an endothermic reaction is spontaneous, then entropy must have decreased.
17. Catalysts act by lowering the activation energy.
18. Catalysts that slow reactions are called inhibitors.
19. In a reaction at equilibrium, the forward and reverse reactions have both stopped.
20. In a reaction at equilibrium, the concentrations of reactants and products are equal.

Matching
For each of the following processes choose the appropriate response from those on the right.
21. a match burns
22. perspiration evaporates
23. melted lead becomes a solid
24. an explosive detonates

a. both entropy and energy increase
b. both entropy and energy decrease
c. entropy increases; energy decreases
d. entropy decreases; energy increases

Three liquid fuels are to be tested. A 1.0 gram sample of each fuel is weighed out and heated to its ignition temperature. When the fuel burns, the total heat liberated is measured. The results of this experiment are given in the table below. For Questions 25-30 choose the answer from the column on the right that best fits each statement on the left.

Fuel	Ignition Temperature	Heat Liberated
X	210°C	1680 cal
Y	110°C	1410 cal
Z	285°C	1206 cal

25. it has the highest activation energy
26. it has the second highest activation energy
27. it has the lowest activation energy
28. it has the smallest energy difference between reactants and products
29. it has the largest energy difference between reactants and products
30. the described reaction is exothermic (exergonic)

a. fuel X
b. fuel Y
c. fuel Z
d. two or more reactant fuels fit this category

Use the following equilibrium expression and match the effects on the equilibrium from the right with the changes made to the equilibrium system listed on the left.

$$\text{heat} + CO + H_2 \rightleftarrows CH_3OH$$

31. add CO
32. remove some H_2
33. add CH_3OH
34. heat the system
35. add a catalyst

a. shift left
b. shift right
c. no effect on equilibrium
d. cannot determine from the information given

Chapter 9: Acids, Bases, and Salts

CHAPTER OUTLINE

9.1 The Arrhenius Theory
9.2 The Brønsted Theory
9.3 Naming Acids
9.4 The Self-Ionization of Water
9.5 The pH Concept
9.6 Properties of Acids
9.7 Properties of Bases
9.8 Salts
9.9 The Strengths of Acids and Bases
9.10 Analyzing Acids and Bases
9.11 Titration Calculations
9.12 Hydrolysis Reactions of Salts
9.13 Buffers

LEARNING OBJECTIVES/ASSESSMENT

When you have completed your study of this chapter, you should be able to:

1. Write reaction equations that illustrate the Arrhenius and Brønsted acid-base behavior. (Sections 9.1 and 9.2; Exercises 9.2 and 9.6)
2. Identify Brønsted acids and bases from written reaction equations. (Section 9.2; Exercise 9.8)
3. Name common acids. (Section 9.3; Exercise 9.22)
4. Do calculations using the concept of the self-ionization of water. (Section 9.4; Exercises 9.28 a & b, and 9.30 a & b)
5. Do calculations using the pH concept. (Section 9.5; Exercises 9.36 and 9.40)
6. Write reaction equations that illustrate the characteristic reactions of acids. (Section 9.6; Exercise 9.50)
7. Write reaction equations that illustrate various ways to prepare salts. (Sections 9.6, 9.7, and 9.8; Exercises 9.54, 9.60, and 9.68)
8. Write reaction equations that represent neutralization reactions between acids and bases. (Section 9.7; Exercise 9.58)
9. Do calculations using the concept of an equivalent of a salt. (Section 9.8; Exercise 9.74)
10. Explain the difference between the words *strong* and *weak* as applied to acids and bases. (Section 9.9; Exercise 9.86)
11. Do calculations related to the analysis of acids and bases by titration. (Sections 9.10 and 9.11; Exercises 9.92 and 9.96)
12. Explain the concept of salt hydrolysis, and write equations to illustrate the concept. (Section 9.12; Exercise 9.108)
13. Explain how buffers work, and write equations to illustrate their action. (Section 9.13; Exercise 9.116)

SOLUTIONS FOR THE END OF CHAPTER EXERCISES

THE ARRHENIUS THEORY (SECTION 9.1)

☑9.2 a. $HBrO_2$ $HBrO_2\,(aq) \rightarrow H^+(aq) + BrO_2^-(aq)$
      b. HS^- $HS^-(aq) \rightarrow H^+(aq) + S^{2-}(aq)$
      c. HBr $HBr\,(aq) \rightarrow H^+(aq) + Br^-(aq)$
      d. $HC_2H_3O_2$ $HC_2H_3O_2\,(aq) \rightarrow H^+(aq) + C_2H_3O_2^-(aq)$

9.4 a. $LiOH$ Arrhenius base $LiOH\,(aq) \rightarrow Li^+(aq) + OH^-(aq)$
      b. $C_2H_5NH_2$ not an Arrhenius base
      c. $Sr(OH)_2$ Arrhenius base $Sr(OH)_2\,(aq) \rightarrow Sr^{2+}(aq) + 2\,OH^-(aq)$
      d. $NaNH_2$ not an Arrhenius base

THE BRØNSTED THEORY (SECTION 9.2)

☑ 9.6
a. $HC_2O_4^- (aq) + H_2O (l) \rightleftharpoons H_3O^+ (aq) + C_2O_4^{2-} (aq)$
b. $HNO_2 (aq) + H_2O (l) \rightleftharpoons H_3O^+ (aq) + NO_2^- (aq)$
c. $PO_4^{3-} (aq) + H_2O (l) \rightleftharpoons HPO_4^{2-} (aq) + OH^- (aq)$
d. $H_2SO_3 (aq) + H_2O (l) \rightleftharpoons HSO_3^- (aq) + H_3O^+ (aq)$
e. $F^- (aq) + H_2O (l) \rightleftharpoons HF (aq) + OH^- (aq)$

	Brønsted Acids	Brønsted Bases
	$HC_2O_4^-$, H_3O^+	H_2O, $C_2O_4^{2-}$
	HNO_2, H_3O^+	H_2O, NO_2^-
	H_2O, HPO_4^{2-}	PO_4^{3-}, OH^-
	H_2SO_3, H_3O^+	H_2O, HSO_3^-
	H_2O, HF	F^-, OH^-

Conjugate Acid-Base Pairs

☑ 9.8
a. $HC_2O_4^-$ and $C_2O_4^{2-}$, H_3O^+ and H_2O
b. HNO_2 and NO_2^-, H_3O^+ and H_2O
c. H_2O and OH^-, HPO_4^{2-} and PO_4^{3-}
d. H_2SO_3 and HSO_3^-, H_3O^+ and H_2O
e. H_2O and OH^-, HF and F^-

9.10
a. $HF (aq) + H_2O (l) \rightleftharpoons F^- (aq) + H_3O^+ (aq)$
b. $HClO_3 (aq) + H_2O (l) \rightleftharpoons ClO_3^- (aq) + H_3O^+ (aq)$
c. $HClO (aq) + H_2O (l) \rightleftharpoons ClO^- (aq) + H_3O^+ (aq)$
d. $HS^- (aq) + H_2O (l) \rightleftharpoons S^{2-} (aq) + H_3O^+ (aq)$

9.12

		Conjugate Base
a.	$HSO_4^- (aq) + H_2O (l) \rightleftharpoons SO_4^{2-} (aq) + H_3O^+ (aq)$	SO_4^{2-}
b.	$CH_3NH_3^+ (aq) + H_2O (l) \rightleftharpoons CH_3NH_2 (aq) + H_3O^+ (aq)$	CH_3NH_2
c.	$HClO_4 (aq) + H_2O (l) \rightleftharpoons ClO_4^- (aq) + H_3O^+ (aq)$	ClO_4^-
d.	$NH_4^+ (aq) + H_2O (l) \rightleftharpoons NH_3 (aq) + H_3O^+ (aq)$	NH_3
e.	$HCl (aq) + H_2O (l) \rightleftharpoons Cl^- (aq) + H_3O^+ (aq)$	Cl^-

9.14

		Conjugate Acid
a.	$HCO_3^- (aq) + H_2O (l) \rightleftharpoons H_2CO_3 (aq) + OH^- (aq)$	H_2CO_3
b.	$S^{2-} (aq) + H_2O (l) \rightleftharpoons HS^- (aq) + OH^- (aq)$	HS^-
c.	$HS^- (aq) + H_2O (l) \rightleftharpoons H_2S (aq) + OH^- (aq)$	H_2S
d.	$HC_2O_4^- (aq) + H_2O (l) \rightleftharpoons H_2C_2O_4 (aq) + OH^- (aq)$	$H_2C_2O_4$
e.	$HN_2O_2^- (aq) + H_2O (l) \rightleftharpoons H_2N_2O_2 (aq) + OH^- (aq)$	$H_2N_2O_2$

9.16

		Missing Formula
a.	$H_2AsO_4^- (aq) + ? \rightarrow NH_4^+ (aq) + HAsO_4^{2-} (aq)$	$NH_3 (aq)$
b.	$C_6H_5NH_2 (aq) + ? \rightarrow C_6H_5NH_3^+ (aq) + OH^- (aq)$	$H_2O (aq)$
c.	$S^{2-} (aq) + H_2O (l) \rightarrow ? + OH^- (aq)$	$HS^- (aq)$
d.	$(CH_3)_2NH (aq) + ? \rightarrow (CH_3)_2NH_2^+ (aq) + Br^- (aq)$	$HBr (aq)$
e.	$H_2PO_4^- (aq) + CH_3NH_2 (aq) \rightarrow HPO_4^{2-} (aq) + ?$	$CH_3NH_3^+ (aq)$

Acids, Bases, and Salts 125

		Acid	Base	Equation
9.18	a.	HS$^-$	NH$_3$	HS$^-$(aq) + NH$_3$(aq) → S^{2-}(aq) + NH$_4^+$(aq)
	b.	H$_2$O	ClO$_3^-$	H$_2$O(l) + ClO$_3^-$(aq) → HClO$_3$(aq) + OH$^-$(aq)
	c.	H$_2$O	NH$_2^-$	H$_2$O(l) + NH$_2^-$(aq) → NH$_3$(aq) + OH$^-$(aq)
	d.	HBO$_3^{2-}$	H$_2$O	HBO$_3^{2-}$(aq) + H$_2$O(l) → BO$_3^{3-}$(aq) + H$_3$O$^+$(aq)
	e.	HNO$_2$	NH$_3$	HNO$_2$(aq) + NH$_3$(aq) → NH$_4^+$(aq) + NO$_2^-$(aq)

NAMING ACIDS (SECTION 9.3)

9.20 HCN (aq) = hydrocyanic acid

☑9.22   a. H$_2$Te (aq)   hydrotelluric acid   c. H$_2$SO$_3$ sulfurous acid
        b. HClO   hypochlorous acid   d. HNO$_2$ nitrous acid

9.24 H$_2$C$_2$H$_4$O$_4$ = succinic acid

9.26 carbonic acid = H$_2$CO$_3$

THE SELF-IONIZATION OF WATER (SECTION 9.4)

9.28		[H$_3$O$^+$]	Calculation	[OH$^-$]
	☑a.	0.044	$[OH^-] = \dfrac{1.0 \times 10^{-14}}{0.044} = 2.\overline{27} \times 10^{-13}$ M	2.3 x 10^{-13} M
	☑b.	1.3 x 10^{-4}	$[OH^-] = \dfrac{1.0 \times 10^{-14}}{1.3 \times 10^{-4}} = 7.6923 \times 10^{-11}$ M	7.7 x 10^{-11} M
	c.	0.0087	$[OH^-] = \dfrac{1.0 \times 10^{-14}}{0.087} = 1.1494 \times 10^{-13}$ M	1.1 x 10^{-13} M
	d.	7.9 x 10^{-10}	$[OH^-] = \dfrac{1.0 \times 10^{-14}}{7.9 \times 10^{-10}} = 1.2658 \times 10^{-5}$ M	1.3 x 10^{-5} M
	e.	3.3 x 10^{-2}	$[OH^-] = \dfrac{1.0 \times 10^{-14}}{3.3 \times 10^{-2}} = 3.\overline{03} \times 10^{-13}$ M	3.0 x 10^{-13} M

9.30		[OH$^-$]	Calculation	[H$_3$O$^+$]
	☑a.	6.9 x 10^{-5}	$[H_3O^+] = \dfrac{1.0 \times 10^{-14}}{6.9 \times 10^{-5}} = 1.449275 \times 10^{-10}$ M	1.4 x 10^{-10} M
	☑b.	0.074	$[H_3O^+] = \dfrac{1.0 \times 10^{-14}}{0.074} = 1.35\overline{135} \times 10^{-13}$ M	1.4 x 10^{-13} M
	c.	4.9	$[H_3O^+] = \dfrac{1.0 \times 10^{-14}}{4.9} = 2.0408 \times 10^{-15}$ M	2.0 x 10^{-15} M
	d.	1.7 x 10^{-3}	$[H_3O^+] = \dfrac{1.0 \times 10^{-14}}{1.7 \times 10^{-3}} = 5.88235 \times 10^{-12}$ M	5.9 x 10^{-12} M
	e.	9.2 x 10^{-9}	$[H_3O^+] = \dfrac{1.0 \times 10^{-14}}{9.2 \times 10^{-9}} = 1.0869565 \times 10^{-6}$ M	1.1 x 10^{-6} M

9.32		9.28			9.30	
	a.	[OH$^-$] = 2.3 x 10^{-13} M	acidic		[H$_3$O$^+$] = 1.4 x 10^{-10} M	basic
	b.	[OH$^-$] = 7.7 x 10^{-11} M	acidic		[H$_3$O$^+$] = 7.4 x 10^{-13} M	basic

126 Chapter 9

9.28
c. [OH⁻] = 1.1 x 10⁻¹³ M   acidic
d. [OH⁻] = 1.3 x 10⁻⁵ M   basic
e. [OH⁻] = 3.0 x 10⁻¹³ M   acidic

9.30
[H₃O⁺] = 2.0 x 10⁻¹⁵ M   basic
[H₃O⁺] = 5.9 x 10⁻¹² M   basic
[H₃O⁺] = 1.1 x 10⁻⁶ M   acidic

THE pH CONCEPT (SECTION 9.5)

9.34   a. pH = 4.7   acidic   d. pH = 11   basic
        b. pH = 8.8   basic   e. pH = 5.9   acidic
        c. pH = 1.5   acidic

☑ 9.36 a. [H⁺] = 4.1 x 10⁻⁹ $pH = -\log[4.1 \times 10^{-9}] = 8.39$ basic

        b. [OH⁻] = 9.4 x 10⁻⁴ $pH = 14 - (-\log[9.4 \times 10^{-4}]) = 10.97$ basic

        c. [OH⁻] = 10 [H⁺] $pH = -\log[\sqrt{\frac{1.0 \times 10^{-14}}{10}}] = 7.50$ basic

        d. [H⁺] = 2.3 x 10⁻² $pH = -\log[2.3 \times 10^{-2}] = 1.64$ acidic

        e. [OH⁻] = 5.1 x 10⁻¹⁰ $pH = 14 - (-\log[5.1 \times 10^{-10}]) = 4.71$ acidic

9.38   a. [H⁺] = 2.2 x 10⁻³ $pH = -\log[2.2 \times 10^{-3}] = 2.66$ acidic

        b. [H⁺] = 3.9 x 10⁻¹² $pH = -\log[3.9 \times 10^{-12}] = 11.41$ basic

        c. [H⁺] = 7.5 x 10⁻⁶ $pH = -\log[7.5 \times 10^{-6}] = 5.12$ acidic

        d. [OH⁻] = 2.5 x 10⁻⁴ $pH = 14 - (-\log[2.5 \times 10^{-4}]) = 10.40$ basic

        e. [OH⁻] = 8.6 x 10⁻¹⁰ $pH = 14 - (-\log[8.6 \times 10^{-10}]) = 4.93$ acidic

☑ 9.40 a. pH = 9.27 $[H^+] = 10^{-9.27} = 5.4 \times 10^{-10}$ M

        b. pH = 2.55 $[H^+] = 10^{-2.55} = 2.8 \times 10^{-3}$ M

        c. pH = 5.42 $[H^+] = 10^{-5.42} = 3.80 \times 10^{-6}$ M

9.42   a. pH = 3.95 $[H^+] = 10^{-3.95} = 1.1 \times 10^{-4}$ M $[OH^-] = \frac{10^{-14}}{10^{-3.95}} = 8.9 \times 10^{-11}$ M

        b. pH = 4.00 $[H^+] = 10^{-4.00} = 1.0 \times 10^{-4}$ M $[OH^-] = \frac{10^{-14}}{10^{-4.00}} = 1.0 \times 10^{-10}$ M

        c. pH = 11.86 $[H^+] = 10^{-11.86} = 1.4 \times 10^{-12}$ M $[OH^-] = \frac{10^{-14}}{10^{-11.86}} = 7.2 \times 10^{-3}$ M

9.44   a. Bile, pH = 8.05 $[H^+] = 10^{-8.05} = 8.9 \times 10^{-9}$ M   basic

        b. Vaginal fluid, pH = 3.93 $[H^+] = 10^{-3.93} = 1.2 \times 10^{-4}$ M   acidic

        c. Semen, pH = 7.38 $[H^+] = 10^{-7.38} = 4.2 \times 10^{-8}$ M   basic

        d. Cerebrospinal fluid, pH = 7.40 $[H^+] = 10^{-7.40} = 4.0 \times 10^{-8}$ M   basic

        e. Perspiration, pH = 6.23 $[H^+] = 10^{-6.23} = 5.9 \times 10^{-7}$ M   acidic

Acids, Bases, and Salts 127

9.46 a. Soft drink, pH = 2.91 $[H^+] = 10^{-2.91} = 1.2 \times 10^{-3}$ M   acidic
      b. Tomato juice, pH = 4.11 $[H^+] = 10^{-4.11} = 7.8 \times 10^{-5}$ M   acidic
      c. Lemon juice, pH = 2.32 $[H^+] = 10^{-2.32} = 4.8 \times 10^{-3}$ M   acidic
      d. Grapefruit juice, pH = 3.07 $[H^+] = 10^{-3.07} = 8.5 \times 10^{-4}$ M   acidic

PROPERTIES OF ACIDS (SECTION 9.6)

9.48 a. $(2\text{ L})(3.0\text{ M}) = (V_c)(6\text{ M}) \Rightarrow V_c = 1\text{ L}$

   I would add approximately 1 L of water to a 2 L graduated cylinder, then dilute up to the 2 L mark with 6 M nitric acid. I would make sure to stir the solution.

   b. $(500\text{ mL})(1.5\text{ M}) = (V_c)(15\text{ M}) \Rightarrow V_c = 50\text{ mL}$

   I would add approximately 450 mL of water to a 500 mL graduated cylinder, then dilute up to the 500 mL mark with 15 M aqueous ammonia. I would make sure to stir the solution.

   c. $(5\text{ L})(0.2\text{ M}) = (V_c)(12\text{ M}) \Rightarrow V_c = 83\text{ mL}$

   I would add approximately 4.9 L of water to a 5 L graduated cylinder, then dilute up to the 5 L mark with 12 M hydrochloric acid. I would make sure to stir the solution.

☑9.50 a. H_2SO_4 (aq) + 2 H_2O (l) → 2 H_3O^+(aq) + SO_4^{2-}(aq)
       b. H_2SO_4 (aq) + CaO (s) → H_2O (l) + $CaSO_4$ (aq)
       c. H_2SO_4 (aq) + $Mg(OH)_2$ (s) → 2 H_2O (l) + $MgSO_4$ (aq)
       d. H_2SO_4 (aq) + $CuCO_3$ (s) → H_2O (l) + $CuSO_4$ (aq) + CO_2 (g)
       e. H_2SO_4 (aq) + 2 $KHCO_3$ (s) → 2 H_2O (l) + K_2SO_4 (aq) + 2 CO_2 (g)
       f. H_2SO_4 (aq) + Mg (s) → H_2 (g) + $MgSO_4$ (aq)

9.52 a. TIE*: 2 H^+ (aq) + SO_4^{2-} (aq) + 2 H_2O (l) → 2 H_3O^+ (aq) + SO_4^{2-}(aq)
         NIE**: H^+ (aq) + H_2O (l) → H_3O^+ (aq)

      b. TIE*: 2 H^+ (aq) + SO_4^{2-} (aq) + CaO (s) → H_2O (l) + Ca^{2+}(aq) + SO_4^{2-}(aq)
         NIE**: 2 H^+ (aq) + CaO (s) → H_2O (l) + Ca^{2+}(aq)

      c. TIE*: 2 H^+ (aq) + SO_4^{2-} (aq) + $Mg(OH)_2$ (s) → H_2O (l) + Mg^{2+}(aq) + SO_4^{2-} (aq)
         NIE**: 2 H^+ (aq) + $Mg(OH)_2$ (s) → 2 H_2O (l) + Mg^{2+}(aq)

      d. TIE*: 2 H^+ (aq) + SO_4^{2-} (aq) + $CuCO_3$ (s) → H_2O (l) + Cu^{2+}(aq) + SO_4^{2-} (aq) + CO_2 (g)
         NIE**: 2 H^+ (aq) + $CuCO_3$ (s) → H_2O (l) + Cu^{2+}(aq) + CO_2 (g)

      e. TIE*: 2 H^+ (aq) + SO_4^{2-} (aq) + 2 $KHCO_3$ (s) → 2 H_2O (l) + 2 K^+(aq) + SO_4^{2-} (aq) + 2 CO_2 (g)
         NIE**: H^+ (aq) + $KHCO_3$ (s) → H_2O (l) + K^+(aq) + CO_2 (g)

*M = molecular equation, **TIE = total ionic equation, ***NIE = net ionic equation

128 Chapter 9

    f.   TIE*: $2\,H^+\,(aq) + SO_4^{2-}\,(aq) + Mg\,(s) \rightarrow H_2\,(g) + Mg^{2+}(aq) + SO_4^{2-}\,(aq)$

         NIE**: $2\,H^+\,(aq) + Mg\,(s) \rightarrow H_2\,(g) + Mg^{2+}(aq)$

☑9.54   a. $2\,HCl\,(aq) + MgO\,(s) \rightarrow H_2O\,(l) + MgCl_2\,(aq)$
            b. $2\,HCl\,(aq) + Mg(OH)_2\,(s) \rightarrow 2\,H_2O\,(l) + MgCl_2\,(aq)$
            c. $2\,HCl\,(aq) + MgCO_3\,(s) \rightarrow H_2O\,(l) + MgCl_2\,(aq) + CO_2\,(g)$
            d. $2\,HCl\,(aq) + Mg(HCO_3)_2\,(s) \rightarrow 2\,H_2O\,(l) + MgCl_2\,(aq) + 2\,CO_2\,(g)$
            e. $2\,HCl\,(aq) + Mg\,(s) \rightarrow H_2\,(g) + MgCl_2\,(aq)$

9.56   a.   M*: $Sn\,(s) + H_2SO_3\,(aq) \rightarrow SnSO_3\,(aq) + H_2\,(g)$
              TIE**: $Sn\,(s) + 2\,H^+(aq) + SO_3^{2-}(aq) \rightarrow Sn^{2+}(aq) + SO_3^{2-}(aq) + H_2\,(g)$
              NIE***: $Sn\,(s) + 2\,H^+(aq) \rightarrow Sn^{2+}(aq) + H_2\,(g)$

        b.   M*: $3\,Mg\,(s) + 2\,H_3PO_4\,(aq) \rightarrow Mg_3(PO_4)_2\,(s) + 3\,H_2\,(g)$
              TIE**: $3\,Mg\,(s) + 6\,H^+\,(aq) + 2\,PO_4^{3-}\,(aq) \rightarrow Mg_3(PO_4)_2\,(s) + 3\,H_2\,(g)$
              NIE***: $3\,Mg\,(s) + 6\,H^+\,(aq) + 2\,PO_4^{3-}\,(aq) \rightarrow Mg_3(PO_4)_2\,(s) + 3\,H_2\,(g)$

        c.   M*: $Ca\,(s) + 2\,HBr\,(aq) \rightarrow CaBr_2\,(aq) + H_2\,(g)$
              TIE**: $Ca\,(s) + 2\,H^+(aq) + 2\,Br^-(aq) \rightarrow Ca^{2+}(aq) + 2\,Br^-(aq) + H_2\,(g)$
              NIE***: $Ca\,(s) + 2\,H^+(aq) \rightarrow Ca^{2+}(aq) + H_2\,(g)$

PROPERTIES OF BASES (SECTION 9.7)

☑9.58   a.   M*: $RbOH\,(aq) + HBr\,(aq) \rightarrow RbBr\,(aq) + H_2O\,(l)$
              TIE**: $Rb^+(aq) + OH^-\,(aq) + H^+(aq) + Br^-(aq) \rightarrow Rb^+(aq) + Br^-(aq) + H_2O\,(l)$
              NIE***: $OH^-\,(aq) + H^+(aq) \rightarrow H_2O\,(l)$

        b.   M*: $2\,RbOH\,(aq) + H_2SO_3\,(aq) \rightarrow Rb_2SO_3\,(aq) + 2\,H_2O\,(l)$
              TIE**: $2\,Rb^+(aq) + 2\,OH^-\,(aq) + 2\,H^+(aq) + SO_3^{2-}(aq) \rightarrow 2\,Rb^+(aq) + SO_3^{2-}(aq) + 2\,H_2O\,(l)$
              NIE***: $OH^-\,(aq) + H^+(aq) \rightarrow H_2O\,(l)$

        c.   M*: $2\,RbOH\,(aq) + H_3PO_3\,(aq) \rightarrow Rb_2HPO_3\,(aq) + 2\,H_2O\,(l)$
              TIE**: $2\,Rb^+(aq) + 2\,OH^-(aq) + 2\,H^+(aq) + HPO_3^{2-}\,(aq) \rightarrow 2\,Rb^+(aq) + HPO_3^{2-}(aq) + 2\,H_2O\,(l)$
              NIE***: $OH^-\,(aq) + H^+(aq) \rightarrow H_2O\,(l)$

☑9.60   a.   M*: $2\,KOH\,(aq) + H_3PO_4\,(aq) \rightarrow K_2HPO_4\,(aq) + 2\,H_2O\,(l)$
              TIE**: $2\,K^+(aq) + 2\,OH^-(aq) + 2\,H^+(aq) + HPO_4^{2-}(aq) \rightarrow 2\,K^+(aq) + HPO_4^{2-}(aq) + 2\,H_2O\,(l)$
              NIE***: $OH^-\,(aq) + H^+(aq) \rightarrow H_2O\,(l)$

        b.   M*: $3\,KOH\,(aq) + H_3PO_4\,(aq) \rightarrow K_3PO_4\,(aq) + 3\,H_2O\,(l)$
              TIE**: $3\,K^+(aq) + 3\,OH^-(aq) + 3\,H^+(aq) + PO_4^{3-}\,(aq) \rightarrow 3\,K^+(aq) + PO_4^{3-}(aq) + 3\,H_2O\,(l)$
              NIE***: $OH^-(aq) + H^+(aq) \rightarrow H_2O\,(l)$

*M = molecular equation, **TIE = total ionic equation, ***NIE = net ionic equation

Acids, Bases, and Salts 129

c. M*: KOH (aq) + H$_2$C$_2$O$_4$ (aq) → KHC$_2$O$_4$ (aq) + H$_2$O (l)

TIE**: K$^+$(aq) + OH$^-$(aq) + H$^+$(aq) + HC$_2$O$_4^-$ (aq) → K$^+$(aq) + HC$_2$O$_4^-$(aq) + H$_2$O (l)

NIE***: OH$^-$ (aq) + H$^+$(aq) → H$_2$O (l)

SALTS (SECTION 9.8)

			Cations	Anions
9.62	a.	CuCl$_2$	Cu^{2+}	2 Cl$^-$
	b.	(NH$_4$)$_2$SO$_4$	2 NH$_4^+$	SO$_4^{2-}$
	c.	Li$_3$PO$_4$	3 Li$^+$	PO$_4^{3-}$
	d.	MgCO$_3$	Mg^{2+}	CO$_3^{2-}$
	e.	Ca(C$_2$H$_3$O$_2$)$_2$	Ca^{2+}	2 C$_2$H$_3$O$_2^-$
	f.	KNO$_3$	K$^+$	NO$_3^-$

			Cations	Base	Anions	Acids
9.64	a.	CuCl$_2$	Cu^{2+}	Cu(OH)$_2$	2 Cl$^-$	HCl
	b.	(NH$_4$)$_2$SO$_4$	2 NH$_4^+$	NH$_3$ or NH$_4$OH	SO$_4^{2-}$	H$_2$SO$_4$
	c.	Li$_3$PO$_4$	3 Li$^+$	LiOH	PO$_4^{3-}$	H$_3$PO$_4$
	d.	MgCO$_3$	Mg^{2+}	Mg(OH)$_2$	CO$_3^{2-}$	H$_2$CO$_3$
	e.	Ca(C$_2$H$_3$O$_2$)$_2$	Ca^{2+}	Ca(OH)$_2$	2 C$_2$H$_3$O$_2^-$	HC$_2$H$_3$O$_2$
	f.	KNO$_3$	K$^+$	KOH	NO$_3^-$	HNO$_3$

9.66 a. Epsom salt = MgSO$_4 \cdot$ 7H$_2$O $\quad$ MgSO$_4 \cdot$ 7 H$_2$O (s) $\xrightarrow{\Delta}$ MgSO$_4$ (s) + 7 H$_2$O (g)

$$1 \text{ mol Epsom salt} \left(\frac{7 \text{ moles H}_2\text{O}}{1 \text{ mole epsom salt}}\right)\left(\frac{18.02 \text{ g H}_2\text{O}}{1 \text{ mole H}_2\text{O}}\right) = 126.1 \text{ g H}_2\text{O}$$

b. Borax = Na$_2$B$_4$O$_7 \cdot$ 10H$_2$O $\quad$ Na$_2$B$_4$O$_7 \cdot$ 10 H$_2$O (s) $\xrightarrow{\Delta}$ Na$_2$B$_4$O$_7$ (s) + 10 H$_2$O (g)

$$1 \text{ mol Borax} \left(\frac{10 \text{ moles H}_2\text{O}}{1 \text{ mole borax}}\right)\left(\frac{18.02 \text{ g H}_2\text{O}}{1 \text{ mole H}_2\text{O}}\right) = 180.2 \text{ g H}_2\text{O}$$

Both of these reactions will produce anhydrous salts as well as water. For every mole of Epsom salt, 7 moles of water are released. For every mole of borax, 10 moles of water are released.

			Acid	Solid
☑9.68	a.	Mg(NO$_3$)$_2$	HNO$_3$	MgCO$_3$
	b.	CaCl$_2$	HCl	CaO
	c.	Rb$_2$SO$_4$	H$_2$SO$_4$	RbHCO$_3$

9.70 a. 2 HNO$_3$ (aq) + MgCO$_3$ (s) → H$_2$O (l) + CO$_2$ (g) + Mg(NO$_3$)$_2$ (aq)

b. 2 HCl (aq) + CaO (s) → H$_2$O (l) + CaCl$_2$ (aq)

c. H$_2$SO$_4$ (aq) + 2 RbHCO$_3$ (s) → 2 H$_2$O (l) + 2 CO$_2$ (g) + Rb$_2$SO$_4$ (aq)

9.72 a. CaCl$_2$ $\qquad$ 1 eq CaCl$_2$ = 0.5 moles CaCl$_2$

b. Cu(NO$_3$)$_2$ $\qquad$ 1 eq Cu(NO$_3$)$_2$ = 0.5 moles Cu(NO$_3$)$_2$

c. CrCl$_3$ $\qquad$ 1 eq CrCl$_3$ = 0.33 moles CrCl$_3$

*M = molecular equation, **TIE = total ionic equation, ***NIE = net ionic equation

130 Chapter 9

☑9.74 a. 0.22 mol ZnCl₂

$$0.22 \text{ mol ZnCl}_2 \left(\frac{2 \text{ eq. ZnCl}_2}{1 \text{ mole ZnCl}_2}\right) = 0.44 \text{ eq ZnCl}_2$$

$$0.22 \text{ mol ZnCl}_2 \left(\frac{2 \text{ eq. ZnCl}_2}{1 \text{ mole ZnCl}_2}\right)\left(\frac{1000 \text{ meq}}{1 \text{ eq.}}\right) = 440 \text{ meq ZnCl}_2$$

b. 0.45 mol CsCl

$$0.45 \text{ mol CsCl}\left(\frac{1 \text{ eq. CsCl}}{1 \text{ mole CsCl}}\right) = 0.45 \text{ eq CsCl}$$

$$0.45 \text{ mol CsCl}\left(\frac{1 \text{ eq. CsCl}}{1 \text{ mole CsCl}}\right)\left(\frac{1000 \text{ meq}}{1 \text{ eq.}}\right) = 450 \text{ meq CsCl}$$

c. 3.12×10^{-2} mol Fe(NO₃)₂

$$3.12 \times 10^{-2} \text{ mol Fe(NO}_3)_2 \left(\frac{2 \text{ eq. Fe(NO}_3)_2}{1 \text{ mole Fe(NO}_3)_2}\right) = 6.24 \times 10^{-2} \text{ eq Fe(NO}_3)_2$$

$$3.12 \times 10^{-2} \text{ mol Fe(NO}_3)_2 \left(\frac{2 \text{ eq. Fe(NO}_3)_2}{1 \text{ mole Fe(NO}_3)_2}\right)\left(\frac{1000 \text{ meq}}{1 \text{ eq.}}\right) = 62.4 \text{ meq Fe(NO}_3)_2$$

9.76 a. $5.00 \text{ g Na}_2\text{CO}_3 \cdot 10 \text{ H}_2\text{O} \left(\frac{1 \text{ mole Na}_2\text{CO}_3 \cdot 10 \text{ H}_2\text{O}}{286.19 \text{ g Na}_2\text{CO}_3 \cdot 10 \text{ H}_2\text{O}}\right)\left(\frac{2 \text{ eq Na}_2\text{CO}_3 \cdot 10 \text{ H}_2\text{O}}{1 \text{ mole Na}_2\text{CO}_3 \cdot 10 \text{ H}_2\text{O}}\right) = 3.49 \times 10^{-2} \text{ eq Na}_2\text{CO}_3 \cdot 10 \text{ H}_2\text{O}$

$5.00 \text{ g Na}_2\text{CO}_3 \cdot 10 \text{ H}_2\text{O} \left(\frac{1 \text{ mole Na}_2\text{CO}_3 \cdot 10 \text{ H}_2\text{O}}{286.19 \text{ g Na}_2\text{CO}_3 \cdot 10 \text{ H}_2\text{O}}\right)\left(\frac{2 \text{ eq Na}_2\text{CO}_3 \cdot 10 \text{ H}_2\text{O}}{1 \text{ mole Na}_2\text{CO}_3 \cdot 10 \text{ H}_2\text{O}}\right)\left(\frac{1000 \text{ meq}}{1 \text{ eq.}}\right) = 34.9 \text{ meq Na}_2\text{CO}_3 \cdot 10 \text{ H}_2\text{O}$

b. $5.00 \text{ g CuSO}_4 \cdot 5 \text{ H}_2\text{O} \left(\frac{1 \text{ mole CuSO}_4 \cdot 5 \text{ H}_2\text{O}}{249.72 \text{ g CuSO}_4 \cdot 5 \text{ H}_2\text{O}}\right)\left(\frac{2 \text{ eq CuSO}_4 \cdot 5 \text{ H}_2\text{O}}{1 \text{ mole CuSO}_4 \cdot 5 \text{ H}_2\text{O}}\right) = 4.00 \times 10^{-2} \text{ eq CuSO}_4 \cdot 5 \text{ H}_2\text{O}$

$5.00 \text{ g CuSO}_4 \cdot 5 \text{ H}_2\text{O} \left(\frac{1 \text{ mole CuSO}_4 \cdot 5 \text{ H}_2\text{O}}{249.72 \text{ g CuSO}_4 \cdot 5 \text{ H}_2\text{O}}\right)\left(\frac{2 \text{ eq CuSO}_4 \cdot 5 \text{ H}_2\text{O}}{1 \text{ mole CuSO}_4 \cdot 5 \text{ H}_2\text{O}}\right)\left(\frac{1000 \text{ meq}}{1 \text{ eq.}}\right) = 40.0 \text{ meq CuSO}_4 \cdot 5 \text{ H}_2\text{O}$

c. $5.00 \text{ g Li}_2\text{CO}_3 \left(\frac{1 \text{ mole Li}_2\text{CO}_3}{73.89 \text{ g Li}_2\text{CO}_3}\right)\left(\frac{1 \text{ eq Li}_2\text{CO}_3}{1 \text{ mole Li}_2\text{CO}_3}\right) = 1.35 \times 10^{-1} \text{ eq Li}_2\text{CO}_3$

$5.00 \text{ g Li}_2\text{CO}_3 \left(\frac{1 \text{ mole Li}_2\text{CO}_3}{73.89 \text{ g Li}_2\text{CO}_3}\right)\left(\frac{1 \text{ eq Li}_2\text{CO}_3}{1 \text{ mole Li}_2\text{CO}_3}\right)\left(\frac{1000 \text{ meq}}{1 \text{ eq.}}\right) = 135 \text{ meq Li}_2\text{CO}_3$

d. $5.00 \text{ g NaH}_2\text{PO}_4 \left(\frac{1 \text{ mole NaH}_2\text{PO}_4}{119.98 \text{ g NaH}_2\text{PO}_4}\right)\left(\frac{1 \text{ eq NaH}_2\text{PO}_4}{1 \text{ mole NaH}_2\text{PO}_4}\right) = 4.17 \times 10^{-2} \text{ eq NaH}_2\text{PO}_4$

$5.00 \text{ g NaH}_2\text{PO}_4 \left(\frac{1 \text{ mole NaH}_2\text{PO}_4}{119.98 \text{ g NaH}_2\text{PO}_4}\right)\left(\frac{1 \text{ eq NaH}_2\text{PO}_4}{1 \text{ mole NaH}_2\text{PO}_4}\right)\left(\frac{1000 \text{ meq}}{1 \text{ eq.}}\right) = 41.7 \text{ meq NaH}_2\text{PO}_4$

9.78 $150 \text{ mL i.c. fluid}\left(\frac{1 \text{ L}}{1000 \text{ mL}}\right)\left(\frac{133 \text{ meq}}{1 \text{ L}}\right)\left(\frac{1 \text{ eq}}{1000 \text{ meq}}\right)\left(\frac{1 \text{ mole K}_2\text{SO}_4}{2 \text{ eq K}_2\text{SO}_4}\right) = 9.98 \times 10^{-3} \text{ moles K}_2\text{SO}_4$

$150 \text{ mL i.c. fluid}\left(\frac{1 \text{ L}}{1000 \text{ mL}}\right)\left(\frac{133 \text{ meq}}{1 \text{ L}}\right)\left(\frac{1 \text{ eq}}{1000 \text{ meq}}\right)\left(\frac{1 \text{ mole K}_2\text{SO}_4}{2 \text{ eq K}_2\text{SO}_4}\right)\left(\frac{174.27 \text{ g K}_2\text{SO}_4}{1 \text{ mole K}_2\text{SO}_4}\right) = 1.74 \text{ g K}_2\text{SO}_4$

THE STRENGTHS OF ACIDS AND BASES (SECTION 9.9)

9.80 (weakest) acid B < acid A < acid C < acid D (strongest)

9.82 a. (weakest) acid B < acid A < acid C < acid D (strongest)
      b. (weakest) base D < base C < base A < base B (strongest)

9.84 a. HNO₂

$$\text{HNO}_2 \text{ (aq)} \rightleftharpoons \text{H}^+\text{(aq)} + \text{NO}_2^-\text{(aq)} \qquad K_a = \frac{[\text{H}^+][\text{NO}_2^-]}{[\text{HNO}_2]}$$

b. HCO₃⁻

$$\text{HCO}_3^-\text{(aq)} \rightleftharpoons \text{H}^+\text{(aq)} + \text{CO}_3^{2-}\text{(aq)} \qquad K_a = \frac{[\text{H}^+][\text{CO}_3^{2-}]}{[\text{HCO}_3^-]}$$

c. H₂PO₄⁻

$$\text{H}_2\text{PO}_4^-\text{(aq)} \rightleftharpoons \text{H}^+\text{(aq)} + \text{HPO}_4^{2-}\text{(aq)} \qquad K_a = \frac{[\text{H}^+][\text{HPO}_4^{2-}]}{[\text{HPO}_4^-]}$$

Acids, Bases, and Salts 131

d. HS⁻ $HS^-(aq) \rightleftharpoons H^+(aq) + S^{2-}(aq)$ $K_a = \dfrac{[H^+][S^{2-}]}{[HS^-]}$

e. HBrO $HBrO(aq) \rightleftharpoons H^+(aq) + BrO^-(aq)$ $K_a = \dfrac{[H^+][BrO^-]}{[HBrO]}$

☑9.86 The 20% acetic acid solution is the weak acid solution. It is a weak acid because acetic acid does not completely dissociate.

The 0.05 M HCl is a dilute strong acid solution. If someone wanted this solution instead, they should use the term "dilute."

ANALYZING ACIDS AND BASES (SECTION 9.10)

9.88 A titration is performed to determine the concentration of a solution.

9.90 a. The endpoint of a titration occurs when an indicator changes color. The equivalence point of a titration is the point at which the unknown solution has completely reacted with the known solution. The resulting mixture from the titration will have a specific pH at the equivalence point. If the indicator selected changes color at the same pH as the pH of the resulting mixture, the endpoint and the equivalence point will be the same.

b. If the indicator selected changes color at a different pH than the pH of the mixture that results from the titration, then the endpoint and the equivalence point will not be the same.

☑9.92 a. $HBr + NaOH \rightarrow H_2O + NaBr$

$250 \text{ mL} \left(\dfrac{1 \text{ L}}{1000 \text{ mL}}\right)\left(\dfrac{0.400 \text{ moles HBr}}{1 \text{ L}}\right)\left(\dfrac{1 \text{ mole NaOH}}{1 \text{ mole HBr}}\right) = 0.100 \text{ moles NaBr}$

b. $HClO_4 + NaOH \rightarrow H_2O + NaClO_4$

$750 \text{ mL} \left(\dfrac{1 \text{ L}}{1000 \text{ mL}}\right)\left(\dfrac{0.300 \text{ moles HClO}_4}{1 \text{ L}}\right)\left(\dfrac{1 \text{ mole NaOH}}{1 \text{ mole HClO}_4}\right) = 0.225 \text{ moles HClO}_4$

TITRATION CALCULATIONS (SECTION 9.11)

9.94 a. $2 \text{ NaOH (aq)} + H_2CrO_4 \text{ (aq)} \rightarrow Na_2CrO_4 \text{ (aq)} + 2 H_2O \text{ (l)}$
      b. $NaOH \text{ (aq)} + HClO_3 \text{ (aq)} \rightarrow NaClO_3 \text{ (aq)} + H_2O \text{ (l)}$
      c. $3 \text{ NaOH (aq)} + H_3AsO_3 \text{ (aq)} \rightarrow Na_3AsO_3 \text{ (aq)} + 3 H_2O \text{ (l)}$

☑9.96 a. $2 \text{ HCl (aq)} + Sr(OH)_2 \text{ (aq)} \rightarrow SrCl_2 \text{ (aq)} + 2 H_2O \text{ (l)}$
       b. $3 \text{ HCl (aq)} + Ni(OH)_3 \text{ (s)} \rightarrow 3 \text{ NaCl (aq)} + 3 H_2O \text{ (l)}$
       c. $2 \text{ HCl (aq)} + Zn(OH)_2 \text{ (s)} \rightarrow ZnCl_2 \text{ (aq)} + 2 H_2O \text{ (l)}$

9.98 $H_2C_2O_4 + 2 \text{ NaOH} \rightarrow 2 H_2O + Na_2C_2O_4$

$\dfrac{43.88 \text{ mL}\left(\dfrac{1 \text{ L}}{1000 \text{ mL}}\right)\left(\dfrac{0.1891 \text{ moles NaOH}}{1 \text{ L}}\right)\left(\dfrac{1 \text{ mole } H_2C_2O_4}{2 \text{ mole NaOH}}\right)}{25.00 \text{ mL}\left(\dfrac{1 \text{ L}}{1000 \text{ mL}}\right)} = 0.1660 \text{ M } H_2C_2O_4$

132 Chapter 9

9.100   a.   NaOH + HClO$_4$ → H$_2$O + NaClO$_4$

$$20.00 \text{ mL} \left(\frac{1\text{L}}{1000\text{ mL}}\right)\left(\frac{0.200 \text{ moles HClO}_4}{1\text{L}}\right)\left(\frac{1 \text{ mole NaOH}}{1 \text{ mole HClO}_4}\right)\left(\frac{1 \text{L NaOH}}{0.120 \text{ moles NaOH}}\right)\left(\frac{1000 \text{ mL}}{1\text{L}}\right) = 33.3 \text{ mL NaOH}$$

b.   2 NaOH + H$_2$SO$_4$ → 2 H$_2$O + Na$_2$SO$_4$

$$20.00 \text{ mL} \left(\frac{1\text{L}}{1000\text{ mL}}\right)\left(\frac{0.125 \text{ moles H}_2\text{SO}_4}{1\text{L}}\right)\left(\frac{2 \text{ moles NaOH}}{1 \text{ mole H}_2\text{SO}_4}\right)\left(\frac{1 \text{L NaOH}}{0.120 \text{ moles NaOH}}\right)\left(\frac{1000 \text{ mL}}{1\text{L}}\right) = 41.7 \text{ mL NaOH}$$

c.   4 NaOH + H$_4$P$_2$O$_6$ → 4 H$_2$O + Na$_4$P$_2$O$_6$

$$20.00 \text{ mL} \left(\frac{1\text{L}}{1000\text{ mL}}\right)\left(\frac{0.150 \text{ moles H}_4\text{P}_2\text{O}_6}{1\text{L}}\right)\left(\frac{4 \text{ moles NaOH}}{1 \text{ mole H}_4\text{P}_2\text{O}_6}\right)\left(\frac{1 \text{L NaOH}}{0.120 \text{ moles NaOH}}\right)\left(\frac{1000 \text{ mL}}{1\text{L}}\right) = 100. \text{ mL NaOH}$$

d.   3 NaOH + H$_3$PO$_4$ → 3 H$_2$O + Na$_3$PO$_4$

$$20.00 \text{ mL} \left(\frac{0.120 \text{ moles H}_3\text{PO}_4}{500 \text{ mL}}\right)\left(\frac{3 \text{ moles NaOH}}{1 \text{ mole H}_3\text{PO}_4}\right)\left(\frac{1 \text{L NaOH}}{0.120 \text{ moles NaOH}}\right)\left(\frac{1000 \text{ mL}}{1\text{L}}\right) = 120. \text{ mL NaOH}$$

e.   2 NaOH + H$_2$SO$_4$ → 2 H$_2$O + Na$_2$SO$_4$

$$20.00 \text{ mL} \left(\frac{6.25 \text{ g H}_2\text{SO}_4}{250 \text{ mL}}\right)\left(\frac{1 \text{ mole H}_2\text{SO}_4}{98.09 \text{ g H}_2\text{SO}_4}\right)\left(\frac{2 \text{ moles NaOH}}{1 \text{ mole H}_2\text{SO}_4}\right)\left(\frac{1 \text{L NaOH}}{0.120 \text{ moles NaOH}}\right)\left(\frac{1000 \text{ mL}}{1\text{L}}\right) = 85.0 \text{ mL NaOH}$$

f.   NaOH + HClO$_3$ → H$_2$O + NaClO$_3$

$$20.00 \text{ mL} \left(\frac{1\text{L}}{1000\text{ mL}}\right)\left(\frac{0.500 \text{ moles HClO}_3}{1\text{L}}\right)\left(\frac{1 \text{ mole NaOH}}{1 \text{ mole HClO}_3}\right)\left(\frac{1 \text{L NaOH}}{0.120 \text{ moles NaOH}}\right)\left(\frac{1000 \text{ mL}}{1\text{L}}\right) = 83.3 \text{ mL NaOH}$$

9.102   a.   H$_2$SO$_4$ + 2 NaOH → 2 H$_2$O + Na$_2$SO$_4$

$$\frac{29.88 \text{ mL} \left(\frac{1\text{L}}{1000\text{ mL}}\right)\left(\frac{1.17 \text{ moles NaOH}}{1\text{L}}\right)\left(\frac{1 \text{ mole H}_2\text{SO}_4}{2 \text{ moles NaOH}}\right)}{5.00 \text{ mL} \left(\frac{1\text{L}}{1000\text{ mL}}\right)} = 3.50 \text{ M H}_2\text{SO}_4$$

b.   HC$_2$H$_3$O$_2$ + KOH → H$_2$O + KC$_2$H$_3$O$_2$

$$\frac{35.62 \text{ mL} \left(\frac{1\text{L}}{1000\text{ mL}}\right)\left(\frac{0.250 \text{ moles KOH}}{1\text{L}}\right)\left(\frac{1 \text{ mole HC}_2\text{H}_3\text{O}_2}{1 \text{ mole KOH}}\right)}{10.00 \text{ mL} \left(\frac{1\text{L}}{1000\text{ mL}}\right)} = 0.891 \text{ M HC}_2\text{H}_3\text{O}_2$$

c.   HCl + NaOH → H$_2$O + NaCl

$$\frac{20.63 \text{ mL} \left(\frac{1\text{L}}{1000\text{ mL}}\right)\left(\frac{6.00 \text{ moles NaOH}}{1\text{L}}\right)\left(\frac{1 \text{ mole HCl}}{1 \text{ mole NaOH}}\right)}{10.00 \text{ mL} \left(\frac{1\text{L}}{1000\text{ mL}}\right)} = 12.4 \text{ M HCl}$$

9.104   benzoic acid + NaOH → H$_2$O + sodium benzoate

$$\frac{0.5823 \text{ g benzoic acid}}{46.75 \text{ mL}\left(\frac{1\text{L}}{1000\text{ mL}}\right)\left(\frac{0.1021 \text{ moles NaOH}}{1\text{L}}\right)\left(\frac{1 \text{ mole benzoic acid}}{1 \text{ mole NaOH}}\right)} = 122.0 \frac{\text{g}}{\text{mole}}$$

HYDROLYSIS REACTIONS OF SALTS (SECTION 9.12)

9.106   Na$_2$CO$_3$ is a salt that could be produced by the reaction of a strong base and a weak acid. The hydrolysis of Na$_2$CO$_3$ produces a alkaline solution.

Acids, Bases, and Salts 133

☑9.108 a. NaOCl   The pH is greater than 7 because NaOH is a strong base and HOCl is a weak acid.

$$OCl^-(aq) + H_2O\ (l) \rightleftharpoons HOCl\ (aq) + OH^-(aq)$$

b. NaCHO$_2$ The pH is greater than 7 because NaOH is a strong base and HCHO$_2$ is a weak acid.

$$CHO_2^-(aq) + H_2O\ (l) \rightleftharpoons HCHO_2\ (aq) + OH^-(aq)$$

c. KNO$_3$ The pH equals 7 because KOH is a strong base and HNO$_3$ is a strong acid.

d. Na$_3$PO$_4$ The pH is greater than 7 because NaOH is a stong base and H$_3$PO$_4$ is a weak acid.

$$PO_4^{3-}(aq) + H_2O\ (l) \rightleftharpoons HPO_4^{2-}\ (aq) + OH^-(aq)$$

9.110   More than one indicator should be available in the laboratory because the pH at the equivalence point of a titration will vary depending on the salt produced by the reaction.

BUFFERS (SECTION 9.13)

9.112 $HPO_4^{2-}(aq) + H_3O^+(aq) \rightleftharpoons H_2PO_4^-(aq) + H_2O\ (l)$

$H_2PO_4^-(aq) + OH^-(aq) \rightleftharpoons HPO_4^{2-}(aq) + H_2O\ (l)$

9.114 $HCO_3^-(aq) + H^+(aq) \rightleftharpoons H_2CO_3\ (aq)$

☑9.116 a. $pH = 3.85 + \log\frac{[0.1]}{[0.1]} = 3.85$

b. $pH = 3.85 + \log\frac{[1]}{[1]} = 3.85$

c. The solution in part b has greater buffer capacity than the solution in part a.

9.118 a. $pH = 4.74 + \log\frac{[0.25]}{[0.40]} = 4.54$

b. $pH = 7.21 + \log\frac{[0.40]}{[0.10]} = 7.81$

c. $pH = 7.00 + \log\frac{[0.20]}{[1.50]} = 6.12$

9.120 $7.65 = 7.21 + \log\frac{[HPO_4^{2-}]}{[H_2PO_4^-]} \Rightarrow 0.44 = \log\frac{[HPO_4^{2-}]}{[H_2PO_4^-]} \Rightarrow \frac{[HPO_4^{2-}]}{[H_2PO_4^-]} = 10^{0.44} = 2.8$

The concentration of Na$_2$HPO$_4$ has to be 2.8 times the concentration of Na$_2$HPO$_4$.

ADDITIONAL ACTIVITIES

Section 9.1 Review:
(1) How many protons, neutrons, and electrons are in this hydrogen ion: $^1_1H^+$?
(2) Write a general formula like Equation 9.1 for an Arrhenius acid (HA) dissolved in water.
(3) Write a general formula like Equation 9.2 for an Arrhenius base (MOH) dissolved in water.

Section 9.2 Review:
(1) Write a general formula like Equation 9.5 for a Brønsted acid (HA) dissolved in water.
(2) Identify the conjugate acid-base pairs in (1).

Section 9.3 Review:
(1) Can H_2O be named as an acid? Explain.
(2) Assume the following are dissolved in water. Name them as acids.

H_2Se	H_2SO_3	HF	H_3P	HClO	$HClO_3$	$HC_2H_3O_2$	$H_2Cr_2O_7$

Section 9.4 Review:
(1) Add the corresponding $[OH^-]$ concentrations to the diagram below.
(2) Label the acidic, basic, and neutral regions of the diagram.
(3) Mark the position of $[H_3O^+]$ = 0.125 M. Is this acidic, basic, or neutral?
(4) Mark the position of $[OH^-]$ = 6.7 x 10^{-7} M. Is this acidic, basic, or neutral?
(5) Mark the position of $[H_3O^+]$ = 2.5 x 10^{-9} M. Is this acidic, basic, or neutral?
(6) Mark the position of $[OH^-]$ = 3.4 x 10^{-3} M. Is this acidic, basic, or neutral?

```
        1 x 10⁰              1 x 10⁻⁷                      1 x 10⁻¹⁴
[H₃O⁺]|                         |                              |
      |_____|_____|
      |                         |                              |
[OH⁻] |                         |                              |
```

Section 9.5 Review:
(1) Calculate the pH for each of the solutions described in (3)-(6) of the Section 9.4 Review.
(2) Classify each of these solutions as acidic, basic, or neutral.
(3) Does the classification from (2) match the classification from the Section 9.4 Review?
(4) What is the $[H_3O^+]$ for a solution with a pH of 4.76? Is this acidic, basic, or neutral?
(5) What is the $[OH^-]$ for a solution with a pH of 8.27? Is this acidic, basic, or neutral?

Section 9.6 Review:
(1) The labels on three containers in the laboratory have fallen off. The containers hold iron (II) oxide, iron (II) carbonate, and iron (II) hydroxide. Can reactions with an acid be performed to tell these compounds apart?
(2) Write the total ionic equations and net ionic equations for any reactions described in (1). Identify spectator ions.
(3) Will iron react with 6.00 M hydrochloric acid? If so, write a molecular equation, a total ionic equation, and a net ionic equation. Also, identify any spectator ions.

Section 9.7 Review:
(1) For water soluble Arrhenius bases and acids, what is the net ionic equation for neutralization?
(2) Why do neutralization reactions occur?

Section 9.8 Review:
(1) Referring back to the net ionic equation in the Section 9.7 Review, identify the reactant ions as cations or anions.
(2) Did the OH^- ion in (1) come from the acid or the base? What was charge on the counterion in this compound?
(3) Did the H^+ ion in (1) come from the acid or the base? What was charge on the counterion in this compound?
(4) Why is the salt excluded from the net ionic equation in the Section 9.7 Review?

Section 9.9 Review:
(1) Does a dilute strong acid have a small value of K_a? Explain.
(2) Does a concentrated weak acid have a small value of K_a? Explain.
(3) If a base is weak, what is the strength of its conjugate acid?
(4) If a base is strong, what is the strength of its conjugate base?

Section 9.10 Review:
(1) Write an equilibrium expression for the dissociation of hydrofluoric acid. The K_a is 6.46×10^{-4}.
(2) Referring back to the seesaw analogy from the Section 8.8 Review, what happens to the seesaw as a weak acid like hydrofluoric acid reacts with a strong base?
(3) How is equilibrium reestablished after the hydrofluoric acid reacts with a strong base?
(4) What is the moment called when all of the acid has reacted with the base?
(5) What is the moment called when the reaction mixture changes color?

Section 9.11 Review:
(1) How can the moles of solute be calculated from the liters and molarity of solution?
(2) How can the liters of solution be calculated from the molarity of the solution and the moles of solute?

Section 9.12 Review:
Consider the following "tug-of-war" competitors. For each pair of reactants, determine the pH of the resulting salts.

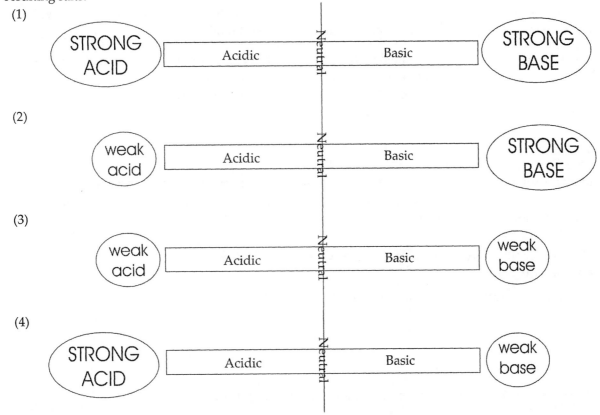

Section 9.13 Review:
Refer back to the dissociation equation written for HF in the Section 9.10 Review.
(1) What ion must be present in a salt added to HF in order to form a buffer?
(2) Should the other ion in the salt come from a strong base or a weak base?
(3) Will the salt completely dissociate or establish an equilibrium of its own?

136 Chapter 9

(4) How will adding the salt affect the seesaw for the HF equilibrium? What will happen to reestablish equilibrium?
(5) For a buffer with pH = pKa, place HF, H⁺, and F⁻ in order of increasing concentration.
(6) Which component of this buffer will react with acids? How will this affect the equilibrium?
(7) Which component of this buffer will react with bases? How will this affect the equilibrium?
(8) What concentration levels ensure a large buffer capacity?

Tying It All Together with a Laboratory Application:

A (1) _____ is performed to determine the concentration of an ammonia (NH₃) solution. Ammonia (2) _____ a base according the Arrhenius theory because (3) _____; however, ammonia will react with water to produce (4) _____ and _____ ions. Ammonia is a(n) (5) _____ according to the Brønsted theory because (6) _____.

An aliquot of 25.00 mL is pipetted from the ammonia solution of unknown concentration into a 125 mL Erlenmeyer flask. A 0.125 M H₂SO₄ solution is added to the (7) _____. The initial volume reading is 5.87 mL. H₂SO₄ is a(n) (8) _____ according to the Arrhenius theory because (9) _____. H₂SO₄ is a(n) (10) _____ according to the Brønsted theory because (11) _____. The name of H₂SO₄ is (12) _____.

The molecular equation for the reaction is (13) _____. The total ionic equation for the reaction is (14) _____. The spectator ion(s) is(are) (15) _____. The net ionic equation for the reaction is (16) _____. This is the reaction between a (17) _____ acid and a (18) _____ base. The solution at the equivalence point will have a pH (19) _____ 7 because of the (20) _____ reaction of the salt with water as shown in (21) _____. The indicator used in this titration should change color at a pH (22) _____ 7.

A few drops of an indicator are added to the Erlenmeyer flask. The titration is performed until the (23) _____ point when the solution changes color. This is also the (24) _____ point because an appropriate indicator was selected. The final reading on the buret is 35.27 mL. The number of moles of H₂SO₄ that reacted were (25) _____. The concentration of the ammonia solution is (26) _____ M.

The [OH⁻] for the ammonia solution is determined to be 2.3 x 10⁻³ M. The [H₃O⁺] is (27) _____ and the pH is (28) _____. The ammonia solution is (29) _____. If ammonium nitrate were added to the ammonia solution, a (30) _____ would form. If the concentration of ammonia and the ammonium nitrate are equal, the pH of the solution (31) _____ the pKa. The amount of H₂SO₄ that could be added to this solution without changing the pH depends on the (32) _____.

SOLUTIONS FOR THE ADDITIONAL ACTIVITIES

Section 9.1 Review:

(1) 1 proton, 0 neutrons, and 0 electrons in $_1^1H^+$; (2) HA (aq) → H⁺(aq) + A⁻(aq);

(3) MOH (aq) → M⁺(aq) + OH⁻(aq)

Section 9.2 Review:

(1) HA (aq) + H₂O (l) → H₃O⁺(aq) + A⁻(aq)
(2) HA = acid, A⁻ = conjugate base; H₂O = base, H₃O⁺ = conjugate acid

Section 9.3 Review:

(1) H₂O cannot be named as an acid because it does not ionize significantly in water.
(2) H₂Se = hydroselenic acid; H₂SO₃ = sulfurous acid; HF = hydrofluoric acid; H₃P = hydrophosphoric acid; HClO = hypochlorous acid; HClO₃ = chloric acid; HC₂H₃O₂ = acetic acid; H₂Cr₂O₇ = dichromic acid

Section 9.4 Review:

(3) acidic; (4) basic; (5) basic; (6) basic

Section 9.5 Review:
(1) $[H_3O^+] = 0.125$ M → pH = 0.903; $[OH^-] = 6.7 \times 10^{-7}$ M → pH = 7.83; $[H_3O^+] = 2.5 \times 10^{-9}$ M → pH = 8.60; $[OH^-] = 3.4 \times 10^{-3}$ M → pH = 11.53; (2) acidic; basic; basic; basic
(3) Yes, the classification from (2) matches the classification from the Section 9.4 Review.
(4) pH = 4.76 → $[H_3O^+] = 1.7 \times 10^{-5}$ M; acidic; (5) pH = 8.27 → $[OH^-] = 1.9 \times 10^{-6}$ M; basic

Section 9.6 Review:
(1) When iron (II) carbonate reacts with an acid, it will produce carbon dioxide gas in addition to salt and water. When iron (II) oxide and iron (II) hydroxide react with acid, they produce a salt and water. The iron (II) carbonate could be identified by a reaction with an acid because bubbles are produced; however, the other two materials cannot be identified unless stoichiometric calculations are performed. Both the iron (II) oxide and the iron (II) hydroxide react in a 1:2 mole ratio with a monoprotic acid. The molecular weights of these compounds are different, though. Consequently, 1 g FeO should react completely with 27.8 mL of a 1 M monoprotic acid and 1 g Fe(OH)$_2$ should react completely with 22.3 mL of a 1 M monoprotic acid. All three solids can be identified.

(2) Molecular = 2 HA (aq) + FeCO$_3$ (s) → CO$_2$ (g) + FeA$_2$ (aq) + H$_2$O (l)
TIE = 2 H$^+$(aq) + 2 A$^-$ (aq) + FeCO$_3$ (s) → CO$_2$ (g) + Fe^{2+} (aq) + 2 A$^-$ (aq) + H$_2$O (l)
NIE = 2 H$^+$(aq) + FeCO$_3$ (s) → CO$_2$ (g) + Fe^{2+} (aq) + H$_2$O (l)
Spectator ions = 2 A$^-$ (aq)

Molecular = 2 HA (aq) + FeO (s) → FeA$_2$ (aq) + H$_2$O (l)
TIE = 2 H$^+$(aq) + 2 A$^-$ (aq) + FeO (s) → Fe^{2+} (aq) + 2 A$^-$ (aq) + H$_2$O (l)
NIE = 2 H$^+$(aq) + FeO (s) → Fe^{2+} (aq) + H$_2$O (l)
Spectator ions = 2 A$^-$ (aq)

Molecular = 2 HA (aq) + Fe(OH)$_2$ (s) → FeA$_2$ (aq) + 2 H$_2$O (l)
TIE = 2 H$^+$(aq) + 2 A$^-$ (aq) + Fe(OH)$_2$ (s) → Fe^{2+} (aq) + 2 A$^-$ (aq) + 2 H$_2$O (l)
NIE = 2 H$^+$(aq) + Fe(OH)$_2$ (s) → Fe^{2+} (aq) + 2 H$_2$O (l)
Spectator ions = 2 A$^-$ (aq)

(3) Yes, iron will react with 6.00 M HCl.
Molecular = 2 HCl (aq) + Fe (s) → FeCl$_2$ (aq) + H$_2$ (g)
TIE = 2 H$^+$(aq) + 2 Cl$^-$ (aq) + Fe (s) → Fe^{2+} (aq) + 2 Cl$^-$ (aq) + H$_2$ (g)
NIE = 2 H$^+$(aq) + Fe (s) → Fe^{2+} (aq) + H$_2$ (g)
Spectator ions = 2 Cl$^-$ (aq)

Section 9.7 Review:
(1) OH$^-$(aq) + H$^+$ (aq) → H$_2$O (l)
(2) Neutralization reactions occur because hydroxide and hydrogen ions (really, hydronium ions) react to form liquid water.

138 Chapter 9

Section 9.8 Review:
(1) OH⁻ = anion; H⁺ = cation (Note: An anion is *a negative ion*. A cation is "paws"itive and cats have paws!)
(2) The OH⁻ came from the base. The charge on the counterion in this compound is positive. It is a cation.
(3) The H⁺ ion came from the acid. The charge on the counterion in this compound is negative. It is an anion.
(4) The positive cation from the base and the negative anion from the acid formed a water soluble salt. Consequently, these spectator ions were removed from the net ionic equation.

Section 9.9 Review:
(1) No, a dilute strong acid does not have a small value of K_a. A strong acid will completely dissociate in water regardless of whether it is concentrated or diluted.
(2) Yes, a concentrated weak acid has a small value of K_a. A weak acid will not completely dissociate in water regardless of whether it is concentrated or diluted.
(3) If a base is weak, it has a strong conjugate acid.
(4) If a base is strong, it has a weak conjugate acid.

Section 9.10 Review:
(1) HF (aq) $\rightleftharpoons$ H⁺(aq) + F⁻ (aq)
(2) During a reaction with a base, the free H⁺ ions will react. The products side of equilibrium between undissociated HF and the hydrogen and fluoride ions will become lighter. More HF will dissociate in order to reestablish equilibrium. Consequently, the concentration of undissociated HF will decrease.

(3) More HF will dissociate into hydrogen and fluoride ions.
(4) equivalence point
(5) endpoint

Section 9.11 Review:
(1) The moles of solute can be calculated by multiplying the liters of solution by the molarity of the solution.
(2) The liters of solution can be calculated by dividing the moles of solute by the molarity of the solution.

Section 9.12 Review:
(1) strong acid + strong base → neutral salt;   (2) weak acid + strong base → basic salt
(3) weak acid + weak base → neutral salt;   (4) strong acid + weak base → acidic salt

Section 9.13 Review:
(1) The salt must contain the fluoride ion.
(2) The other ion (cation) should come from a strong base. This ensures that the cation will not undergo hydrolysis and establish an additional equilibrium system in the buffer.
(3) The salt will completely dissociate in water.
(4) The addition of the salt to water will increase the concentration of the F⁻ ion. Because the F⁻ ion is a product, the seesaw will tip toward the products. In order to reestablish equilibrium, some of the additional F⁻ will react with the H⁺ to produce more HF. Equilibrium will shift to the left.

(5) H⁺ << F⁻ = HF; In a buffer with pH = pKa, the concentrations of the undissociated acid and its conjugate base are equal. Both of these chemical species are in much higher concentration than the hydrogen (hydronium) ion.
(6) The F⁻ ion will react with acids. This will remove some of the F⁻ from solution and create more HF. To reestablish equilibrium, some of the HF will dissociate and equilibrium will shift to the right.
(7) The HF will react with bases. This will remove some of the HF from solution. To reestablish equilibrium, some of the H⁺ and F⁻ will combine to form HF and equilibrium will shift to the left.
(8) High concentrations of both the HF and salt containing F⁻ will ensure a large buffer capacity.

Tying It All Together with a Laboratory Application:
(1) titration (2) is not (3) it does not dissociate to produce OH⁻ ions in water
(4) OH⁻, NH$_4^+$ (5) base (6) it is a proton acceptor
(7) buret (8) acid (9) it dissociates to produces H⁺ ions in water
(10) acid (11) it is a proton donor (12) sulfuric acid
(13) H_2SO_4 (aq) + 2 NH_3 (aq) → $(NH_4)_2SO_4$ (aq)
(14) 2 H⁺(aq) + SO_4^{2-} (aq) + 2 NH_3 (aq) → 2 NH_4^+(aq) + SO_4^{2-} (aq)
(15) SO_4^{2-} (aq) (16) H⁺ (aq) + NH_3 (aq) → NH_4^+(aq) (17) strong
(18) weak (19) less than (20) hydrolysis
(21) NH_4^+(aq) + H_2O (l) ⇌ NH_3 (aq) + H_3O^+(aq)

(22) less than (23) endpoint (24) equivalence (25) 3.68 x 10⁻³ moles
(26) 0.294 M (27) 4.3 x 10⁻¹² M (28) 11.36 (29) basic
(30) buffer (31) equals (32) buffer capacity

SELF-TEST QUESTIONS
Multiple Choice
1. The Arrhenius definition of a base focuses on
   a. the acceptance of H⁺
   b. the production of OH⁻
   c. the formation of covalent bonds
   d. more than one response is correct

2. The pH of solution that is 0.100 M in OH⁻.
   a. 0.794   b. 1.000   c. 1.259   d. 13.000

3. A beaker contains 100 mL of a liquid with a pH of 7.0. When 0.5 mL of 0.2 M acid is added, the pH changes to 6.88. When 0.5 mL of 0.1 M base is added to another 100 mL sample of the liquid, the pH changes to 7.20. The liquid in the beaker is
   a. water   b. an acid solution   c. a base solution   d. a buffer solution

140 Chapter 9

Refer to the following reaction for Questions 4 through 7: $HCl + NaOH \rightarrow NaCl + H_2O$

4. The above reaction between hydrochloric acid and sodium hydroxide is correctly classified as
   a. combustion
   b. dehydration
   c. neutralization
   d. more than one response is correct

5. The HCl solution is prepared to be 0.120 M. A 20.00 mL sample requires 18.50 mL of NaOH solution for complete reaction. What is the molarity of the NaOH solution?
   a. 0.0800
   b. 0.110
   c. 0.130
   d. 0.200

6. How many moles of HCl would be contained in the 20.00 mL sample used in Question 5?
   a. 0.204
   b. 2.40
   c. 0.120
   d. 0.0024

7. How many grams of HCl would be contained in the 20.00 mL sample used in Question 5?
   a. 0.0875
   b. 4.38
   c. 8.75
   d. 36.5

8. A 25.00 mL sample of monoprotic acid is titrated with a standard 0.100 M base. Exactly 20.00 mL of base is required to titrate to the proper end point. What is the molarity of the acid?
   a. 0.0500
   b. 0.100
   c. 0.0800
   d. 0.125

9. A certain solution has a pH of 1. This solution is best described as
   a. very basic
   b. neutral
   c. slightly acidic
   d. very acidic

10. If the pH of an aqueous solution cannot be changed significantly by adding small amounts of strong acid or strong base, the solution contains
    a. an indicator
    b. a buffer
    c. a protective colloid
    d. a strong acid and a strong base

True-False

11. The terms *weak acid* and *dilute acid* can be used interchangeably.
12. Some H_3O^+ ions are present in pure water.
13. One of the products formed in the titration of a strong acid by a strong base is water.
14. A solution with a pH of 3 is correctly classified as acidic.
15. A solution with a pH of 6.00 has a concentration of 6.00 M H^+.
16. Three different sodium salts of phosphoric acid, H_3PO_4, are possible.
17. The anion produced by the first step in the dissociation of sulfurous acid, H_2SO_3, is SO_3^{2-}.
18. The second step in the dissociation of H_3PO_4 produces $H_2PO_4^-$.
19. In pure water, $[H_3O^+] = [OH^-]$.
20. In an acid-base titration, the point at which the acid and base have exactly reacted is called the equivalence point.
21. The pH is 7 at the equivalence point of all acid-base titrations.

Matching
Choose the response that best completes each reaction below. The acid involved in each reaction is represented by HA.

22. $2\ HA + ? \rightarrow H_2 + MgA_2$ a. NaA
23. $HA + H_2O \rightarrow ? + A^-$ b. Na_2CO_3
24. $HA + NaOH \rightarrow H_2O + ?$ c. Mg
25. $2\ HA + ? \rightarrow CO_2 + H_2O + 2\ NaA$ d. H_3O^+

Match the classifications given on the right to the species listed on the left. The species are involved in the following reversible reaction. Responses can be used more than once.

$$Na^+ (aq) + C_2H_3O_2^- (aq) + H_2O (l) \rightleftharpoons Na^+ (aq) + HC_2H_3O_2 (aq) + OH^- (aq)$$

26. Na^+ a. behaves as a Brønsted acid
27. $C_2H_3O_2^-$ b. behaves as a Brønsted base
28. H_2O c. behaves as neither a Brønsted acid nor base
29. $HC_2H_3O_2$ d. behaves as a both Brønsted acid and base

Choose a silver compound formula from the right to complete each of the reactions used to prepare silver salts. In each reaction, HA represents an acid.

30. $2\ HA + ? \rightarrow 2\ AgA + H_2O$ a. AgOH
31. $2\ HA + ? \rightarrow H_2O + 2\ AgA + CO_2$ b. Ag_2O
32. $HA + ? \rightarrow H_2O + AgA$ c. $AgHCO_3$
33. $HA + ? \rightarrow CO_2 + H_2O + AgA$ d. Ag_2CO_3

Classify the systems described on the left into one of the pH ranges given as responses.

34. the $[OH^-]=[H_3O^+]$ in pure H_2O a. the pH is much lower than 7
35. oven cleaners are strongly basic   b. the pH is much higher than 7
36. the active ingredient in the stomach, digestive juice, is 0.01 M hydrochloric acid   c. the pH is near 7
37. a carbonated soft drink has a tart taste   d. the pH is exactly 7

Choose a description from the right that best characterizes the solution made by dissolving in water each of the salts indicated on the left.

38. Na_2SO_4 a. it is acidic because hydrolysis occurs
39. Na_3PO_4 b. it is basic because hydrolysis occurs
40. NH_4Cl c. it is neutral because no hydrolysis occurs
    d. more than one of the above is correct

Chapter 10: Radioactivity and Nuclear Processes

CHAPTER OUTLINE
10.1 Radioactive Nuclei
10.2 Equations for Nuclear Reactions
10.3 Isotope Half-Life
10.4 The Health Effects of Radiation
10.5 Measurement Units of Radiation
10.6 Medical Uses of Radioisotopes
10.7 Nonmedical Uses of Radioisotopes
10.8 Induced Nuclear Reactions
10.9 Nuclear Energy

LEARNING OBJECTIVES/ASSESSMENT
When you have completed your study of this chapter, you should be able to:
1. Describe and characterize the common forms of radiation emitted during radioactive decay and other nuclear processes. (Section 10.1; Exercise 10.2)
2. Write balanced equations for nuclear reactions. (Section 10.2; Exercise 10.12)
3. Solve problems using the half-life concept. (Section 10.3; Exercise 10.16)
4. Describe the effects of radiation on health. (Section 10.4; Exercise 10.22)
5. Describe and compare the units used to measure quantities of radiation. (Section 10.5; Exercise 10.24)
6. Describe with examples medical and nonmedical uses of radioisotopes. (Sections 10.6 and 10.7; Exercises 10.30 and 10.36)
7. Show that you understand the concept of induced nuclear reactions. (Section 10.8; Exercise 10.38)
8. Describe the differences between nuclear fission and nuclear fusion reactions. (Section 10.9; Exercise 10.46)

SOLUTIONS FOR THE END OF CHAPTER EXERCISES
RADIOACTIVE NUCLEI (SECTION 10.1)
☑10.2
   a. mass number = 0: beta, gamma, positron
   b. positive charge: alpha, positron
   c. charge = 0: gamma, neutron

10.4
   a. A beta particle = an electron
   b. An alpha particle = 2 protons and 2 neutrons
   c. A positron = positive electron

EQUATIONS FOR NUCLEAR REACTIONS (SECTION 10.2)

10.6		Atomic number change	Mass number change
	a. An alpha particle is emitted.	decrease by 2	decrease by 4
	b. A beta particle is emitted.	increase by 1	no change
	c. An electron is captured.	decrease by 1	no change
	d. A gamma ray is emitted.	no change	no change
	e. A positron is emitted.	decrease by 1	no change

10.8		
	a. An iron-54 nucleus	$^{54}_{26}\text{Fe}$
	b. A nucleus of element number 48 with a mass number of 110	$^{110}_{48}\text{Cd}$
	c. A nucleus of the titanium isotope that contains 26 neutrons	$^{48}_{22}\text{Ti}$

10.10 a. $^{234}_{90}\text{Th} \rightarrow ? + ^{234}_{91}\text{Pa}$ $? = ^{0}_{-1}\beta$ (beta)   d. $^{226}_{88}\text{Ra} \rightarrow ? + ^{4}_{2}\alpha$ $? = ^{222}_{86}\text{Rn}$
       b. $^{15}_{8}\text{O} \rightarrow ? + ^{15}_{7}\text{N}$ $? = ^{0}_{+1}\beta$ (positron)   e. $^{113}_{47}\text{Ag} \rightarrow ^{113}_{48}\text{Cd} + ?$ $? = ^{0}_{-1}\beta$ (beta)
       c. $? \rightarrow ^{4}_{2}\alpha + ^{244}_{98}\text{Cf}$ $? = ^{248}_{100}\text{Fm}$ f. $^{41}_{18}\text{Ar} + ^{0}_{-1}e \rightarrow ?$ $? = ^{41}_{17}\text{Cl}$

☑10.12 a. $^{121}_{50}\text{Sn}$ (beta emission) $^{121}_{50}\text{Sn} \rightarrow ^{0}_{-1}\beta + ^{121}_{51}\text{Sb}$
        b. $^{190}_{78}\text{Pt}$ (daughter = osmium-186) $^{190}_{78}\text{Pt} \rightarrow ^{4}_{2}\alpha + ^{186}_{76}\text{Os}$
        c. $^{54}_{26}\text{Fe}$ (electron capture) $^{54}_{26}\text{Fe} + ^{0}_{-1}e \rightarrow ^{54}_{25}\text{Mn}$
        d. $^{72}_{31}\text{Ga}$ (daughter Ge-72) $^{72}_{31}\text{Ga} \rightarrow ^{0}_{-1}\beta + ^{72}_{32}\text{Ge}$
        e. $^{10}_{4}\text{Be}$ (beta emission) $^{10}_{4}\text{Be} \rightarrow ^{0}_{-1}\beta + ^{10}_{5}\text{B}$
        f. $^{238}_{92}\text{U}$ (alpha emission) $^{238}_{92}\text{U} \rightarrow ^{4}_{2}\alpha + ^{234}_{90}\text{Th}$

ISOTOPE HALF-LIFE (SECTION 10.3)

10.14   Half-life is the amount of time required for half of a sample to undergo a specific process. For example, if the half-life of a cake is one day, then half of a cake will be eaten the first day, the next day half of the remaining cake (¼ of the original cake) would be eaten, the following day half of the remaining cake (⅛ of the original cake) would be eaten, etc.

☑10.16 Amount remaining $= 9.0 \text{ ng} \left(\frac{1}{2}\right)^{24 \text{ hours}\left(\frac{1 \text{ half-life}}{6 \text{ hours}}\right)} = 0.56$ ng

10.18 $12.5\% = 100\% \left(\frac{1}{2}\right)^{\text{half-lives}} \Rightarrow \log(0.125) = \log\left(\left(\frac{1}{2}\right)^{\text{half-lives}}\right) \Rightarrow \log(0.125) = \text{half-lives}\left(\log\left(\frac{1}{2}\right)\right)$

        half – lives = 3.000

        Time elapsed = 3 half – lives $\left(\dfrac{5600 \text{ years}}{1 \text{ half-life}}\right) = 16800$ years $= 1.68 \times 10^4$ years

10.20 $\left(\dfrac{1}{2}\right)^{\text{half-lives}} = \dfrac{1}{16} \Rightarrow \log\left(\left(\dfrac{1}{2}\right)^{\text{half-lives}}\right) = \log\left(\dfrac{1}{16}\right) \Rightarrow \text{half-lives}\left(\log\left(\dfrac{1}{2}\right)\right) = \log\left(\dfrac{1}{16}\right)$

        half – lives = 4

        Time elapsed = 4 half – lives $\left(\dfrac{5600 \text{ years}}{1 \text{ half-life}}\right) = 22400$ years $= 2.24 \times 10^4$ years

THE HEALTH EFFECTS OF RADIATION (SECTION 10.4)

☑10.22 Long-term, low-level exposure to radiation may lead to genetic mutations because ionizing radiation can produce free radicals in exposed tissues. Short-term exposure to intense radiation destroys tissue rapidly and causes radiation sickness. Both forms of exposure have negative effects on health.

MEASUREMENT UNITS FOR RADIATION (SECTION 10.5)

☑10.24 Physical units of radiation indicate the activity of a source of radiation; whereas, biological units of radiation indicate the damage caused by radiation in living tissue. Examples of physical units of radiation include the Curie and the Becquerel. Examples of biological units of radiation include the Roentgen, the Rad, the Gray, and the Rem.

Radioactivity and Nuclear Processes

10.26 $$3.1 \text{ rads beta radiation}\left(\frac{1 \text{ roentgen}}{0.96 \text{ rad}}\right) = 3.2 \text{ roentgen}$$

10.28 $$4.6 \text{ μCi}\left(\frac{1 \text{ Ci}}{10^6 \text{ μCi}}\right)\left(\frac{3.7 \times 10^{10} \frac{\text{disintegrations}}{\text{second}}}{1 \text{ Ci}}\right) = 1.7 \times 10^5 \frac{\text{disintegrations}}{\text{second}}$$

MEDICAL USES OF RADIOISOTOPES (SECTION 10.6)

☑**10.30** Radioactive isotopes can be used for diagnostic work. When the radioactive isotope concentrates in a tissue under observation, the location is called a hot spot. When the radioactive isotope is excluded or rejected by a tissue under observation, the location is called a cold spot. Both hot spots and cold spots can be used for diagnostic work.

10.32 $^{51}_{24}\text{Cr} + ^{0}_{-1}\text{e} \rightarrow ^{51}_{23}\text{V}$; daughter nucleus = vanadium-51

NONMEDICAL USES OF RADIOISOTOPES (SECTION 10.7)

10.34 By using water that contains a radioactive isotope of oxygen, the oxygen gas produced could be analyzed to see if it contains the radioactive isotope of oxygen from the water or a nonradioactive isotope of oxygen from the hydrogen peroxide.

☑**10.36** The radioactivity of gallon of water that contains a radioisotope with a long half-life could be measured with the Geiger-Müller counter. The water could then be added to the pool and given time to circulate. A sample of pool water could then be taken and the level of radioactivity measured. The dilution formula could then be used to determine the volume of the pool based on the relative levels of radioactivity.

INDUCED NUCLEAR REACTIONS (SECTION 10.8)

☑**10.38** $^{24}_{12}\text{Mg} + ^{4}_{2}\alpha \rightarrow ^{27}_{14}\text{Si} + ^{1}_{0}\text{n}$

10.40 $^{238}_{92}\text{U} + ^{1}_{0}\text{n} \rightarrow ^{239}_{94}\text{Pu} + 2\,^{0}_{-1}\beta$

10.42 A moderator is a material that can be placed in the path of a neutron in order to slow down the neutrons as they pass through the moderator. As the neutrons slow, their kinetic energy decreases and the nuclear forces allow the neutrons to be captured.

10.44 $^{66}_{30}\text{Zn} + ^{1}_{1}\text{p} \rightarrow ^{67}_{31}\text{Ga}$

NUCLEAR ENERGY (SECTION 10.9)

☑**10.46** $^{1}_{1}\text{H} + ^{2}_{1}\text{H} \rightarrow ^{3}_{2}\text{He} + \gamma$ or another reaction that indicates two nuclei come together to form a single larger nucleus.

10.48 Critical mass is the minimum amount of fissionable material needed to sustain a critical chain reaction at a constant rate. Neutrons released during a fission process might escape from a inside the fissionable material without colliding and splitting another atom; however, once the critical mass of the fissionable material is obtained, the chance for neutrons to escape without causing another fission process decreases and a chain reaction can occur.

10.50 $^{238}_{94}\text{Pu} \rightarrow ^{4}_{2}\alpha + ^{234}_{92}\text{U}$

146 Chapter 10

ADDITIONAL ACTIVITIES

Section 10.1 Review:
(1) Draw representations of the types of radiation in Table 10.1.
(2) Rank the types of radiation from Table 10.1 in order of increasing mass.
(3) Place the types of radiation on the "continuum of charge" shown below.

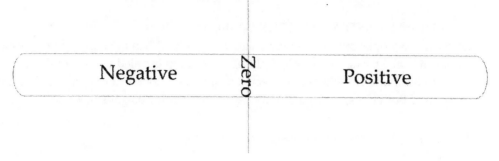

Section 10.2 Review:
Consider whether hydrogen-3 could theoretically undergo the following radioactive processes. Write an equation to represent each process or explain why the process is not possible.
(1) alpha decay
(2) beta decay*
(3) gamma emission
(4) positron emission
(5) electron capture

*Note: Beta decay is the actual decay process that hydrogen-3 undergoes.

Section 10.3 Review:
All of the following questions refer to the picture at the beginning of this chapter of the study guide of the 60-minute stopwatch.
(1) What is the half-life for the element represented in the picture?
(2) How many half-lives have occurred at 7.5 seconds?
(3) Refer to Figure 10.3, predict what fraction of the sample will remain after 7.5 seconds.
(4) If each atom in the picture represents a mole of atoms, how many moles of atoms would remain after 7.5 seconds?
(5) Do answers (3) and (4) agree? Explain.

Section 10.4 Review:
(1) The inverse square law of radiation is mathematically similar to Graham's Law for gases. Rewrite Equation 10.2 in the same form as Equation 6.13.
(2) The inverse square law for radiation relates intensity of radiation to the distance from the radiation source. Graham's law relates the rate of effusion to the molecular mass of a gas. Based on (1), which variables in the two equations are similar?
(3) Why does an X-ray technician step behind the wall when taking dental X-rays?
(4) Why do dental patients wear a lead apron when having X-rays taken?
(5) A dental X-ray typically exposes patients to 2-3 mrem. Classify a dental X-ray as low-level or intense radiation.

Section 10.5 Review:
(1) Rank the physical units of radiation in order of increasing size.
(2) Divide the biological units into two categories: units for absorbed doses and units for effective doses.
(3) Why would one rad of alpha radiation have a more severe health effect than one rad of gamma radiation? (Note: 1 rad of alpha radiation has the same health effect as 10 rad of gamma radiation.)

Radioactivity and Nuclear Processes 147

Section 10.6 Review:
(1) What are the similarities in the properties desired in radioisotopes used as tracers and in therapy?
(2) What are the differences in the properties desire in radioisotopes used as tracers and in therapy?

Section 10.7 Review:
(1) Are the properties desired in radioisotopes used as nonmedical tracers the same as the properties required in medical tracers?
(2) Which type of radioactive decay would be the most effective for determining the thickness of thick metal sheets?
(3) Would radiocarbon dating work well for metal artifacts from a primitive civilization? Explain.

Section 10.8 Review:
(1) What are the four possible outcomes from bombarding a nucleus with high-energy particles?
(2) How many transuranium elements have been synthesized? Where are they located on the periodic table?

Section 10.9 Review:
(1) What is the major difference between fission and fusion?
(2) Why are fission reactors currently used to generate electricity, but fusion reactors are not?

Tying It All Together with a Laboratory Application:
Radioisotopes of chlorine are under investigation for use in nuclear medicine. Approximately 0.12% of the human body is made of the chloride ion. Stomach acid is hydrochloric acid. The following chart provides information for the radioisotopes of chlorine. (1) Complete the chart.

Isotope	Half-Life	Mode of Decay	Daughter Nuclei	Daughter Nuclei's Mode of Decay or Granddaughter identity
^{36}Cl	3.01 x 10^5 yrs	beta		stable
^{38}Cl	37.2 min		^{38}Ar	stable
^{39}Cl	55.6 min	beta		beta (268 yrs)
^{40}Cl	1.38 min		^{40}Ar	stable
^{41}Cl	34 sec	beta		beta (1.82 hr)
^{42}Cl	6.8 sec		^{42}Ar	^{42}K (33 yrs)
^{43}Cl	3.3 sec	beta		^{43}K (5.4 min)

On the basis of the daughter nuclei produced, the three best radioisotopes of chlorine for nuclear medicine are (2) _____ because they produce (3) _____. The (4) _____ of the daughter nuclei in the body should also be investigated. On the basis of a half-life that is long enough to prepare and administer a sample, the best radioisotopes of chlorine for nuclear medicine are (5) _____. On the basis of having a short enough half-life that the radioisotopes will decay during the diagnosis but emit little radiation after the procedure, the best radioisotopes of chlorine for nuclear medicine are (6) _____. Combining all of the qualifications for using a radioisotope of chlorine in nuclear medicine, (7) _____ is(are) the best choice(s). If the chlorine radioisotopes were used as tracers for stomach illnesses, the radioisotopes would form a (8) _____ spot. The isotopes of chlorine than emit beta rays would be better used in (9) _____ rather than as (10) _____.

The activity of a sample of radioactive chlorine could be measured using (11) _____, _____, or _____. The units to report the disintegrations per second are (12) _____ units, like the curie or (13) _____. The units to report the health effects of the radioisotopes of chlorine are (14) _____ units, like the rad. To compare the health effects of chlorine-41 to the health effects of an X-ray source, the best units to use are (15) _____.

148 Chapter 10

The granddaughter isotope of chlorine-39 is (16) _____. The mode of decay for the daughter nuclei of chlorine-43 is (17) _____. If chlorine-36 was bombarded with an alpha particle and a new nucleus formed, the product would be (18) _____. If chlorine-36 were bombarded with a neutron, chlorine-(19) _____ could form which would be (20) _____ (stable or unstable) or nuclear (21) _____ or _____ could occur. Bombarding radioisotopes with energetic particles leads to (22) _____ nuclear reactions.

The percentage of chlorine-40 that will remain after 2.5 minutes is (23) _____. The amount of time required for only 2.5 mg to remain from a 40 mg sample of chlorine-40 is (24) _____. If the distance from a sample of chlorine-40 is tripled, the measurement of the intensity of radiation will (25) _____ (increase or decrease) by a factor of (26) _____.

Hospital personnel who are administering radioisotopes should be aware of symptoms like nausea, fatigue, vomiting, and general malaise. These are symptoms of (27) _____ which can occur from (28) _____ exposure to radiation.

SOLUTIONS FOR THE ADDITIONAL ACTIVITIES
Section 10.1 Review:
(1) alpha = [image]; beta = ⊖; gamma = ∿; neutron = ⓞ; positron = ⊕

(2) (least massive) gamma < beta = positron < neutron < alpha (most massive)

(3)
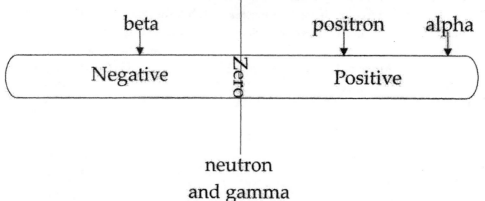

Section 10.2 Review:
(1) Hydrogen-3 could not undergo alpha decay because hydrogen-3 only contains 1 proton and two neutrons. An alpha particle contains two protons and two neutrons.

(2) $^3_1H \rightarrow \,^0_{-1}\beta + \,^3_2He$; (3) $^3_1H \rightarrow \,^0_0\gamma + \,^3_1H$; (4) $^3_1H \rightarrow \,^0_{+1}\beta + 3\,^1_0n$; (5) $^3_1H + \,^0_{-1}e \rightarrow 3\,^1_0n$

Section 10.3 Review:
(1) half-life = 15 seconds; (2) 0.50 half-lives at 7.5 seconds; (3) less than ¾ of the original sample will remain; (4) 5.7 moles of atoms would remain; (5) Yes, 5.7 moles out of 8 moles is 71% (from 4) which is less than 75% or ¾ (from (3)).

Section 10.4 Review:
(1) $\dfrac{I_x}{I_y} = \dfrac{d_y^2}{d_x^2} \Rightarrow \sqrt{\dfrac{I_x}{I_y}} = \dfrac{d_y}{d_x}$

(2) The distance from the radiation source in the inverse square law for radiation is equivalent to the rate of effusion in Graham's law. The intensity of radiation in the inverse square law for radiation is analogous to the molecular mass from Graham's law.

(3) The X-ray technician steps behind the wall when taking dental X-rays to increase his or her distance from the radiation source as well as to insert a barrier between himself or herself and the radiation source.
(4) Dental patients wear a lead apron when having X-rays taken to minimize the exposure of other tissues to the X-ray radiation.
(5) A dental X-ray is a low-level exposure to radiation.

Section 10.5 Review:
(1) becquerel < curie
(2) units for absorbed doses: roentgen, rad, gray; units for effective doses: rem
(3) Gamma rays have high energy; however, they do not have any mass. Alpha particles, on the other hand, have less energy than gamma rays; however, they have mass and are more easily stopped by tissues than gamma rays are. Consequently, alpha particles do not penetrate tissue as deeply as gamma rays. In order for alpha particles to transfer the energy associated with one rad (1×10^{-2} J of energy to 1 kg of tissue), more alpha particles are required to interact with the tissue than for gamma rays. Both alpha particles and gamma rays are forms of ionizing radiation. If more alpha particles strike the tissue, then more free radicals will likely be produced from the alpha particles than the number of free radicals that the gamma rays could produce.

Section 10.6 Review:
(1) Both tracer and therapy radioisotopes should have half-lives that should be long enough for the radioisotope to be prepared and used and the decay products should be nontoxic and give off little or no radiation. The half-life of the radioisotope should be as short as is consistent with use in order to minimize the radioactive waste products in the environment.
(2) Tracer radioisotopes should have half-lives that are short enough that they will decay while the diagnosis is performed, emit gamma rays which can be detected outside the body, and have chemical properties that will allow it to concentrate or be excluded from diseased areas. Therapy radioisotopes should emit alpha or beta radiation to restrict the extent of damage to the desired tissue and chemical properties that will allow it to concentrate in the diseased area.

Section 10.7 Review:
(1) Radioisotopes used as tracers in nonmedical applications do need to have chemical compatibility with the application, depending on the size of the system being traced the type of radioactive decay could be alpha, beta, or gamma, and the half-life needs to be sufficiently long that the study can be completed. The requirement of nontoxic, stable decay products is not as important for nonmedical applications; however, waste products that are generated will need to be disposable. Stable, nontoxic decay products are much safer (and cheaper) for waste disposal.
(2) For thick metal sheets, gamma rays would be the most effective at penetrating the material.
(3) Metal artifacts contain little very little carbon, so radiocarbon dating would not be effective. Looking for another radioisotope might prove to be more successful.

Section 10.8 Review:
(1) produce a stable nucleus, produce an unstable nucleus, nuclear fission occurs, nuclear fusion occurs
(2) Twenty-one transuranium elements have been synthesized. They occupy sites in the d, f, and p blocks of the seventh period.

Section 10.9 Review:
(1) Fission is the process of heavier nuclei breaking apart into smaller nuclei. Fusion is the process of smaller nuclei coming together to form larger nuclei.

150 Chapter 10

(2) Fission can occur at low temperatures; however, fusion only occurs at extremely high temperatures. Scientists are still working to determine how to contain the high temperatures needed to maintain a fusion reactor.

Tying It All Together with a Laboratory Application:

(1)
Isotope	Mode of Decay	Daughter Nuclei
^{36}Cl	beta	^{36}Ar
^{38}Cl	**beta**	^{38}Ar
^{39}Cl	beta	^{39}Ar
^{40}Cl	**beta**	^{40}Ar
^{41}Cl	beta	^{41}Ar
^{42}Cl	**beta**	^{42}Ar
^{43}Cl	beta	^{43}Ar

(2) $^{36}Cl, ^{38}Cl, ^{40}Cl$
(3) stable daughter nuclei
(4) toxicity
(5) $^{36}Cl, ^{38}Cl, ^{39}Cl, ^{40}Cl$
(6) $^{38}Cl, ^{39}Cl, ^{40}Cl, ^{41}Cl, ^{42}Cl, ^{43}Cl$
(7) $^{38}Cl, ^{40}Cl$
(8) hot
(9) therapy
(10) tracers
(11) Geiger-Müller tube, scintillation counter, film badge
(12) physical
(13) becquerel
(14) biological
(15) rem
(16) potassium-39
(17) beta
(18) potassium-40
(19) 37
(20) stable
(21) fission, fusion
(22) induced
(23) 28%
(24) 5.5 minutes
(25) decrease
(26) 9
(27) radiation sickness
(28) intense, short-term

SELF-TEST QUESTIONS

Multiple Choice

1. The three common types of radiation emitted by naturally radioactive elements are
   a. electrons, protons, and neutrons
   b. alpha rays, beta rays, and neutrons
   c. x-rays, gamma rays, and protons
   d. alpha particles, beta particles, and gamma rays

2. Which of the following types of radiation is composed of particles which carry a +2 charge?
   a. alpha
   b. beta
   c. gamma
   d. neutrons

3. Which of the following types of radiation is not composed of particles?
   a. alpha
   b. beta
   c. gamma
   d. neutrons

4. After four half-lives have elapsed, the amount of a radioactive sample which has not decayed is
   a. 40% of the original sample
   b. ¼ of the original sample
   c. ⅛ of the original sample
   d. ¹⁄₁₆ of the original sample

5. If ⅛ of an isotope sample is present after 22 days, what is the half-life of the isotope?
   a. 10 days
   b. 5 days
   c. 2¾ days
   d. 7⅓ days

6. By doubling the distance between yourself and a source of radiation, the intensity of the radiation
   a. is ½ as great
   b. is ⅓ as great
   c. is ¼ as great
   d. is ⅛ as great

7. Which of the following would be the most convenient unit to use when determining the total dose of radiation received by an individual who was exposed to several different types of radiation?
   a. Roentgen
   b. rad
   c. gray
   d. rem

8. Which type of radiation has the lowest penetration?
   a. α   b. β   c. γ   d. X

9. The reaction, $^{24}_{11}Na \rightarrow \,^{24}_{12}Mg + \,^{0}_{-1}\beta$, is an example of
   a. alpha decay   b. beta decay   c. positron decay   d. fission

10. The reaction, $^{235}_{92}U + \,^{1}_{0}n \rightarrow \,^{103}_{41}Nb + \,^{131}_{51}Sb + 2\,^{1}_{0}n$, is an example of
    a. alpha decay   b. beta decay   c. positron decay   d. fission

True-False

11. Radioactive tracers are useful in both medical and nonmedical applications.
12. Radioactive isotopes are not taken into the body during medical uses of radioisotopes.
13. A Curie is a measurement technique used to determine the amount of radiation produced by a sample.
14. The rem is a biological radiation measurement unit.
15. Fission and fusion processes produce energy.
16. The mass number of potassium-40 is 19.
17. Radiation sickness is most commonly associated with long-term, low-level exposure to radiation.

Matching

18. $^{13}_{7}N \rightarrow \,^{13}_{6}C + ?$ a. beta, $^{0}_{-1}\beta$
19. $^{27}_{13}Al + \,^{2}_{1}H \rightarrow \,^{25}_{12}Mg + ?$ b. neutron, $^{1}_{0}n$
20. $^{9}_{4}Be + \,^{4}_{2}He \rightarrow \,^{12}_{6}C + ?$ c. positron, $^{0}_{1}\beta$
     d. alpha, $^{4}_{2}\alpha$

Chapter 11: Organic Compounds: Alkanes

CHAPTER OUTLINE

11.1 Carbon: The Element of Organic Compounds
11.2 Organic and Inorganic Compounds Compared
11.3 Bonding Characteristics and Isomerism
11.4 Functional Groups: The Organization of Organic Chemistry
11.5 Alkane Structures
11.6 Conformations of Alkanes
11.7 Alkane Nomenclature
11.8 Cycloalkanes
11.9 The Shape of Cycloalkanes
11.10 Physical Properties of Alkanes
11.11 Alkane Reactions

LEARNING OBJECTIVES/ASSESSMENT

When you have completed your study of this chapter, you should be able to:

1. Show that you understand the general importance of organic chemical compounds. (Section 11.1; Exercise 11.2)
2. Be able to recognize the molecular formulas of organic and inorganic compounds. (Section 11.1; Exercise 11.4)
3. Explain some general differences between inorganic and organic compounds. (Section 11.2; Exercise 11.8)
4. Be able to use structural formulas to identify compounds that are isomers of each other. (Section 11.3; Exercise 11.20)
5. Write condensed or expanded structural formulas for compounds. (Section 11.4; Exercise 11.24)
6. Classify alkanes as normal or branched. (Section 11.5; Exercise 11.28)
7. Use structural formulas to determine whether compounds are structural isomers. (Section 11.6; Exercise 11.30)
8. Assign IUPAC names and draw structural formulas for alkanes. (Section 11.7; Exercise 11.34)
9. Assign IUPAC names and draw structural formulas for cycloalkanes. (Section 11.8; Exercise 11.44)
10. Name and draw structural formulas for geometric isomers of cycloalkanes. (Section 11.9; Exercise 11.54)
11. Describe the key physical properties of alkanes. (Section 11.10; Exercise 11.56)
12. Write alkane combustion reactions. (Section 11.11; Exercise 11.60)

SOLUTIONS FOR THE END OF CHAPTER EXERCISES

CARBON: THE ELEMENT OF ORGANIC COMPOUNDS (SECTION 11.1)

☑11.2 Fruits and vegetables, the family pet, plastics, sugar, cotton, and wood are a few of the many items composed of organic compounds.

☑11.4 All organic compounds contain carbon atoms.

ORGANIC AND INORGANIC COMPOUNDS COMPARED (SECTION 11.2)

11.6 Covalent bonding is the most prevalent type of bonding in organic compounds.

☑11.8 a. A liquid that readily burns is most likely an **organic compound** because organic compounds can exist in any of the states of matter and are often flammable.
b. A white solid with a melting point of 735°C is most likely an **inorganic compound** because inorganic compounds are usually high melting point solids.
c. A liquid that floats on the surface of water and does not dissolve is most likely an **organic compound** because organic compounds can exist in any of the states of matter and often have low water solubility.

d. A compound that exists as a gas at room temperature and ignites easily is most likely an **organic compound** because organic compounds can exist in any of the states of matter and are often flammable.

e. A solid substance that melts at 65°C is most likely an **organic compound** because organic compounds can exist as low melting point solids.

11.10 Organic compounds are nonconductors of electricity because they do not form ions readily on their own or in water. In fact, most organic compounds do not even dissolve in water, let alone, dissociate. Without ions to carry the charge, electricity cannot flow.

BONDING CHARACTERISTICS AND ISOMERISM (SECTION 11.3)

11.12 Numerous organic compounds exist because each carbon atom can form four covalent bonds, including bonds to other carbon atoms, and carbon containing molecules can exhibit isomerism.

11.14 The carbon-hydrogen bond in CH_4 results from the overlap of the sp^3 hybrid orbital on the carbon atom overlaps with the $1s$ orbital on the hydrogen atom.

11.16 An unhybridized p orbital has two lobes of equal size separated by a node; however, a hybridized sp^3 orbital has two lobes of unequal size separated by a node.

11.18 a. [structure]   c. [structure]
     b. [structure]   d. [structure]

11.20 a. not isomers (C_4H_8 and C_4H_{10})   d. isomers (C_3H_6O)
      b. isomers (C_5H_{12})   e. isomers (C_3H_9N)
      c. not isomers ($C_4H_{10}O$ and C_4H_8O)

11.22 a. [structure] This structure is incorrect because the circled hydrogen atom is making 2 bonds. Hydrogen atoms are only able to make 1 bond.

b. [structure] This structure is correct. Each hydrogen has 1 bond. Each carbon has 4 bonds. The nitrogen has 3 bonds.

c. [structure] This structure is incorrect because the circled carbon atom is making 5 bonds. Carbon atoms are only able to make 4 bonds.

d. [structure] This structure is incorrect because the circled carbon atoms are only making 3 bonds each. Each carbon atom must make 4 bonds.

e. [structure] This structure is correct. Each carbon atom has 4 bonds. Each hydrogen atom has 1 bond. The oxygen atom has two bonds.

Organic Compounds: Alkanes

FUNCTIONAL GROUPS: THE ORGANIZATION OF ORGANIC CHEMISTRY (SECTION 11.4)

☑11.24
    Expanded   Condensed

a. [expanded structure of dipropyl ether] $CH_3CH_2CH_2-O-CH_2CH_2CH_3$

b. [expanded structure of 3-methyl-2-butanone] $CH_3-\underset{\underset{O}{\|}}{C}-CH(CH_3)-CH_3$

11.26
    Condensed   Expanded

a. $CH_3-CH_2-CH_2-NH_2$ [expanded structure of propylamine]

b. $CH_3-CH_2-\underset{\underset{O}{\|}}{C}-CH_3$ [expanded structure of 2-butanone]

ALKANE STRUCTURES (SECTION 11.5)

☑11.28
a. branched (3 C longest chain)
b. normal (5 C longest chain)
c. normal (4 C longest chain)
d. normal (5 C longest chain)
e. branched (5 C longest chain)
f. branched (4 C longest chain)

CONFORMATIONS OF ALKANES (SECTION 11.6)

☑11.30
a. same compound (C_4H_{10} normal)
b. same compound (C_6H_{14} normal)
c. isomers (C_4H_{10} branched and normal)
d. same compound (C_5H_{12} same branched)

ALKANE NOMENCLATURE (SECTION 11.7)

11.32
a. [structure with C1–C6 numbering]   6 C atoms
b. [structure with C1–C6 numbering]   6 C atoms
c. [structure with C1–C8 numbering]   8 C atoms

☑11.34
a. [structure] 3-methylpentane
b. [structure] 2-methylpropane
c. [structure] 3,6-diethyl-4-methylnonane

156 Chapter 11

d.

4,7-diethyl-5-methyldecane

e.

3-ethyl-5-methylnonane

11.36   a.   3-ethylpentane

CH₃CH₂—CH(CH₂CH₃)—CH₂CH₃

b.   2,2-dimethylbutane

CH₃—C(CH₃)(CH₃)—CH₂CH₃

c.   4-ethyl-3,3-dimethyl-5-propyldecane

CH₃CH₂—C(CH₃)(CH₃)—CH(CH₂CH₃)—CH(CH₂CH₂CH₃)—CH₂CH₂CH₂CH₂CH₃

d.   5-sec-butyldecane

CH₃CH₂CH₂CH₂—CH[CH(CH₃)CH₂CH₃]—CH₂CH₂CH₂CH₂CH₃

11.38

CH₃CH₂CH₂CH₂CH₃
pentane

CH₃—CH(CH₃)—CH₂CH₃
2-methylbutane

CH₃—C(CH₃)(CH₃)—CH₃
2,2-dimethylpropane

11.40

CH₃CH₂CH₂CH₂CH₂CH₃
hexane

CH₃—CH(CH₃)—CH₂CH₂CH₃
2-methylpentane

CH₃CH₂—CH(CH₃)—CH₂CH₃
3-methylpentane

CH₃—C(CH₃)(CH₃)—CH₂CH₃
2,2-dimethylbutane

CH₃—CH(CH₃)—CH(CH₃)—CH₃
2,3-dimethylbutane

11.42   a.   incorrect = 1,2-dimethylpropane

H₂C=CH(CH₃)—CH₃ with CH₃ branch
corrected = 2-methylbutane

The longest chain (4 C) was not identified as the parent chain.

b. incorrect = 3,4-dimethylpentane

H₃C—CH₂—CH(CH₃)—CH(CH₃)—CH₃
 5   1
←
corrected = 2,3-dimethylpentane

The parent chain was not numbered so that the attached groups were on the lowest numbered carbon atoms.

c. incorrect = 2-ethyl-4-methylpentane

CH₃—CH₂   CH₃
 | |
CH₃—CH—CH₂—CH—CH₃

corrected = 2,4-dimethylhexane

The longest chain (6 C) was not identified as the parent chain.

d. incorrect = 2-bromo-3-ethylbutane

       Br
       |
H₃C—CH—CH—CH₃
         |
       H₂C—CH₃

corrected = 2-bromo-3-methylpentane

The longest chain (5 C) was not identified as the parent chain.

CYCLOALKANES (SECTION 11.8)

☑ 11.44 a. cyclopentane   c. 1,1-dimethylcyclohexane

b. 1,2-dimethylcyclobutane   d. 1,2,3-trimethylcyclobutane

11.46 a. isopropylcyclopentane   c. 1-isobutyl-3-isopropylcyclohexane

b. 1,1-dimethylcyclobutane

11.48 a. same compound (C₆H₁₂; 1,2-dimethylcyclobutane)
b. structural isomers (C₈H₁₆; 1-ethyl-2-methylcyclopentane & 1-ethyl-3-methylcyclopentane)
c. structural isomers (C₅H₁₀; ethylcyclopropane & 1,2-dimethylcyclopropane)
d. structural isomers (C₈H₁₆; 1,1-dimethylcyclohexane & 1,2-dimethylcyclohexane)

THE SHAPE OF CYCLOALKANES (SECTION 11.9)

11.50 Each carbon atom in cyclohexane has 4 single bonds and assumes a tetrahedral shape. The only way for six carbon atoms in a ring to each have a tetrahedral shape is for the overall shape to be a chair rather than a planar hexagon.

158 Chapter 11

11.52 a. no geometric isomers

b.
trans-1,2-dimethylcyclohexane   cis-1,2-dimethylcyclohexane

c.
trans-1,3-dimethylcyclobutane   cis-1,3-dimethylcyclobutane

d.
trans-1-ethyl-2-methylcyclopropane   cis-1-ethyl-2-methylcyclopropane

☑11.54 a.
trans-1-ethyl-2-methylcyclopropane

c.
trans-1-methyl-2-propylcyclobutane

b.
cis-1-bromo-2-chlorocyclopentane

d.
trans-1,3-dimethylcyclohexane

PHYSICAL PROPERTIES OF ALKANES (SECTION 11.10)

☑11.56 a. Decane is a liquid at room temperature.   c. Decane is soluble in hexane.
b. Decane is not soluble in water.   d. Decane is less dense than water.

11.58 The sample with the higher melting point and boiling point is 2-methylheptane. It has a higher molecular mass than 2-methylhexane. Melting point and boiling point both increase with molecular mass as long as the intermolecular forces within each sample are identical.

ALKANE REACTIONS (SECTION 11.11)

☑11.60 a. propane $C_3H_8 + 5\ O_2\ (g) \rightarrow 3\ CO_2\ (g) + 4\ H_2O\ (l)$

b. $2\ C_4H_{10} + 13\ O_2\ (g) \rightarrow 8\ CO_2\ (g) + 10\ H_2O\ (l)$

c. $2\ C_5H_{10} + 15\ O_2\ (g) \rightarrow 10\ CO_2\ (g) + 10\ H_2O\ (l)$

11.62   2 C₆H₁₄ + 13 O₂ (g) → 12 CO (g) + 14 H₂O (l)

CHEMISTRY AND YOUR HEALTH, CHEMISTRY AROUND US, AND OVER THE COUNTER

11.64   Organic foods may have more flavor than conventionally grown produce because organic foods are allowed to ripen naturally on the tree or vine and are generally fresher than conventionally grown produce.

11.66   Mineral oil and petroleum jelly are petroleum products sometimes found in skin moisturizers.

11.68   More than half of all industrial synthetic organic compounds are made from petroleum.

11.70   The antidote for CO poisoning is fresh air. Victims of CO poisoning need oxygen because CO occupies the oxygen binding site on the hemoglobin of red cells.

ADDITIONAL ACTIVITIES

Section 11.1 Review: Link the ideas in the rectangles to the appropriate category.

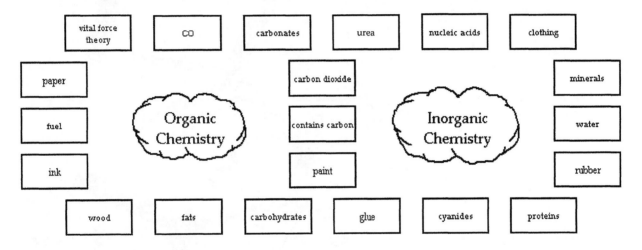

Section 11.2 Review: Identify the facts/properties from each of the following categories for "typical" organic compounds.

Category	Possibility A	Possibility B
Number of known compounds	250,000	> 6 million
Bonding within molecules	ionic	covalent
Interparticle forces	weak	strong
Flammability	flammable	nonflammable
Water solubility	high	low
Conductivity of water solutions	conductor	nonconductor
Rate of chemical reactions	slow	fast
Melting point	high	low

160 Chapter 11

Section 11.3 Review: Copy this page onto another sheet of paper or a transparency, then cut out the shapes. For the large triangles, fold the smaller triangles up from the small triangular base and tape the edges of the small triangles together to form a "closed" tetrahedron. For the bows, fold along the black-gray borders and tape the two separate skinny gray rectangles together to form 3 legs. These legs can be stabilized by adding pieces of wooden toothpicks behind them. To form an "open" tetrahedron, align two of the legs on two of the "bows" and tape. This structure will be stabilized by adding the other two "bows" in such a manner that all of the resulting endpoints have three connected bows. The open tetrahedron will fit inside the closed tetrahedron.

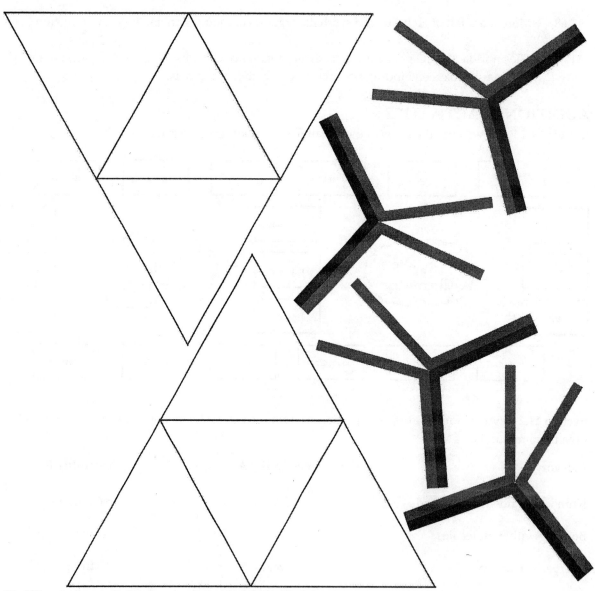

(1) What is the angle between unhybridized p orbitals?
(2) What type of hybridized orbital is represented by the legs on the bows?
(3) What is the angle between the legs on the bows?
(4) Why is this angle different from (1)?
(5) If carbon atoms can be represented as tetrahedrons, how many points of the tetrahedrons will touch for a carbon-carbon single bond?
(6) If carbon atoms can be represented as tetrahedrons, how many points of the tetrahedrons will touch for a carbon-carbon double bond?

(7) If carbon atoms can be represented as tetrahedrons, how many points of the tetrahedrons will touch for a carbon-carbon triple bond?

(8) Can carbon atoms make a quadruple bond? Explain using the image of carbon as a tetrahedron.

Section 11.4 Review:

 Study Tip: Make flashcards of the class name and the corresponding key functional groups from Table 11.2. Practice with the flashcards until you can identify the key functional group by seeing the class name and the class name by seeing the key functional group.

Section 11.5 Review:

(1) List the similarities between methane, ethane, propane, and butane.
(2) What is similar in the structures of methane, ethane, and propane? How is butane different?
(3) Pentane is the fifth member of this series. What is its formula? Can it have a normal and/or a branched structure?

Section 11.6 Review:

Place a toothpick into the top point of one of your closed tetrahedrons from the Section 11.3 Review. Place the other end of the toothpick into the top point of the other closed tetrahedron from the Section 11.3 Review. Color one of the "free" points on each tetrahedron. These will represent CH_3– groups.

(1) Hold the tetrahedrons in such a way that both of the colored points are oriented toward the ceiling and you are looking at one end of the toothpick. Can you see the other tetrahedron?
(2) Tip the structure so that the "front" tetrahedron is lower than the "rear" tetrahedron. Are the CH_3– groups aligned?
(3) Rotate the "front" tetrahedron 90° clockwise. (The colored point should move from the "12 o'clock" position to the "3 o'clock" position). Are the CH_3–groups aligned? Are the points on the tetrahedrons aligned?
(4) Rotate the "front tetrahedron another 30° clockwise. (The colored point should move to the "4 o'clock" position.) Are the CH_3–groups aligned? Are the points on the tetrahedrons aligned?
(5) What are the different orientations seen in (2), (3), and (4) called?
(6) Has the bonding order changed in (2), (3), or (4)?
(7) Are (2), (3), and (4) structural isomers of each other?

Section 11.7 Review:

 Study Tip: (1) Make flashcards for the number of carbon atoms in the longest chain and their corresponding stem names. Practice with the flashcards until you can name the longest carbon chain based on the number of carbon atoms in the chain and determine how many carbon atoms are in the longest chain based on the name.

 (2) Make flashcards of the names and structures of the alkyl groups as well as the common nonalkyl groups. Practice with the flashcards until you can draw the structures from the names and name the structures within a compound without having to refer to your notes.

(1) Suggest a reason alkanes have an –*ane* ending.
(2) Suggest a reason that unbranched carbon chains are given the prefix *n-*.
(3) What is the common structural element for all *iso-* compounds?
(4) Identify the following alkyl groups by their names.

162 Chapter 11

[Structural formulas shown at top of page:
- CH₂CH₃ attached to H₃C-CH₂- group
- A branched structure with CH, H₃C, H₂C—CH₃, CH₃, CH₃—CH
- CH₂CH₂CH₂CH₃ with vertical bond and CH₃CHCH₂CH₃
- CH₃, CH₂—, CH₃—CH—CH₃]

Section 11.8 Review:
(1) Is the prefix *cyclo-* appropriate to describe compounds that contain a ring? Explain.
(2) What is the general formula for cycloalkanes?

Section 11.9 Review:
(1) Can the two linked tetrahedrons from the Section 11.6 Review be placed on one of the small triangles from the closed tetrahedron template in the Section 11.3 Review in order to represent cyclopropane? If so, what is the bond angle between the carbon atoms? If not, can cyclopropane be represented if the tetrahedrons are removed from the toothpick? What is the associated bond angle?
(2) What are the angles in a planar square?
(3) What are the angles in a planar pentagon?
(4) What is the preferred bond angle for adjacent carbon-carbon bonds in a cycloalkane?
(5) What must the atoms in cyclobutane and cyclopentane do to obtain a more desirable bond angle?
(6) Can the atoms in cyclopropane make the same change as in (5)?
(7) What do the prefixes *cis-* and *trans-* mean?

Section 11.10 Review:
Box the following properties that apply to alkanes. Correct the properties that do not apply to alkanes as they are currently written.

unpleasant odor   high boiling point   low melting point   polar   dispersion forces

soluble in water   less dense than water   hydrophobic

Section 11.11 Review:
(1) What is the major type of reaction that alkanes undergo?
(2) Is the reaction endothermic or exothermic?
(3) What is an application for this type of alkane reaction?
(4) What is the limiting reactant when CO is a product of this type of reaction?
(5) What is the limiting reactant when only carbon dioxide and water are the products of this type of reaction?

Tying It All Together with a Laboratory Application:
The expanded structural formula of pentane is (1)_____. The condensed structural formula of pentane is (2)_____. Rotating the carbon-carbon bonds in the pentane will produce different (3)_____. The molecular formula of pentane is (4)_____. The relationship between pentane and 2-methylbutane is called (5)_____. The condensed structural formula of cyclopentane is (6)_____. The molecular formula of cyclopentane is (7)_____. Pentane and cyclopentane (8)_____ isomers; however, they both contain only (9)_____ or _____ bonds. If two of the hydrogen atoms in pentane were replaced with chlorine atoms, several (10)_____ isomers could be formed; however, no (11)_____ isomers could be formed. If two hydrogen atoms in cyclopentane

were replaced with chlorine atoms, several (12)_____ isomers could be formed as long as the two chlorine atoms were attached to (13)_____ carbon atom(s).

When 25 mL of pentane is poured into a beaker that contains 25 mL of water, a (14)_____ mixture forms. Pentane is (15)_____ dense than water and occupies the (16)_____ portion of the mixture. Pentane is a (17)_____ (polar, nonpolar, or ionic) molecule and experiences (18)_____ (dispersion, dipole-dipole, or hydrogen bonding) intermolecular forces. Water is a (19)_____ (polar, nonpolar, or ionic) molecule and experiences predominately (20)_____ (dispersion, dipole-dipole, or hydrogen bonding) intermolecular forces. These compounds do not fit the guideline for solubility which is (21)_____. Pentane is also said to be (22)_____ because it (23)_____ dissolve in water.

The beaker is placed in the fume hood and a smoldering splint is placed into the beaker above the liquid level. What happens? (24)_____

        A. Nothing, because the pentane and water are not flammable.
        B. The pentane begins to burn because it is flammable.
        C. The pentane and water begin to burn because they are both flammable.

Flammability is a property of most (25)_____ compounds. The process of burning a compound in the presence of oxygen is called (26)_____. When oxygen is the limiting reactant, the reaction is classified as (27)_____. When oxygen is the excess reactant, the reaction is classified as (28)_____. Burning is an (29)_____ process because heat is released.

SOLUTIONS FOR THE ADDITIONAL ACTIVITIES
Section 11.1 Review:

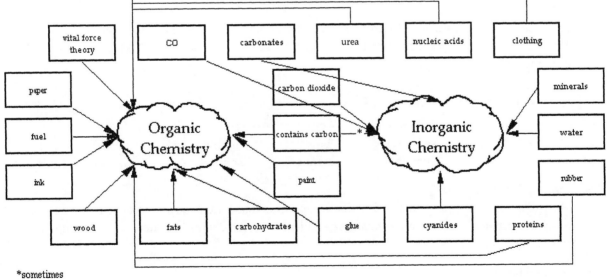

*sometimes

Section 11.2 Review:

Number of known compounds	250,000	> 6 million
Bonding within molecules	ionic	covalent
Interparticle forces	weak	strong
Flammability	flammable	nonflammable
Water solubility	high	low
Conductivity of water solutions	conductor	nonconductor
Rate of chemical reactions	slow	fast
Melting point	high	low

164 Chapter 11

Section 11.3 Review:
(1) The angle between unhybridized p orbitals is 90°.
(2) The legs on the bows represent the larger lobe of the sp^3 hybridized orbitals.
(3) The angle between the legs on the bows is 109.5°.
(4) When a s orbital blends with 3 p orbitals to make 4 new sp^3 hybrid orbitals on a carbon atom, the four new orbitals share the same energy level. Each hybrid orbital contains one electron. These orbitals repel each other and establish a new angle of 109.5°. The reorganization of the orbitals results in a new angle between the orbitals.

(5) single bond – 1 point   (6) double bond – 2 points   (7) triple bond – 3 points

(8) Carbon atoms cannot make a quadruple bond because it is impossible to have all four corners of the tetrahedrons touch.

Section 11.5 Review:
(1) Methane, ethane, propane, and butane are saturated hydrocarbons with the general formula C_nH_{2n+2}. They have bond angles of 109.5°. They can be burned as fuels.
(2) Methane, ethane, and propane cannot not have branched structures. They are normal alkanes. Butane can have a branched or a normal structure.
(3) Pentane, C_5H_{12}, can have either a branched or a normal structure.

Section 11.6 Review:
(1) no; (2) yes; (3) no, no; (4) no, yes; (5) conformations; (6) no; (7) no

Section 11.7 Review:
(1) The *–ane* ending for saturated hydrocarbons is also contained within the word "*alkane.*"
(2) The *n-* prefix means unbranched and begins with the same letter as "*normal*" which means unbranched.

(3) CH₃—CH—
       |
       CH₃

(4)
isopropyl	isobutyl	*t*-butyl	propyl
ethyl	*sec*-butyl	butyl	methyl
ethyl	isopropyl	*sec*-butyl	isobutyl

Section 11.8 Review:
(1) The prefix *cyclo-* is appropriate to describe compounds that contain a ring because it means of a circle or wheel, circular. These compounds are considered "closed chains" because no beginning or end exists. (2) C_nH_{2n}

Section 11.9 Review:
(1) No, the two linked tetrahedrons from the Section 11.6 Review cannot be placed on one of the small triangles in order to represent cyclopropane. The distance between the tetrahedron points is too large. By removing the tetrahedrons from the toothpick and adjusting the adjacent carbon-carbon bond angles to 60°, a representation of cyclopropane can be made.
(2) 90°; (3) 120°; (4) 109.5°; (5) Cyclobutane and cyclopentane can both deviate from a planar arrangement of carbon atoms in order to obtain a more desirable bond angle.
(6) No, cyclopropane cannot deviate from a planar arrangement of carbon atoms because the three carbon atoms automatically form a plane.
(7) *cis-* = on this side of; *trans-* = on the other side of, to the other side of, over, across, through

Organic Compounds: Alkanes 165

Section 11.10 Review:

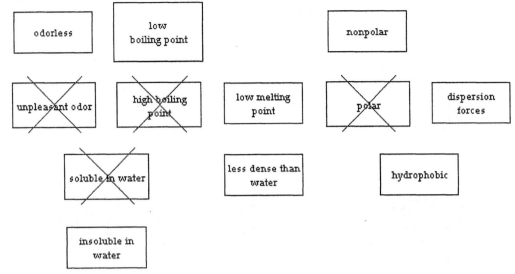

Section 11.11 Review:
(1) Alkanes undergo rapid oxidation (combustion) reactions.
(2) Combustion reactions are exothermic.
(3) Alkane combustion can be used as a source of heat, for cooking, or for running an engine.
(4) Oxygen is the limiting reactant when CO is produced. This is incomplete combustion.
(5) The alkane is the limiting reactant when only carbon dioxide and water are the products of combustion. This is complete combustion.

Tying It All Together with a Laboratory Application:

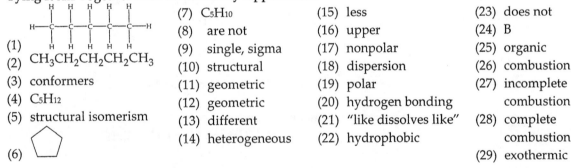

(2) $CH_3CH_2CH_2CH_2CH_3$
(3) conformers
(4) C_5H_{12}
(5) structural isomerism
(6)

(7) C_5H_{10}
(8) are not
(9) single, sigma
(10) structural
(11) geometric
(12) geometric
(13) different
(14) heterogeneous

(15) less
(16) upper
(17) nonpolar
(18) dispersion
(19) polar
(20) hydrogen bonding
(21) "like dissolves like"
(22) hydrophobic

(23) does not
(24) B
(25) organic
(26) combustion
(27) incomplete combustion
(28) complete combustion
(29) exothermic

SELF-TEST QUESTIONS
Multiple Choice
1. The maximum number of covalent bonds which carbon can form is
   a. 1   b. 2   c. 3   d. 4

2. _____ is considered an organic compound.
   a. CH_4 b. NaOH   c. Na_2CO_3 d. KCN

3. How many hydrogen atoms are needed to complete the following structure:

   $$C-\overset{\overset{O}{\|}}{C}-C$$

   a. 2   b. 4   c. 6   d. 8

4. A C—H bond in CH₄ is formed by the overlap of what orbitals?
   a. sp^3 and $1s$
   b. $1s$ and $1s$
   c. p and $1s$
   d. sp and $1s$

5. Which of the following are the same compound?

   I. CH₃—CH₂—CH₂—CH₃

   IV. CH₃—CH₂—CH—CH₃
          |
          CH₃

   II.
     CH₃
     |
   CH₃—CH—CH₂—CH₂—CH₃

   V.
      CH₃
      |
   CH₃—C—CH₂—CH₃
      |
      CH₃

   III.
    CH₃
    |
   CH₂—CH₂—CH₃

   a. I and II
   b. I and III
   c. II and III
   d. IV and V

6. The molecular formula for [cyclobutane with CH₃ substituent] is
   a. C₅H₁₂
   b. C₅H₁₀
   c. C₅H₉
   d. C₄H₉

7. Which of the following compounds is a structural isomer of
       CH₃
       |
   CH₃—CH₂—CH—CH₃ ?

   a. CH₃—CH₂—CH₂—CH₃

   b.
      CH₃
      |
   CH₃—C—CH₃
      |
      CH₃

   c. CH₃—CH—CH₂—CH₃
      |
      CH₃

   d.
      CH₃ CH₃
      | |
   CH₃—CH—CH—CH₃

8. Which of the following compounds is a structural isomer of [methylcyclopropane] ?
   a. [cyclobutane]
   b. [methylcyclopropane with CH₃ below]
   c. CH₃—CH₂—CH₂—CH₃
   d.
      CH₃
      |
   CH₃—CH—CH₃

9. The unbranched alkane that contains eight carbon atoms is called
   a. hexane
   b. heptane
   c. octane
   d. nonane

10. How many structural isomers have the formula C₄H₁₀?
    a. 2
    b. 3
    c. 4
    d. 5

Organic Compounds: Alkanes 167

11. The number of carbon atoms in the longest chain of

    CH₃
    |
    CH₃—CH—CH₂
              \
               CH₂—CH₃   is

    a. 3   b. 4   c. 5   d. 6

12. The correct IUPAC name for CH₃—CH—CH₂—CH—CH₃   is
                                    | |
                                    CH₂   CH₃
                                    |
                                    CH₃

    a. 2-ethyl-4-methylpentane   c. 3,5-dimethylhexane
    b. 2-methyl-4-ethylpentane   d. 2,4-dimethylhexane

13. The correct IUPAC name for is

    a. 1,3-dimethylhexane   c. 1,5-dimethylcyclohexane
    b. 1,3-methylcyclohexane   d. 1,3-dimethylcyclohexane

14. The bromine in (cyclopentane with CH₃ and Br) is located at position

    a. 1   b. 2   c. 3   d. 5

15. A 12-carbon alkane should be a _____ at room temperature.
    a. solid   b. liquid   c. gas   d. none of these

Matching
Match an alkyl group name to each structure on the left.
16. CH₃—CH₂—   a. propyl
17. CH₃—CH₂—CH₂—   b. sec-butyl
18. CH₃—CH—CH₂—   c. isobutyl
         | d. ethyl
         CH₃

Match the structures on the left to the descriptions on the right.

19.

a. a *cis* compound
b. a *trans* compound
c. neither *cis* nor *trans*

20.

21.

22.

True-False

23. Most organic compounds are very soluble in water
24. Covalent bonds are more prevalent in inorganic compounds than in organic compounds.
25. Solutions of inorganic compounds are better electrical conductors than solutions of organic compounds.
26. There are more known inorganic compounds than organic compounds.
27. A few compounds of carbon, such as CO_2, are classified as inorganic.
28. Structural isomers always have the same molecular formula.
29. Structural isomers always have the same functional groups.
30. A condensed structural formula may show some bonds.
31. An expanded structural formula may not show all the bonds.
32. A molecule may have more than one functional group.
33. Pentane and cyclopentane are isomers of each other.
34. Alkanes have lower boiling points than other organic compounds.
35. The main component of natural gas is butane.
36. Complete combustion of pentane produces H_2O and CO_2.
37. Alkanes are polar molecules.
38. Cycloalkanes experience dispersion forces.
39. The general formula for cycloalkanes is C_nH_{2n+2}.
40. Alkanes are also known as paraffins.

Chapter 12: Unsaturated Hydrocarbons

CHAPTER OUTLINE
12.1 The Nomenclature of Alkenes
12.2 The Geometry of Alkenes
12.3 Properties of Alkenes
12.4 Addition Polymers
12.5 Alkynes
12.6 Aromatic Compounds and the Benzene Structure
12.7 The Nomenclature of Benzene Derivatives
12.8 Properties and Uses of Aromatic Compounds

LEARNING OBJECTIVES/ASSESSMENT
When you have completed your study of this chapter, you should be able to:
1. Classify unsaturated hydrocarbons as alkenes, alkynes, or aromatics. (Section 12.1; Exercise 12.2)
2. Write the IUPAC names of alkenes from their molecular structures. (Section 12.1; Exercise 12.4)
3. Predict the existence of geometric (cis-trans) isomers from formulas of compounds. (Section 12.2; Exercise 12.18)
4. Write the names and structural formulas for geometric isomers. (Section 12.2; Exercise 12.20)
5. Write equations for addition reactions of alkenes, and use Markovnikov's rule to predict the major products of certain reactions. (Section 12.3; Exercise 12.26)
6. Write equations for addition polymerization, and list uses for addition polymers. (Section 12.4; Exercise 12.36)
7. Write the IUPAC names of alkynes from their molecular structures. (Section 12.5; Exercise 12.44)
8. Classify organic compounds as aliphatic or aromatic. (Section 12.6; Exercise 12.48)
9. Name and draw structural formulas for aromatic compounds. (Section 12.7; Exercises 12.52 and 12.54)
10. Recognize uses for specific aromatic compounds. (Section 12.8; Exercise 12.66)

SOLUTIONS FOR THE END OF CHAPTER EXERCISES
THE NOMENCLATURE OF ALKENES (SECTION 12.1) AND ALKYNES (SECTION 12.5)

☑12.2 An alkene is a hydrocarbon that contains at least one carbon-carbon double bond.
An alkyne is a hydrocarbon that contains at least one carbon-carbon triple bond.
An aromatic hydrocarbon is a compound that contains a benzene ring or other similar feature.

☑12.4
a. $CH_3CH=CHCH_3$
2-butene

b. 3-ethyl-2-pentene

c. 4,4-dimethyl-2-hexyne

d. 4-methylcyclopentene

e. 6-bromo-2-methyl-3-heptyne

f. 1-ethyl-2,3-dimethylcyclopropene

g. 6-methyl-1,4-heptadiene

12.6
a. 4-ethyl-2-heptene

b. 4,4-diethyl-1-hexyne

c. 3-ethyl-1,4-pentadiene

d. 2,3-dimethylcyclohexene

e. 1-methylcyclopropene

12.8 a. C₅H₈ alkyne

HC≡C—CH₂—CH₂—CH₃   H₃C—C≡C—CH₂—CH₃
1-pentyne   2-pentyne

$$\text{HC} \equiv \text{C} - \underset{\underset{\text{CH}_3}{|}}{\text{CH}} - \text{CH}_3$$

3-methyl-1-butyne

b. C₅H₈ diene

H₂C=C=CH—CH₂—CH₃   H₃C—CH=C=CH—CH₃
1,2-pentadiene   2,3-pentadiene

H₂C=CH—CH=CH—CH₃   H₂C=CH—CH₂—CH=CH₂
1,3-pentadiene   1,4-pentadiene

$$\text{H}_2\text{C} = \text{C} = \underset{\underset{\text{CH}_3}{|}}{\text{C}} - \text{CH}_3$$

3-methyl-1,2-butadiene

c. C₅H₈ cyclic alkene

1-ethylcyclopropene 3-ethylcyclopropene 1,2-dimethylcyclopropene

1,3-dimethylcyclopropene 3,3-dimethylcyclopropene 1-methylcyclobutene

3-methylcyclobutene cyclopentene

Unsaturated Hydrocarbons 171

12.10 CH₃C=CHCH₂CH₂C=CHCH₂CH=CCH=CH₂   3,7,11-trimethyl-1,3,6,10-dodecatetraene
       (with CH₃ branches)

12.12 a. incorrect = 2-ethyl-2-pentene
          corrected = 3-methyl-3-hexene

          H_2C—CH_3
          |
          H_3C—C=CH—CH_2—CH_3

       b. incorrect = 3,6-octadiene
          corrected = 2,5-octadiene

          H_3C—CH_2—CH=CH—CH_2—CH=CH—CH_3
          8   1

       c. incorrect = 5-ethylcyclopentene
          corrected = 3-ethylcyclopentene

          (cyclopentene with CH₂CH₃ at position 3, double bond at 1)

THE GEOMETRY OF ALKENES (SECTION 12.2)

12.14 Pi bonds are formed when unhybridized *p* orbitals overlap. The pi bond is represented by π. The pi bond contains 2 electrons.

12.16 Structural isomers have a different order of linkage of atoms. Geometric isomers have the same order of linkage of atoms; however, the three dimensional structures are unique because of restricted rotation as a result of a ring or a double bond.

☑12.18 a. CH₃CH₂CH₂CH₂CH=CH₂; no geometric isomers

        b. H   H   CH₃CH₂   H
             \\C=C/   \\C=C/
            /   \\   /   \\
        CH₃CH₂   CH₂CH₃   H   CH₂CH₃
            cis-3-hexene   *trans*-3-hexene

        c.   CH₃
              |
           CH₃C=CHCH₂CH₃; no geometric isomers

☑12.20 a. H   H   b. CH₃   H
             \\C=C/   \\C=C/
            /   \\   /   \\
          CH₃   CH₂CH₃   H   CH₂CH₂CH₃
            cis-2-pentene   *trans*-2-hexene

PROPERTIES OF ALKENES (SECTION 12.3)

12.22 The alkenes and alkanes are both nonpolar molecules that have low solubility in water and high solubility in nonpolar solvents. They are also less dense than water.

12.24 Markovnikov's rule states that when a heteroatomic compound containing hydrogen is added to a multiple bond, the hydrogen will attach to the carbon atom in the multiple bond that is directly bonded to more hydrogen atoms. An example is shown on the next page.

172 Chapter 12

$$H_2C=CH\text{-}CH_3(\text{with CH}_3) + HX \longrightarrow H_2C\text{-}CH(\text{with CH}_3)\text{ with H and X}$$

☑12.26 a. (cyclohexyl-CH=CH-CH₃) + Br₂ ⟶ (cyclohexyl-CHBr-CHBr-CH₃)

b. (cyclohexene) + H₂O —H₂SO₄→ (cyclohexanol, with H and OH)

c. CH₂=C(CH₃)-CH=CH₂ + 2 H₂ —Pt→ CH₂(H)-C(CH₃)(H)-CH(H)-CH₂(H)

d. (methylenecyclohexane) + HCl ⟶ (1-chloro-1-methylcyclohexane)

12.28   a. To prepare 1,2-dichlorocyclohexane from cyclohexene, I would use Cl₂.
          b. To prepare hexane from cyclohexene, I would use hydrogen gas and a platinum catalyst.
          c. To prepare chlorocyclohexane from cyclohexene, I would use HCl.
          d. To prepare hydroxycyclohexane (cyclohexanol) from cyclohexene, I would use water and a sulfuric acid catalyst.

12.30   Adding reddish-brown bromine to cyclohexane will produce a reddish-brown solution. Adding reddish-brown bromine to 2-hexene will produce a clear solution. The differences in the chemical reactivity of these substances will allow them to be differentiated.

ADDITION POLYMERS (SECTION 12.4)

12.32   A monomer is the starting material for a polymer. It can be a small molecule.
A polymer is a large molecule that may contain thousands of repeating units.
An addition polymer is a polymer formed by the reaction of monomers that contained multiple bonds to form the repeating units of a polymer.
A copolymer is a polymer formed by the reaction of at least two types of monomers.

12.34   All of the monomers in Table 12.3 contain one carbon-carbon double bond.

☑12.36 n CH$_2$=CH(Cl) →(polymerization) [—CH$_2$—CH(Cl)—]$_n$

ALKYNES (SECTION 12.5)

12.38 Each carbon atom bonded with a triple bond has two *sp* hybrid orbitals.

12.40 The geometry of a triple bond is linear.

12.42 Acetylene is the simplest alkyne and it is used in torches for welding steel and in making plastics and synthetic fibers.

☑12.44   HC≡C—CH$_2$—CH$_2$—CH$_3$ H$_3$C—C≡C—CH$_2$—CH$_3$
              1-pentyne   2-pentyne

HC≡C—CH(CH$_3$)—CH$_3$
3-methyl-1-butyne

AROMATIC COMPOUNDS AND THE BENZENE STRUCTURE (SECTION 12.6)

12.46 The pi bonding in a benzene ring is the result of unhybridized *p* orbitals overlapping.

☑12.48 Aromatic means a molecule contains a benzene ring or one of its structural relatives. Aliphatic means a molecule does not contain a benzene ring or one of its structural relatives.

12.50 a. *cis*-1,2-dibromohexane   *trans*-1,2-dibromohexane

Dibromocyclohexane is not a planar molecule. Each carbon atom has two attached groups (H or another atom) in addition to two positions of attachment in the ring. In order to maintain a tetrahedral geometry around the carbon atoms, the attached groups can be "above" or "below" the ring. Consequently, the molecule can exhibit cis-trans isomerism when it has two attached groups (other than hydrogen) on two different carbon atoms in the ring.

b. 1,2-dibromobenzene is a planar molecule because each carbon atom in the ring is only bonded to 3 other atoms. Consequently, the attached bromine atoms cannot be "above" or "below" the plane of the molecule. This in turn means the molecule cannot exhibit cis-trans isomerism.

THE NOMENCLATURE OF BENZENE DERIVATIVES (SECTION 12.7)

☑12.52 a. 1,3,5-trimethylbenzene

b. p-diethylbenzene
1,4-diethylbenzene

☑12.54 a. 3-phenyl-1-pentene

b. 2-phenylhexane

12.56 a. p-ethylphenol

b. o-chlorophenol

12.58 a. 2-ethyl-3-propyltoluene
(1-ethyl-2-methyl-6-propylbenzene)

b. 3-bromo-5-chlorobenzoic acid

12.60 a. *o*-ethylphenol (structure: HO and CH₃CH₂ groups on benzene ring, ortho)

c. 3-methyl-3-phenylpentane (structure: benzene ring attached to C with CH₂CH₃, CH₃, CH₂CH₃)

b. *m*-chlorobenzoic acid (structure: Cl and C(=O)OH groups on benzene ring, meta)

PROPERTIES AND USES OF AROMATIC COMPOUNDS (SECTION 12.8)

12.62 Aromatic hydrocarbons are nonpolar molecules that are insoluble in water and soluble in nonpolar solvents. They are also less dense than water.

12.64 Benzene is less reactive than cyclohexene because benzene cannot undergo the addition reactions that cyclohexene does. Both benzene and cyclohexene can undergo combustion.

☑12.66 a. Used in the production of Formica®   phenol
        b. A starting material for polystyrene   styrene
        c. Used to manufacture drugs   aniline
        d. A starting material for Bakelite®   phenol

CHEMISTRY AROUND US, CHEMISTRY AND YOUR HEALTH, OVER THE COUNTER, AND HOW REACTIONS OCCUR

12.68 Tomatoes, guava, and pink grapefruit are good sources of lycopene in addition to watermelon.

12.70 Cooking meat over an open flame may cause cancer-promoting compounds to form in the meat. Baked meat is cooked at a lower temperature which does not allow the cancer-promoting compounds to form.

12.72 The active ingredient in patches, gum, and sprays designed to help break the smoking habit is nicotine.

ADDITIONAL ACTIVITIES

Section 12.1 Review:
(1) What is the relationship between the general formulas of alkenes and alkanes?
(2) What is the relationship between the general formulas alkenes and cycloalkanes?
(3) Is –*ene* a fitting ending for compounds that contain a carbon-carbon double bond? Explain.
(4) What is the significance of the ending –*diene*?
(5) In terms of numbering, what takes priority in the alkene longest chain? Explain.

176 Chapter 12

Section 12.2 Review: Copy this page onto another sheet of paper or a transparency, then cut out the shapes. The circle is included for additional stability, but does not have any real physical meaning. Insert a pipe cleaner into the center of each circle. Center the structure on the pipe cleaner.

(1) What type of orbitals are represented by the straight black lines in the above structures?
(2) What type of orbital is represented by the pipe cleaner?
(3) What is the angle between the straight black lines in the above structures?
(4) What is the angle between the straight black lines and the pipe cleaner?
(5) Bring the two structures close enough for the sp^2 hybrid orbitals on the two carbon atoms to overlap. What type of bond does this form? How many bonds does this form?
(6) Bend the tops and bottoms of the pipe cleaner so they can also overlap. What type of bond does this form? How many bonds does this form?
(7) How does a pi bond differ from a sigma bond?

Section 12.3 Review:

Box the following properties that apply to alkenes. Correct the properties that do not apply to alkenes as they are currently written.

unpleasant odor   high boiling point   low melting point   polar   dispersion forces

soluble in water   less dense than water   hydrophobic

(1) List the four types of addition reactions alkenes can undergo.
(2) Divide these types of reactions into homoatomic and heteroatomic reactants based on the material added to the alkene.
(3) Which group from (2) requires knowledge and use of Markovnikov's rule?
(4) Divide the types of addition reactions from (1) into two groups: those that require a catalyst and those that do not.
(5) Are the groups the same in (2) and (4)?
(6) Divide the type of addition reactions from (1) into groups by product formed. For the types of reactions grouped together, can any distinction be made between their products?
(7) Which of the groupings (2), (4), or (6) will be most useful for remembering these reactions?

Section 12.4 Review:
(1) Write a polymerization reaction using the monomer labeled (1) on the next page.
(2) Draw the chemical structures of the monomers necessary to form the copolymer labeled (2).

(3) What are the differences between the polymers in (1) and (2)?

Section 12.5 Review:
(1) Based on the activities in the Section 11.3 Review and the Section 12.2 Review, describe how to model a carbon atom that has undergone *sp* hybridization.
(2) What is the angle between the *sp* hybrid orbitals?
(3) What is the angle between the *sp* hybrid orbitals and the unhybridized *p* orbitals?
(4) What is the angle between the unhybridized *p* orbitals?
(5) How many *sp* hybrid orbitals are formed?
(6) How many *p* orbitals are unhybridized?
(7) What type of bond is formed when the *sp* hybrid orbitals on two carbon atoms overlap?
(8) How many pi bonds will these carbon atoms form?
(9) Explain why molecules that contain a carbon-carbon triple bond have a –*yne* ending.
(10) Box the following properties that apply to alkynes. Correct the properties that do not apply to alkynes as they are currently written.

combustible   unpleasant odor   high boiling point   low melting point   polar   dispersion forces

do not undergo addition reactions   soluble in water   less dense than water   hydrophobic

Section 12.6 Review:
(1) Which class of hydrocarbons contains carbon atoms with the same type of hybridized orbitals as are found in benzene?
(2) What is the major difference in the bonding within the hydrocarbon class from (1) and benzene?
(3) What differences does (2) present in terms of structure?

Section 12.7 Review:

Study Tip: Make flashcards of the benzene derivatives known by common names that are also IUPAC accepted. Practice the flashcards until you can correctly identify benzene derivative by name or name the structure of a given benzene derivative without referring to your notes.

(1) What are the numbers associated with the attached groups in a benzene structure with the prefix *o-*?
(2) What are the numbers associated with the attached groups in a benzene structure with the prefix *m-*?
(3) What are the numbers associated with the attached groups in a benzene structure with the prefix *p-*?

178 Chapter 12

Section 12.8 Review:
(1) What physical properties do all hydrocarbons share?
(2) Which class of hydrocarbons has the most similar reactivity to the aromatic compounds?
(3) Could the difference identified in (2) of the Section 12.6 Review account for the difference in reactivity between alkenes and aromatic compounds? Explain.
(4) What is the major structural difference between benzopyrene and riboflavin?
(5) What is the difference between the interaction of benzopyrene and riboflavin with the human body?

Tying It All Together with a Laboratory Application:
A laboratory student has five unknown liquids to identify. The unknowns are labeled A, B, C, D, and E. The identities of the unknowns (not in order) are benzene, 1-hexene, 1-hexyne, cyclohexene, and cyclohexane. The student also has the following information:

	benzene	1-hexene	1-hexyne	cyclohexene	cyclohexane	hexane
Color	colorless	colorless	colorless	colorless	colorless	colorless
Odor	paint thinner like	?	?	sweetish	mild sweet	gasoline like
Boiling Point (°C)	80.1	63	71	83	80.7	69
Freezing Point (°C)	5.5	-139.8	-131.9	-104	6.6	-95
Density (g/mL)	0.8786	0.673	0.716	0.811	0.779	0.6548

(1) Draw the structural formulas for each of the unknowns.
(2) Write the molecular formulas for each of the unknowns.
(3) Are any of the compounds structural isomers? If so, which ones?
(4) Are any of the compounds geometric isomers? If so, which ones?

The student wants to obtain 5.00 mL of each unknown and find the mass of each sample in order to determine the (5) _____ of the unknowns. The unknowns in order of increasing mass (for 5.00 mL samples) are:

(6) _____ < (7) _____ < (8) _____ < (9) _____ < (10) _____

(11) Comment on this order. The student takes the test tubes over to the balance and discovers that the balance is out of order. Another procedure is needed to distinguish the compounds.

All of the unknowns are (12) _____ because they contain only carbon and hydrogen atoms. They are also (13) _____ molecules that (14) _____ dissolve in water. All of the compounds are (15) _____ dense than water. The student decides that adding a small quantity of each unknown to separate beakers of water (16) _____ allow the unknowns to be identified by density.

(17) Write the reactions for the complete combustion of each of the unique molecular formulas. Aromatic compounds generally burn with more soot than (18) _____ compounds.
(19) Write the reactions for the addition of bromine to each of the compounds.

The student decides to add a bromine solution to 0.5 mL samples of each of the unknowns. The student determined the number of drops required to obtain a reddish color for each unknown:

unknown	A	B	C	D	E
drops	43	77	2	36	1

Based on this data, unknowns C and E (20) _____ react with bromine. The student considers distinguishing these compounds by smell, but unfortunately, this student has a cold. The student decides to combust a few drops of each of these compounds in watch glasses in the fume hood. The student observes that unknown C produces much more smoke than unknown E as it burns. A black residue is evident on both watch glasses; however, it is much thicker on the watch glass for unknown C. The identity of unknown C is (21) _____. The identity of unknown E is (22) _____.

Based on the remaining data obtained from the bromine reaction, the student identifies unknown B as (23) _____ because (24) _____. To double check this identification, the student decides to check the boiling points of unknowns A, B, and D. The student sets up a beaker of water on a hot plate in the fume hood and uses clamps and a ring stand to secure the three test tubes of unknowns in the water bath. The student also adds a secured thermometer to each test tube. The student does not use a Bunsen burner to check the boiling points of these compounds because they are (25) _____. The boiling temperatures the student obtains are:

unknown	A	B	D
boiling point	~80°C	~70°C	~60°C

The identity of unknown A is (26) _____, unknown B is (27) _____, and unknown D is (28) _____.

SOLUTIONS FOR THE ADDITIONAL ACTIVITIES

Section 12.1 Review:
(1) alkene = C_nH_{2n}; alkane = C_nH_{2n+2}; If an alkene and an alkane contain the same number of carbon atoms, the alkene will contain 2 fewer hydrogen atoms than the alkane because it contains a double bond.
(2) alkene = C_nH_{2n}; cycloalkane = C_nH_{2n}; If an alkene and a cycloalkane contain the same number of carbon atoms, they will also contain the same number of hydrogen atoms. Forming a double bond or a ring, requires two hydrogen atoms to be removed from the corresponding alkane structure.
(3) Compounds that contain a carbon-carbon double bond are alkenes. The ending of the group is also the ending for the IUPAC names of these compounds.
(4) The ending *–diene* means that a compound contains two double bonds. It is an alkene twice over.
(5) The double bond takes priority in the alkene longest chain. The carbon atoms in the longest chain will be numbered so that the double bond is associated with the carbon atom with the lowest possible number.

Section 12.2 Review:
(1) sp^2 hybrid orbitals; (2) unhybridized *p* orbital; (3) 120°; (4) 90°; (5) sigma bond, one bond;
(6) pi bond, one bond; (7) A sigma bond has one area of overlap along the internuclear axis. A pi bond has two areas of overlap, one above and one below the internuclear axis.

Section 12.3 Review:

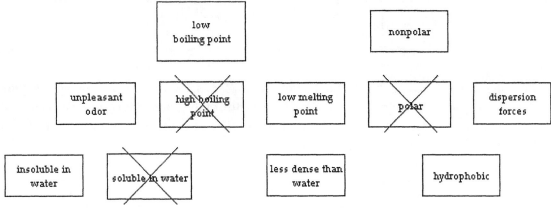

180 Chapter 12

(1) halogenation, hydrogenation, addition of hydrogen halides, hydration
(2) homoatomic reactants: halogenation, hydrogenation
heteroatomic reactants: addition of hydrogen halides, hydration
(3) Heteroatomic reactants require knowledge and use of Markovnikov's rule.
(4) Require catalyst: hydrogenation, hydration
Do not require catalyst: halogenation, addition of hydrogen halides
(5) No, the groups are different in (2) and (4).
(6) haloalkane (alkyl halide): halogenation, addition of hydrogen halides
alkane: hydrogenation
alcohol: hydration
While both halogenation and the addition of hydrogen halides both produce haloalkanes, halogenation produces disubstituted haloalkanes (contain 2 halogen atoms), while the addition of hydrogen halides produces monosubstituted haloalkanes (contain 1 halogen atom).
(7) The "best" grouping is a matter of personal study style. Markovnikov's rule is an important part of this chapter, so the grouping from (2) is useful. The need for a catalyst is important both in recording the details of a reaction and in performing reactions in the laboratory, so the grouping from (4) is useful. The ability to predict products for a reaction is also important, so the grouping from (6) is also useful.

Section 12.4 Review:

(1)

(2)

(3) The polymer in (2) is a copolymer and the polymer in (1) is not a copolymer. The repeating unit for (2) is larger than the repeating unit for (1). The structure of the polymer in (2) contains two HCCl goups in a row then two HC-phenyl groups in a row. The structure of the polymer in (1) contains alternating HCCl groups and HC-phenyl groups.

Section 12.5 Review:
(1) The *sp* hybrid orbitals could be represented as two different colored straws joined to make a straight line. The two unhybridized *p* orbitals could be represented using two different colored pipe cleaners at 90° angles to each other and the straws.
(2) 180°; (3) 90°; (4) 90°; (5) 2; (6) 2; (7) sigma; (8) 2
(9) Molecules that contain a carbon-carbon triple bond key functional group are alkynes. The molecules in this class are given *-yne* endings in the IUPAC nomenclature system to indicate they are alkynes.

(10)
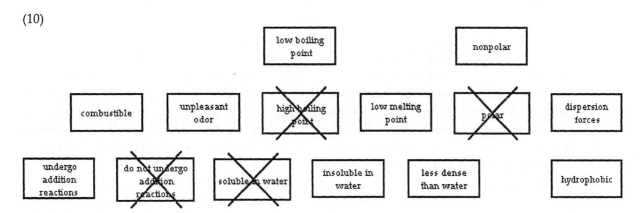

Section 12.6 Review:
(1) Alkenes and aromatic compounds contain sp^2 hybridized carbon atoms.
(2) The major difference in bonding between alkenes and benzene is that alkenes contain 1 localized pi bond, while benzene forms 2 delocalized pi lobes.
(3) The carbon atoms in the double bond of an alkene have two attached groups, one above the double bond, one below the double bond. Consequently, some alkene molecules have geometric isomers. The carbon atoms in a benzene ring have one attached group other than the carbon atoms in the benzene ring. Consequently, aromatic molecules cannot have geometric isomers.

Section 12.7 Review:
(1) 1 and 2; (2) 1 and 3; (3) 1 and 4

Section 12.8 Review:
(1) They are nonpolar molecules that are less dense than water and do not dissolve in water. They dissolve in nonpolar solvents.
(2) Aromatic compounds react similarly to alkanes. They do not undergo addition reactions like alkenes and alkynes.
(3) Yes, the pi cloud in the aromatic compounds provides stability that a lone carbon-carbon double bond does not have. Consequently, aromatic compounds do not undergo addition reactions, but alkenes do.
(4) Benzopyrene contains only carbon and hydrogen. It is completely nonpolar. Riboflavin contains nitrogen and oxygen in addition to carbon and hydrogen. It is polar.
(5) Benzopyrene is a carcinogen and riboflavin is a vitamin. Riboflavin is important to maintaining the health of the human body, while benzopyrene can cause cancer within the human body.

182 Chapter 12

Tying It All Together with a Laboratory Application:

(1)

(2) C₆H₆   C₆H₁₂   C₆H₁₀   C₆H₁₀   C₆H₁₂

(3) structural isomers = 1-hexene and cyclohexane (C₆H₁₂); 1-hexyne and cyclohexene (C₆H₁₀)

(4) no geometric isomers (5) density (7) 1-hexyne (9) cyclohexene
                          (6) 1-hexene (8) cyclohexane (10) benzene

(11) The compound with the lowest density only has a double bond, the next compound has a triple bond, the three highest density compounds are all cyclic. The lowest of these three does not contain any multiple bonds, the next compound contains one double bond, and the highest density compound is benzene which is aromatic. Based on this group of compounds, density increases with degree of unsaturation and with cyclic compounds.

(12) hydrocarbons   (13) nonpolar   (14) will not   (15) less   (16) will not

(17) 2 C₆H₆ + 15 O₂ → 12 CO₂ + 6 H₂O; C₆H₁₂ + 9 O₂ → 6 CO₂ + 6 H₂O; 2 C₆H₁₀ + 17 O₂ → 12 CO₂ + 10 H₂O

(18) aliphatic

(19)

(20) did not   (22) cyclohexane   (24) it reacted with   (25) flammable   (27) 1-hexyne
(21) benzene   (23) 1-hexyne   the most Br₂   (26) cyclohexene (28) 1-hexene

SELF-TEST QUESTIONS

Multiple Choice

1. What is the correct IUPAC name for this compound?

    CH₃—CH=CH—CH₂—CH₂—CH₂
                                    |
                                    Cl

    a. 1-chloro-2-hexene   c. 1-chloro-4-hexene
    b. 6-chloro-2-hexene   d. 1-chlorohexene

2. What is the correct IUPAC name for this compound?

    a. 1-methylcyclopentene   c. 3-methylcyclopentene
    b. 2-methylcyclopentene   d. 1-methyl-2-cyclopentene

3. What is the correct IUPAC name for this compound?

   a. 3-chloro-2-butene
   b. cis-2-chloro-2-butene
   c. trans-3-chloro-2-butene
   d. trans-2-chloro-2-butene

4. Which of the following can exhibit *cis-trans* isomerism?
   a. 2-methyl-1-butene
   b. 2,3-dimethyl-2-butene
   c. 2-methyl-2-butene
   d. 2,3-dichloro-2-butene

5. Markovnikov's rule is useful in predicting the product of a reaction between an alkene and
   a. H₂   b. Br₂   c. HBr   d. O₂

6. What is the product of the following reaction?

   $CH_3-C(CH_3)=CH_2 + HCl \longrightarrow$

   a. $CH_3-CH(CH_3)-CH_2-Cl$
   b. $CH_3-C(CH_3)(Cl)-CH_2-Cl$
   c. $CH_3-C(CH_3)(Cl)-CH_3$
   d. $CH_3-CH(CH_3)-CH_3$

7. Which of the following is the correct representation of the polymer produced from the polymerization of $H_2C=CHCl$?

   a. $\mathrm{-[CH_2-CH_2-CH(Cl)]-}_n$
   b. $\mathrm{-[CH_2-CH(Cl)]-}_n$
   c. $\mathrm{-[CH_2-C(Cl)_2]-}_n$
   d. $\mathrm{-[CH(Cl)-CH(Cl)]-}_n$

8. A correct name of 1,2,6-trichlorobenzene is
   a. 1,2,5-trichlorobenzene
   b. 1,3,5-trichlorobenzene
   c. 1,2,3-trichlorobenzene
   d. 1,2,4-trichlorobenzene

9. What is the correct IUPAC name for this compound?

   a. 2-phenyl-1-butene
   b. 3-phenyl-3-butene
   c. 2-phenyl-3-butene
   d. 3-phenyl-1-butene

10. Addition of H₂O to CH₃—CH=C(CH₃)—CH₃ produces

a. CH₃—CH(OH)—CH(CH₃)—CH₃

b. HO—CH₂—CH₂—CH(CH₃)—CH₃

c. CH₃—CH₂—C(OH)(CH₃)—CH₃

d. CH₃—CH₂—CH(CH₃)—CH₂—OH

11. What reactant and catalyst are necessary to hydrogenate an alkene?
    a. H⁺ and Pt
    b. HCl and Pt
    c. H₂ and Pt
    d. H₂O and Pt

12. What is the IUPAC name for HC≡C—CH₂—C(CH₃)₂—CH₃ ?
    a. 2,2-dimethyl-2-pentyne
    b. 4,4-dimethyl-4-pentyne
    c. 2,2-dimethyl-4-pentyne
    d. 4,4-dimethyl-1-pentyne

Matching
For each aromatic compound on the left, select the correct use or derived product from the responses on the right.
13. phenol
14. toluene
15. aniline
16. riboflavin

a. a solvent
b. a vitamin
c. formica
d. dyes

For each description on the left, select a correct polymer from the responses on the right.
17. used for insulation
18. a copolymer
19. use in airplane windows

a. saran wrap
b. plexiglass
c. polystyrene
d. PVC

True-False
20. Carbon-carbon double bonds do not rotate as freely as carbon-carbon single bonds.
21. The compound BrCH=CHCH₃ can exhibit *cis-trans* isomerism.
22. Alkenes are polar substances.
23. All aromatic compounds are cyclic.
24. Two hydrogens may be bonded to the same benzene carbon atom.
25. Benzene is a completely planar molecule.
26. A *sp* hybrid orbital is obtained by hybridizing an *s* orbital and two *p* orbitals.
27. A copolymer is formed from two different monomers.
28. All the atoms in ethene lie in the same plane.
29. Acetylene is the common name of the simplest alkene.
30. A triple bond contains 1 sigma bond and 2 pi bonds.

Chapter 13: Alcohols, Phenols, and Ethers

CHAPTER OUTLINE
13.1 The Nomenclature of Alcohols and Phenols
13.2 Classification of Alcohols
13.3 Physical Properties of Alcohols
13.4 Reactions of Alcohols
13.5 Important Alcohols
13.6 Characteristics and Uses of Phenols
13.7 Ethers
13.8 Properties of Ethers
13.9 Thiols
13.10 Polyfunctional Compounds

LEARNING OBJECTIVES/ASSESSMENT
When you have completed your study of this chapter, you should be able to:
1. Name and draw structural formulas for alcohols and phenols. (Section 13.1; Exercises 13.4 and 13.10)
2. Classify alcohols as primary, secondary, or tertiary on the basis of their structural formulas. (Section 13.2; Exercise 13.14)
3. Discuss how hydrogen bonding influences the physical properties of alcohols. (Section 13.3; Exercise 13.18)
4. Write equations for alcohol dehydration and oxidation reactions. (Section 13.4; Exercises 13.22 and 13.26)
5. Recognize uses for specific alcohols. (Section 13.5; Exercise 13.36)
6. Recognize uses for specific phenols. (Section 13.6; Exercise 13.38)
7. Name and draw structural formulas for ethers. (Section 13.7; Exercise 13.42)
8. Describe the key physical and chemical properties of ethers. (Section 13.8; Exercise 13.46)
9. Write equations for a thiol reaction with heavy metal ions and production of disulfides. (Section 13.9; Exercise 13.52)
10. Identify functional groups in polyfunctional compounds. (Section 13.10; Exercise 13.56)

SOLUTIONS FOR THE END OF CHAPTER EXERCISES
THE NOMENCLATURE OF ALCOHOLS AND PHENOLS (SECTION 13.1)

13.2 R—O—R'
where the R and the R' are hydrocarbon groups

13.4
a. $CH_3CH_2CH_2CH_2-OH$
1-butanol

b.
$$\begin{array}{ccc} Br & Br & OH \\ | & | & | \\ CH_2 & CH & CH & CH_3 \end{array}$$
3,4-dibromo-2-butanol

c.
$$\begin{array}{c} CH_2CH_3 \\ | \\ CH_3CH_2C-OH \\ | \\ CH_2CH_3 \end{array}$$
3-ethyl-3-pentanol

d. (phenyl ring)—CH_2CH_2-OH
2-phenyl-1-ethanol

e. 2-bromo-6-methylcyclohexanol

f. OH OH
CH₃CHCH₂CHCH₃
2,4-pentanediol

13.6 a. CH₃—OH
methanol

c. CH₃CH₂—OH
ethanol

e. OH OH OH
CH₂—CH—CH₂
1,2,3-propanetriol

b. OH
CH₃CHCH₃
2-propanol

d. OH OH
CH₂—CH₂
1,2-ethanediol

13.8 a. 2-methyl-2-pentanol

CH₃
CH₃CCH₂CH₂CH₃
OH

b. 1,3-butanediol
CH₂CH₂CHCH₃
OH   OH

c. 1-ethylcyclopentanol
(cyclopentane with OH and CH₂CH₃)

13.10 a. 2-ethylphenol
o-ethylphenol
(benzene with CH₂CH₃ and OH ortho)

b. 2-ethyl-4,6-dimethylphenol
(benzene with OH, H₃C, CH₂CH₃, CH₃)

13.12 a. o-bromophenol
(benzene with OH and Br ortho)

b. 2,3,5-triethylphenol
(benzene with OH, CH₂CH₃ groups, CH₃CH₂)

CLASSIFICATION OF ALCOHOLS (SECTION 13.2)

☑13.14 a. CH_3CHCH_3 with OH — 2°
  b. CH_3CH_2-OH — 1°
  c. CH_3CHCH_2-OH with CH_3 — 1°

13.16
  $CH_2CH_2CH_2CH_3$ with OH — 1-butanol/1°
  $CH_3CHCH_2CH_3$ with OH — 2-butanol/2°
  CH_3CCH_3 with CH_3 and OH — 2-methyl-2-propanol/3°
  CH_2CHCH_3 with CH_3 and OH — 2-methyl-1-propanol/1°

PHYSICAL PROPERTIES OF ALCOHOLS (SECTION 13.3)

☑13.18 a. methanol < ethanol < 1-propanol
  b. butane < 1-propanol < ethylene glycol

13.20 a. [hydrogen bonding between ethanol and 1-butanol]
  b. [hydrogen bonding between water and cyclohexanol]

REACTIONS OF ALCOHOLS (SECTION 13.4)

	Reactant	Product
☑13.22 a.	$CH_3CHCH_2CH_3$ with OH	$CH_3CH=CHCH_3$
b.	$CH_3CHCHCH_3$ with CH_3, OH, OH	$CH_3CH=CCH_3$ with CH_3

	Reactant	Product
13.24 a.	CH_3-OH	CH_3-O-CH_3
b.	$CH_3CH_2CH_2CH_2-OH$	$CH_3CH_2CH_2CH_2-O-CH_2CH_2CH_2CH_3$
c.	phenyl-CH_2-OH	phenyl-CH_2-O-CH_2-phenyl

188 Chapter 13

☑13.26
           Product   Reactant

a. cyclopentanone   cyclopentanol

b. $CH_3CH_2\overset{O}{\overset{\|}{C}}CH{-}CH_3$ $CH_3CH_2\overset{OH}{\overset{|}{C}H}CH{-}CH_3$
                    | |
                   CH_3 CH_3

c. $CH_3CH_2CH_2{-}\overset{O}{\overset{\|}{C}}{-}OH$ $CH_3CH_2CH_2{-}\overset{OH}{\overset{|}{C}H_2}$

13.28 a. 2-methyl-2-butanol is subjected to controlled oxidation → no reaction

$$CH_3{-}\underset{\underset{OH}{|}}{\overset{\overset{CH_3}{|}}{C}}{-}CH_2CH_3 \xrightarrow{(O)} \text{no reaction}$$

      b. 1-propanol is heated to 140°C in the presence of sulfuric acid → dipropyl ether, water

$$2\ CH_2CH_2CH_3\ \overset{|}{OH} \xrightarrow[140°C]{H_2SO_4} CH_2CH_2CH_3\ \overset{|}{O{-}CH_2CH_2CH_3} + H_2O$$

      c. 3-pentanol is subjected to controlled oxidation → 3-pentanone

$$CH_3CH_2{-}\underset{\underset{}{}}{\overset{\overset{OH}{|}}{C}H}{-}CH_2CH_3 \xrightarrow{(O)} CH_3CH_2{-}\overset{\overset{O}{\|}}{C}{-}CH_2CH_3 + H_2O$$

      d. 3-pentanol is heated to 180°C in the presence of sulfuric acid → 2-pentene, water

$$CH_3CH_2{-}\overset{\overset{OH}{|}}{C}H{-}CH_2CH_3 \xrightarrow[180°C]{H_2SO_4} CH_3CH{=}CH{-}CH_2CH_3 + H_2O$$

      e. 1-hexanol is subjected to an excess of oxidizing agent → hexanoic acid, water

$$\overset{\overset{OH}{|}}{C}H_2CH_2CH_2CH_2CH_2CH_3 \xrightarrow{(O)} HO{-}\overset{\overset{O}{\|}}{C}CH_2CH_2CH_2CH_2CH_3 + H_2O$$

13.30 a.

$$CH_3CH_2CH{=}CH_2 + H_2O \xrightarrow{H_2SO_4} CH_3CH_2\overset{\overset{OH}{|}}{C}H{-}CH_3$$

$$CH_3CH_2\overset{\overset{OH}{|}}{C}H{-}CH_3 \xrightarrow{(O)} CH_3CH_2\overset{\overset{O}{\|}}{C}CH_3 + H_2O$$

Alcohols, Phenols, and Ethers 189

b. [2-methylcyclopentanol] $\xrightarrow{H_2SO_4}_{180°C}$ [1-methylcyclopentene] + H_2O

[1-methylcyclopentene] + H_2O $\xrightarrow{H_2SO_4}$ [1-methylcyclopentanol]

c. $CH_3CH_2CH_2CH_2OH$ $\xrightarrow{H_2SO_4}_{180°C}$ $CH_3CH_2CH=CH_2 + H_2O$

$CH_3CH_2CH=CH_2 + H_2O \xrightarrow{H_2SO_4} CH_3CH_2CH(OH)CH_3$

$CH_3CH_2CH(OH)CH_3 \xrightarrow{(O)} CH_3CH_2C(=O)CH_3 + H_2O$

13.32 pyruvic acid

$$CH_3-\underset{\underset{O}{\|}}{C}-\underset{\underset{O}{\|}}{C}-OH$$

IMPORTANT ALCOHOLS (SECTION 13.5)

13.34 Methanol is an important industrial chemical because its oxidation product is formaldehyde, which is a major starting material for the production of plastics.

☑13.36
a. a moistening agent in many cosmetics — 1,2,3-propanetriol (glycerin, glycerol)
b. the solvent in solutions called tinctures — ethanol (ethyl alcohol)
c. automobile antifreeze — 1,2-ethanediol (ethylene glycol)
d. rubbing alcohol — 2-propanol (isopropyl alcohol)
e. a flavoring in cough drops — menthol
f. present in gasohol — ethanol (ethyl alcohol)

CHARACTERISTICS AND USES OF PHENOLS (SECTION 13.6)

☑13.38
a. a disinfectant used for cleaning walls — o-phenylphenol, 2-benzyl-4-chlorophenol
b. an antiseptic found in some mouthwashes — 4-hexylresorcinol
c. an antioxidant used to prevent rancidity in foods — BHA, BHT

ETHERS (SECTION 13.7)

13.40 a. $CH_3CH_2-O-CH(CH_3)CH_3$
ethyl isopropyl ether

b. $CH_3-O-CH_2CH_2CH_2CH_3$
butyl methyl ether

c. diphenyl ether

13.42 a. CH₃CH₂—O—CH₂CH₂CH₃
1-ethoxypropane

c. CH₃CH₂—O—⟨phenyl⟩
ethoxybenzene

b. CH₃CH₂—O—CHCH₃
          |
          CH₃
2-ethoxypropane

d.
OCH₃
 |
⟨cyclopentane⟩—OCH₃
1,2-dimethoxycyclopentane

13.44 a. isopropyl methyl ether
                CH₃
               /
CH₃—O—CH
               \
                CH₃

b. phenyl propyl ether
⟨phenyl⟩—O—CH₂CH₂CH₃

c. 2-methoxypentane
     OCH₃
      |
CH₃—CH—CH₂CH₂CH₃

d. 1,2-diethoxycyclopentane
OCH₂CH₃
 |
⟨cyclopentane⟩—OCH₂CH₃

e. 2-phenoxy-2-butene
                CH₃
                 |
⟨phenyl⟩—O—C=CHCH₃

PROPERTIES OF ETHERS (SECTION 13.8)

☑13.46 The chief chemical property of ethers is they are inert (relatively unreactive) with most reagents.

13.48 CH₃CH₂CH₂—OH > CH₃CH₂—O—CH₃ > CH₃CH₂CH₂CH₃
All of the compounds have approximately the same molecular mass. The major differences between these molecules are their key functional groups and the strength of their interparticle forces. The 1-propanol will have the strongest hydrogen bonds to water because of its hydroxy group; therefore, 1-propanol is the most soluble compound. The ethyl methyl ether (methoxyethane) can also hydrogen bond to water; therefore, it is more soluble than the butane, but less soluble than the 1-propanol. The butane cannot form hydrogen bonds with water; therefore, it is the least soluble.

13.50
CH₃CH₂
       \
        O------H—O
       /   \
CH₃CH₂   H

Alcohols, Phenols, and Ethers

THIOLS (SECTION 13.9)

☑13.52 a. $2\ CH_3CH_2CH_2-SH + Hg^{2+} \rightarrow CH_3CH_2CH_2-S-Hg-S-CH_2CH_2CH_3 + 2\ H^+$

b. $2\ CH_3CH_2CH_2-SH + (O) \rightarrow CH_3CH_2CH_2-S-S-CH_2CH_2CH_3 + H_2O$

c.

13.54 When alcohols are oxidized in a controlled way, they produce aldehydes or carboxylic acids (1° alcohols), ketones (2° alcohols), or no reaction (3° alcohols). When thiols are oxidized in a controlled way they produce disulfides.

POLYFUNCTIONAL COMPOUNDS (SECTION 13.10)

☑13.56

The key functional groups in this molecule include a benzene ring (aromatic), a carbon-carbon double bond (alkene), and a carbonyl group (aldehyde).

CHEMISTRY AROUND US, CHEMISTRY AND YOUR HEALTH, OVER THE COUNTER, AND HOW REACTIONS OCCUR

13.58

13.60 The oil in the poison ivy plant is urushiol.

13.62 Vitamin E is an antioxidant that is believed to help prevent cellular damage that leads to degenerative diseases.

ADDITIONAL ACTIVITIES

Section 13.1 Review:

(1) Is the *-ol* ending appropriate for compounds that contain a hydroxy group attached to an aliphatic hydrocarbon chain?

(2) What is the priority for numbering the longest carbon chain in a compound that contains a hydroxy group?

(3) What is the meaning of a compound name that ends in -*triol*?
(4) Which other class of compounds studied so far can have an ending similar to (3)?
(5) Phenols are studied here as their own compound class; however, this is not the first appearance of phenol in the textbook. Where else in the textbook were phenols introduced?

Section 13.2 Review:
(1) Draw and name the simplest secondary alcohol.
(2) Draw and name the simplest tertiary alcohol.

Section 13.3 Review:
Box the following properties that apply to low molecular weight alcohols. Correct the properties that do not apply to low molecular weight alcohols as they are currently written.

high boiling point   low melting point   polar   dispersion forces

soluble in water   less dense than water   hydrophobic

High molecular weight alcohols do not dissolve in water. Explain this observation.

Section 13.4 Review:
(1) Alcohol dehydration to produce an alkene is the opposite process of a previously studied chemical reaction. What is the process?
(2) How do the conditions differ between alcohol dehydration and the other process from (1)?
(3) Should alcohol dehydration to produce an alkene and the process from (1) be represented with equilibrium arrows?
(4) Can Markovnikov's rule be applied to predicting the major product of alcohol dehydration to an alkene?
(5) How do the conditions for alcohol dehydration to an alkene differ from alcohol dehydration to an ether?
(6) In general, what is the relationship between addition and elimination reactions?
(7) Write the reaction for the oxidation of isopropyl alcohol.
(8) Calculate the oxidation numbers for each atom in the reactant and product molecules of (7).
(9) Which atom(s) experience(s) a change in oxidation number? Is it (are they) oxidized or reduced?
(10) Classify primary, secondary, and tertiary alcohols in order of increasing reactivity.

Section 13.5 Review:
(1) Classify the reaction for the formation of methanol.
(2) How does the reaction to form glucose from ethylene differ from the reaction to form ethanol from glucose?
(3) What is different in the reaction conditions of the hydration of an alkene as studied in Chapter 12 and the industrial production of ethanol from ethylene?
(4) Classify the important alcohols in Section 13.5 as toxic or nontoxic.

Section 13.6 Review:
(1) Phenol is also known as carbolic acid. Is the use of the term "acid" appropriate?
(2) What are some of the major uses of phenol derivatives?

Section 13.7 Review:
(1) Draw and label the following alkoxy groups: methoxy, ethoxy, proproxy, isoproproxy, butoxy, isobutoxy, *sec*-butoxy, *t*-butoxy.

(2) Which atom from the alkoxy groups in (1) would connect to a carbon chain in a compound?
(3) Propose an explanation for why the IUPAC system of nomenclature does not utilize an -*er* or -*ther* ending for compounds that belong to the compound class called ethers.
(4) Why are furan and pyran called heterocyclic rings?
(5) Give an example of a homocyclic ring.

Section 13.8 Review:
Box the following properties that apply to ethers. Correct the properties that do not apply to ethers as they are currently written.

high boiling point   polar   low melting point

nonflammable   soluble in water

dispersion forces   hydrophobic   high volatility

Section 13.9 Review:
(1) Which element in the sulfhydryl group is responsible for the odors associated with thiols?
(2) Which of the following best corresponds to the odors associated with thiols: roses, baby powder, rotten eggs?
(3) Propose a reaction cycle to convert methanethiol to its corresponding disulfide and to return the disulfide to methanethiol.
(4) Are the reactions in (2) redox reactions?
(5) What happens to the methanethiol as it is converted to its corresponding disulfide?
(6) What happens to the corresponding disulfide as it is converted to methanethiol?
(7) Is the reaction of heavy metal ions with thiols as easily reversed as the reaction to convert methanethiol to its corresponding disulfide?

Section 13.10 Review:
(1) Does the term "polyfunctional compounds" adequately describe the molecules within this class?
(2) Most biological compounds are polyfunctional compounds. How can your current study of the individual functional groups benefit your future understanding of biological compounds?

Tying It All Together with a Laboratory Application:
A student must distinguish between five unknown compounds. The unknowns are labeled A, B, C, D, and E. The possible identities of the unknowns (not in order) are phenol, cyclohexanol, methoxycylcopentane, 2,3-dimethyl-2-butanol, and 1-hexanol.

(1) Draw the structural formulas for all of the unknowns.
(2) Write the molecular formulas for each of the unknowns.
(3) Are any of the compounds structural isomers? If so, which ones?
(4) Are any of the compounds geometric isomers? If so, which ones?
(5) Identify the compound(s) that is(are) alcohol(s).
(6) Classify the compound(s) from (5) as primary, secondary, or tertiary alcohol(s).
(7) Identify the compound(s) that is(are) ether(s).
(8) Identify the compound(s) that contain(s) a benzene ring.

194 Chapter 13

The student is allowed to use "solutions" of the unknown compounds in water or cyclohexane solvents. All of the solutions are colorless. The unknown A in water mixture appears to have two layers; however, unknown A in cyclohexane is completely dissolved. This suggests that of the two regions within unknown A, the (9) _____ (polar or nonpolar) region of the molecule dominates. The student suspects that unknown A may be (10) _____ (phenol, cyclohexanol, methoxycylcopentane, 2,3-dimethyl-2-butanol, or 1-hexanol) because it has the (11) _____ (strongest or weakest) intermolecular forces of all of the compounds.

The student tests the pH of all of the aqueous solutions and determines unknown E has a pH of 6.2. The other solutions are closer to a pH of 7. The student suspects that unknown E is (12) _____ because it has a(n) (13) _____ pH which indicates that the compound (14) _____ in water. The student also notices that this compound has a (15) _____ odor, like Chloroseptic® throat spray.

The student reacts orange potassium dichromate with small samples of all of the unknowns dissolved in water and records the following observations:

Unknown	Observations
A	the reaction mixture turned orange
B	the reaction mixture turned gray-green immediately and felt hot to the touch
C	the reaction mixture turned gray-green immediately and felt hot to the touch
D	the reaction mixture turned orange
E	the reaction mixture turned orange

Unknowns (16) _____ reacted with potassium dichromate; therefore, they must be (17) _____ because they were (18) _____ by the potassium dichromate. The student filters the resulting mixtures from the potassium dichromate experiment and isolates the organic materials from the filtrates. The student then dissolves the organic materials from the filtrates in water and tests the pH of the resulting solutions. The pH of solution C and E are less than 7. The other compounds have pH values of approximately 7. This suggests to the student that unknown B is (19) _____ and unknown C is (20) _____. Therefore, unknown D is (21) _____. To confirm the identities of unknowns A and D, the student could check the boiling points of pure samples of the compounds. If the student is correct, unknown (22) _____ will have the higher boiling point.

SOLUTIONS FOR THE ADDITIONAL ACTIVITIES
Section 13.1 Review:
(1) The -ol ending is appropriate for compounds that contain a hydroxy group attached to an aliphatic hydrocarbon chain because the hydroxy group is the key functional group for the class of compounds known as alcohols. The compound names and compound class both have the same ending.
(2) When numbering the longest carbon chain in a compound that contains a hydroxy group, the chain is numbered in such a manner that the hydroxy group is attached to the carbon atom with the smallest possible number.
(3) A compound name that ends in -triol suggests that the compound contains three hydroxy groups.
(4) Alkenes with multiple double bonds (dienes, trienes, tetraenes) have endings showing multiple substitutions of the same type, as do alcohols with multiple hydroxy groups (diols, triols, tetraols).
(5) Phenol was first introduced in this textbook as a benzene derivative. The structure of phenol contains a benzene ring.

Section 13.2 Review:

(1)
```
     OH
     |
CH₃—CH—CH₃
```
isopropyl alcohol
2-propanol

(2)
```
      OH
      |
CH₃—C—CH₃
      |
      CH₃
```
t-butyl alcohol
2-methyl-2-propanol

Section 13.3 Review:

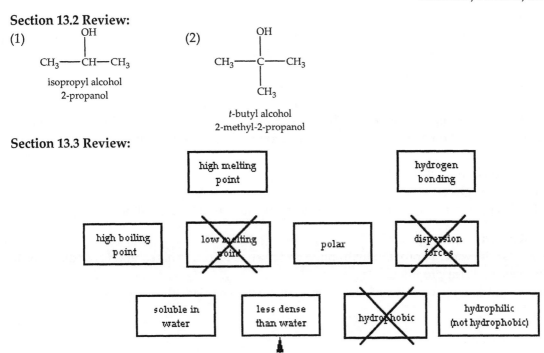

This is difficult to qualitatively observe because the alcohol will dissolve in water.

High molecular weight alcohols do not dissolve in water because the hydrogen bonds between the hydroxy group and the water is not sufficient to overcome the repulsion between the hydrophobic aliphatic tail and the polar water molecules. Water and high molecular weight alcohols are not enough like for dissolving to occur. High molecular weight alcohols behave more like hydrocarbons than like water.

Section 13.4 Review:

(1) hydration of an alkene
(2) Dehydration of an alcohol to form an alkene requires a sulfuric acid catalyst and temperatures of 180°C. Hydration of an alkene to form an alcohol requires water and a sulfuric acid catalyst.
(3) These processes do not require equilibrium arrows because of the differences in reaction conditions.
(4) The opposite of Markovnikov's rule can be applied to predicting the major product of alcohol dehydration to an alkene. The hydrogen atom will be removed from the carbon atom (adjacent to the carbon with the hydroxy group) that contains the fewest number of hydrogen atoms.
(5) Alcohol dehydration to an alkene requires a sulfuric acid catalyst and 180°C. Alcohol dehydration to an ether requires a sulfuric acid catalyst and 140°C. The temperatures of these reactions differ.
(6) Addition and elimination reactions are opposite processes. In an addition reaction, two reactants produce one product. In an elimination reaction, one reactant produces two products.

(7)
```
      OH                    O
      |                     ||
CH₃—CH—CH₃  —(O)→   CH₃—C—CH₃  + H₂O
```

(8)
```
   O = -2  OH                    O = -2  O                   O = -2
   H = +1  |                     C = 0   ||                  H = +1
           |                             ||                  H₂O
      CH₃—CH—CH₃             CH₃—C—CH₃
      C=-2  C=-2  C=-2          C=-2    C=-2                 2 H + O =
      H=+1  H=+1  H=+1          H=+1    H=+1                 2 + (-2) = 0

      3 C + 8 H + O =           3 C + 6 H + O =
      (-6) + 8 + (-2) = 0       (-4) + 6 + (-2) = 0
```

196 Chapter 13

(9) The carbon atom, attached to the hydroxy group as a reactant and the double bonded oxygen as a product, begins with an oxidation number of -2 and ends with an oxidation number of 0. This carbon atom has been oxidized. (10) tertiary < secondary < primary

Section 13.5 Review:
(1) redox, addition; (2) ethylene = redox, addition; glucose = fermentation, redox, elimination
(3) The hydration of an alkene as studied in Chapter 12 required the presence of a sulfuric acid catalyst. The industrial production of ethanol from ethylene requires high pressure (70 atm) and high temperature (300°C).
(4) toxic: methanol, isopropyl alcohol
    nontoxic: ethanol, glycerol
    cannot be classified from the reading: ethylene glycol (toxic), propylene glycol (nontoxic)

Section 13.6 Review:
(1) Phenol reacts with water to produce hydronium and phenolate ions. This solution is acidic, thus the name carbolic acid is appropriate.
(2) Phenol derivatives are used as disinfectants, antiseptics, and antioxidants as well as many other industrial uses.

Section 13.7 Review:

(1)

O—CH_3	O—$CH_2CH_2CH_3$	O—$CH_2CH_2CH_2CH_3$	O—$CHCH_2CH_3$ with CH_3
methoxy	proproxy	butoxy	sec-butoxy

O—CH_2CH_3	O—$CHCH_3$ with CH_3	O—CH_2CHCH_3 with CH_3	O—C(CH_3)$_3$
ethoxy	isoproproxy	isobutoxy	t-butoxy

(2) The oxygen atom from the alkoxy groups in (1) connects to a carbon chain in a compound.
(3) The key functional group in ethers is unique from any of the other key functional groups studied in depth up to this point. The ether key functional group is a linkage of two carbon based groups through an oxygen atom. Numbering the longest carbon chain in such a way that the oxygen atom is attached to the smallest numbered carbon atom could be used identify the location of the oxygen atom, but not the identity of the other carbon group. An ending of -er or -ther would not contain enough information to be useful. Consequently, the IUPAC system names ethers using the hydrocarbon name for the longest chain connected to the oxygen atom and an alkoxy name to describe the location and structure of other atoms bonded to the oxygen atom.
(4) Furan and pyran are heterocyclic rings because they contain more than one type of atom in the ring structure.
(5) Any cycloalkane, cycloalkene, or benzene derivative is an example of a homocyclic ring. Only carbon atoms are bonded into the ring structure.

Alcohols, Phenols, and Ethers 197

Section 13.8 Review:

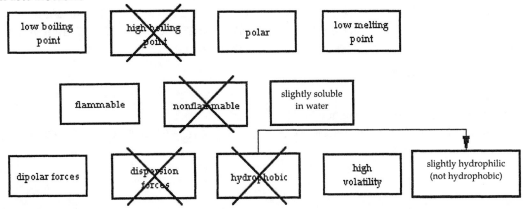

Section 13.9 Review:
(1) sulfur; (2) rotten eggs;
(4) yes; (5) The methanethiol is oxidized. (6) The corresponding disulfide is reduced.
(7) No, the reaction of heavy metal ions with thiols is not as easily reversed as the reaction to convert methanethiol to its corresponding

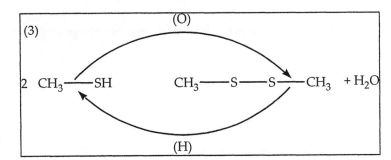

disulfide because this reaction is based on the activity series. When the heavy metal ions replace the hydrogen atoms, they gain electrons and form a more stable compound than the original thiol. Consequently, the hydrogen ions produced will not be able to "steal" electrons back from the metals in their reduced oxidation state.

Section 13.10 Review:
(1) Yes, the term "polyfunctional compounds" adequately describes the molecules within this class because they are compounds that contain more than one key functional group. The prefix *poly-* means many.
(2) The physical and chemical properties of polyfunctional compounds are similar in many ways to the physical and chemical properties of compounds containing only one of the functional groups in the polyfunctional compounds. Understanding the physical and chemical principles associated with the individual parts of a polyfunctional compound will enable better understanding of complex compounds.

Tying It All Together with a Laboratory Application:
(1)
(2) phenol C_6H_6O; cyclohexanol $C_6H_{12}O$; methoxycyclopentane $C_6H_{12}O$; 2,3-dimethyl-2-butanol $C_6H_{14}O$; 1-hexanol $C_6H_{14}O$
(3) cyclohexanol and methoxycyclopentane ($C_6H_{12}O$); 2,3-dimethyl-2-butanol and 1-hexanol ($C_6H_{14}O$)
(4) no geometric isomers
(5) alcohols: cyclohexanol, 2,3-dimethyl-2-butanol, 1-hexanol
(6) 1° = 1-hexanol, 2° = cyclohexanol, 3° = 2,3-dimethyl-2-butanol

(7) ether: methoxycyclopentane (11) weakest   (15) medicinal   (19) cyclohexanol
(8) benzene ring: phenol   (12) phenol   (16) B and C   (20) 1-hexanol
(9) nonpolar   (13) acidic   (17) 1° or 2° alcohols (21) 2,3-dimethyl-2-butanol
(10) methoxycyclopentane   (14) dissociated   (18) oxidized   (22) D

SELF-TEST QUESTIONS
Multiple Choice

1. The structure of 1-propoxypropane is
   a. CH₃—CH₂—O——CH₂—CH₂—CH₃
   b. CH₃—CH₂—O——CH(CH₃)—CH₃
   c. CH₃—CH₂—CH₂—O——CH₂—CH₂—CH₃
   d. CH₃—CH₂—CH₂—O——CH(CH₃)—CH₃

2. What is the correct IUPAC name for this compound?

   CH₃—CH₂—CH₂—C(CH₃)(CH₂CH₃)—OH

   a. 1-ethyl-1-methylbutanol
   b. 2-ethyl-2-pentanol
   c. 3-methyl-3-hexanol
   d. 3-methylheptanol

3. What is the correct IUPAC name for this compound?

   HO—⟨benzene ring⟩—CH₃

   a. 4-methylphenol
   b. 1-methylphenol
   c. 2-methyl-4-phenol
   d. 4-methyl-2-phenol

4. The alcohol is
   a. primary
   b. secondary
   c. tertiary
   d. quaternary

5. Which of the following is a primary alcohol?
   a. 1-butanol
   b. 2-butanol
   c. 2-propanol
   d. 2-methyl-2-propanol

6. Which of the following is the most soluble in water?
   a. CH₃CH₂CH₂—O—CH₃
   b. CH₃CH₂CH₂CH₂—OH
   c. CH₃CH₂—O—CH₂CH₃
   d. CH₃CH₂CH₂CH₂CH₂—OH

7. What reagent is required to carry out the reaction
   2 CH₃—SH + ? → CH₃—S—S—CH₃ + H₂O
   a. NaOH
   b. [O]
   c. H⁺
   d. H₂SO₄

8. The chemical reactivity of ethers is closest to the
   a. alkanes
   b. alkenes
   c. aromatics
   d. alcohols

9. A common name for ethoxybenzene is
   a. ethyl benzene ether
   b. ethyl benzyl ether
   c. ethyl phenyl ether
   d. ethoxy benzyl ether

10. The IUPAC name of is
    a. ethoxycyclobutane
    b. 1-ethyloxycyclobutane
    c. 1-ethylcyclobutane
    d. ethylcyclobutoxy

Matching
For each description on the left, select the correct alcohol from the responses on the right.
11. automobile antifreeze
12. moistening agent in cosmetics
13. rubbing alcohol
14. present in alcoholic beverages

a. glycerol
b. ethylene glycol
c. isopropyl alcohol
d. ethyl alcohol

Select the correct phenol for each description on the left.
15. antioxidant used in food packaging
16. used in throat lozenges
17. present in Lysol

a. o-phenylphenol
b. BHA
c. 4-hexylresorcinol
d. eugenol

For each reaction on the left, select the correct product from the responses on the right.
18. an alcohol is dehydrated at 140°C
19. an alcohol is dehydrated at 180°C
20. a secondary alcohol is oxidized

a. a ketone
b. an ether
c. an alkene
d. an aldehyde

True-False
21. Methyl alcohol is a product of the fermentation of sugars and starch.
22. Vitamin E is a natural antioxidant.
23. Tertiary alcohols undergo oxidation to produce ketones.
24. Hydrogen bonding accounts for the water solubility of certain alcohols.
25. Alcohols have a higher boiling point than ethers of similar molecular weight.
26. Oxidation of a thiol produces a disulfide.
27. Two Hg^{2+} ions are required for the reaction with one thiol.
28. Ethers can form hydrogen bonds with water molecules.
29. Thiols are responsible for the pleasant fragrance of many flowers.
30. Ethylene glycol is used as a moisturizer in foods.

Chapter 14: Aldehydes and Ketones

CHAPTER OUTLINE
14.1 The Nomenclature of Aldehydes and Ketones
14.2 Physical Properties
14.3 Chemical Properties
14.4 Important Aldehydes and Ketones

LEARNING OBJECTIVES/ASSESSMENT
When you have completed your study of this chapter, you should be able to:
1. Recognize the carbonyl group in compounds and classify the compounds as aldehydes or ketones. (Section 14.1; Exercise 14.4)
2. Assign IUPAC names to aldehydes and ketones. (Section 14.1; Exercise 14.6)
3. Compare the physical properties of aldehydes and ketones with those of compounds in other classes. (Section 14.2; Exercise 14.16)
4. Write key reactions for aldehydes and ketones. (Section 14.3; Exercise 14.42)
5. Give specific uses for aldehydes and ketones. (Section 14.4; Exercise 14.52)

SOLUTIONS FOR THE END OF CHAPTER EXERCISES
THE NOMENCLATURE OF ALDEHYDES AND KETONES (SECTION 14.1)

14.2
H—C(=O)—H aldehyde
H_3C—C(=O)—CH_3 ketone

14.4
a. CH_3CH_2—C(=O)—H aldehyde
b. CH_3—C(=O)—O—CH_3 neither (ester)
c. cyclopentanone ketone
d. $CH_3CH_2CH_2$—C(=O)—OH neither (carboxylic acid)
e. CH_3CH_2—C(=O)—CH_3 ketone
f. CH_3CH_2—C(=O)—NH_2 neither (amide)

14.6
a. CH_3CH_2—C(=O)—H   propanal
b. $CH_3CHBrCH_2$—C(=O)—H   3-bromobutanal
c. phenyl–CH_2CH_2—C(=O)—H   3-phenylpropanal
d. $CH_3CH(CH_3)CH_2$—C(=O)—CH_3 4-methyl-2-pentanone
e. 2-isopropylcyclopentanone

201

202 Chapter 14

14.8 a. butanal

$$CH_3CH_2CH_2-\overset{\overset{O}{\|}}{C}-H$$

c. 2-ethyl-4-propylcyclohexanone

(structure: cyclohexanone with CH₂CH₃ at position 2 and CH₂CH₂CH₃ at position 4)

b. 2-methyl-3-pentanone

$$CH_3\overset{\overset{CH_3}{|}}{CH}-\overset{\overset{O}{\|}}{C}-CH_2CH_3$$

d. 4-chloro-2-phenylpentanal

$$CH_3\overset{\overset{Cl}{|}}{CH}CH_2\overset{\overset{}{}}{CH}(Ph)-\overset{\overset{O}{\|}}{C}-H$$

14.10

$$H_3C-\overset{\overset{O}{\|}}{C}-CH_3 \qquad H-\overset{\overset{O}{\|}}{C}-CH_2CH_3$$

   2-propanone   propanal

14.12 a. incorrect = 3-ethyl-2-methylbutanal
          correct = 2,3-dimethylpentanal

   b. incorrect = 2-methyl-4-butanone
      correct = 3-methylbutanal

   c. incorrect = 4,5-dibromocyclopentanone
      correct = 2,3-dibromocyclopentanone

(structures showing 2,3-dimethylpentanal, 3-methylbutanal, and 2,3-dibromocyclopentanone with numbering 1, 3 on ring)

PHYSICAL PROPERTIES (SECTION 14.2)

14.14 The acetone dissolves the remaining water in the glassware and the bulk of the mixture is discarded. The remaining traces of acetone evaporates quickly because it is much more volatile than water.

☑14.16 Propane is nonpolar and only experiences dispersion forces with other propane molecules. Ethanal is polar and experiences dipolar forces with other ethanal molecules. The stronger interparticle forces in ethanal cause it to have a higher boiling point than propane.

14.18 a. [structure: cyclopentyl-C(=O)-CH₃ with H-bonded H-O-H above carbonyl oxygen]

b. [structure: benzaldehyde (Ph-CHO) with H-bonded H-O-H above carbonyl oxygen]

14.20 [structures of menthone (cyclohexanone with methyl and isopropyl substituents) and menthol (cyclohexanol with methyl and isopropyl substituents)]

menthone   menthol

Menthol experiences hydrogen bonding and menthone experiences dipolar interparticle forces. Menthol is likely the solid because hydrogen bonding is a stronger interparticle force than dipolar forces. Menthone is likely the liquid.

CHEMICAL PROPERTIES (SECTION 14.3)

14.22 a.
$$CH_3-\underset{OH}{\underset{|}{CH}}-CH_2\underset{CH_3}{\underset{|}{CH}}CH_3 \xrightarrow{[O]} CH_3-\underset{O}{\underset{\|}{C}}-CH_2\underset{CH_3}{\underset{|}{CH}}CH_3 + H_2O$$

b. [cyclopentyl-CH₂-OH] $\xrightarrow{[O]}$ [cyclopentyl-CHO] + H₂O

c. [2-ethylcyclopentanol] $\xrightarrow{[O]}$ [2-ethylcyclopentanone] + H₂O

14.24 a.
$$CH_3-\underset{H}{\underset{|}{\overset{OH}{\underset{|}{C}}}}-OCH_3$$
hemiacetal

b.
$$CH_3CH_2-\underset{CH_3}{\underset{|}{\overset{OCH_3}{\underset{|}{C}}}}-OH$$
hemiketal

c.
$$CH_3\underset{}{CH}CH_2-OH \quad \overset{OCH_3}{|}$$
neither
(alcohol and ether functional groups)

d. [cyclopentane with OH and OCH₂CH₃ on same carbon]
hemiketal

14.26 a. [cyclohexane with OCH₃ and OCH₂CH₃ on adjacent carbons]
neither
(2 ether functional groups)

b. [cyclohexyl-O-C(OCH₃)(CH₂CH₃)(CH₃)]
ketal

c.
$$\underset{}{CH_2}\begin{matrix}OCH_3\\|\\-OCH_3\end{matrix}$$
acetal

204 Chapter 14

14.28 a. ketal b. none of these (alcohol) c. hemiacetal

14.30 Hemiacetals are potential aldehydes because hemiacetals are not stable products. The hemiacetal can decompose into an alcohol and an aldehyde.

14.32 A positive Tollen's test is the appearance of a silver mirror inside a test tube.

14.34 a. no reaction   (This is a ketone, not an aldehyde.)

b. [benzaldehyde] →(Tollen's Test) [benzoate $O^-NH_4^+$]

c. no reaction   (This is a hemiacetal, not an aldehyde.)

d. CH_3CH_2—C(=O)—H →(Tollen's Test) CH_3CH_2—C(=O)—$O^-NH_4^+$

e. no reaction   (This is a ketone, not an aldehyde.)

14.36
a. CH_3CH_2—C(=O)—H   no reaction
b. [benzaldehyde]   no reaction
c. [cyclopentyl-C(=O)-C(=O)-H]   reacts
d. $CH_3CH(OH)$—C(=O)—H   reacts
e. CH_3—O—CH_2—C(=O)—H   no reaction

14.38 OH OH OH OH ⟨OH O⟩
CH_2—CH—CH—CH—CH—C—H   (an aldehyde with an adjacent alcohol group)

14.40 OH OH OH ⟨OH O⟩ OH
CH_2—CH—CH—CH—C—CH_2 (a ketone with an adjacent alcohol group)

☑**14.42** a. [dicyclohexyl ketone] + H_2 →(Pt) [dicyclohexyl carbinol with OH and H]

Aldehydes and Ketones 205

b.

$$CH_3CH_2-\underset{\underset{CH_3}{|}}{\overset{\overset{OH}{|}}{C}}-OCH_3 + CH_3CH_2CH_2-OH \underset{}{\overset{H^+}{\rightleftharpoons}} CH_3CH_2-\underset{\underset{CH_3}{|}}{\overset{\overset{OCH_2CH_2CH_3}{|}}{C}}-OCH_3 + H_2O$$

c.

$$CH_3CH_2-\overset{\overset{O}{\|}}{C}-H + 2\ CH_3\underset{\underset{}{|}}{\overset{\overset{OH}{|}}{CH}}CH_3 \overset{H^+}{\rightleftharpoons} CH_3CH_2-\underset{\underset{H_3C}{\overset{|}{\underset{CH-O}{}}}}{\overset{\overset{O-CH\diagdown CH_3}{|}}{C}}-H + H_2O$$

d.

cyclopentyl-$CH_2-\overset{\overset{O}{\|}}{C}-H$ + (O) $\longrightarrow$ cyclopentyl-$CH_2-\overset{\overset{O}{\|}}{C}-OH$

e.

$$CH_3CH_2\underset{\underset{}{|}}{\overset{\overset{CH_3}{|}}{CH}}-\overset{\overset{O}{\|}}{C}-H + H_2 \overset{Pt}{\longrightarrow} CH_3CH_2\underset{\underset{H}{|}}{\overset{\overset{CH_3}{|}}{CH}}-\underset{\underset{H}{|}}{\overset{\overset{OH}{|}}{C}}-H$$

14.44   When hydrogen is added to an alkene, an alkane results.
When hydrogen is added to an alkyne, an alkene or alkane results.
When hydrogen is added to an aldehyde, a primary alcohol results.
When hydrogen is added to a ketone, a secondary alcohol results.

14.46 a.

cyclohexane with two OCH_2CH_3 groups + $H_2O \overset{H^+}{\rightleftharpoons}$ cyclohexanone + 2 $HOCH_2CH_3$

b.

$$CH_3CH_2-\underset{\underset{CH_2CH_3}{|}}{\overset{\overset{OCH_3}{|}}{C}}-OCH_3 + H_2O \overset{H^+}{\rightleftharpoons} CH_3CH_2-\underset{\underset{CH_2CH_3}{|}}{\overset{\overset{O}{\|}}{C}} + 2\ HOCH_3$$

c. $CH_3CH_2CH_2O-CH_2-OCH_2CH_2CH_3 + H_2O \overset{H^+}{\rightleftharpoons} \overset{\overset{O}{\|}}{\underset{CH_2}{}} + 2\ HOCH_2CH_3CH_3$

d. $CH_3O-\underset{\underset{OCH_3}{|}}{CH}-CH_2CH_3 + H_2O \overset{H^+}{\rightleftharpoons} HC-CH_2CH_3 + 2\ HOCH_3$ (with =O on HC)

14.48 a. [cyclic hemiketal structure: 2-methyl-2-ethoxytetrahydrofuran] + H₂O ⇌ (H⁺) [HOCH₂CH₂CH₂C(OH)(CH₃)... open form with C=O and CH₃, and HOCH₂ group] + HOCH₂CH₃

b. [tetrahydropyran with OCH₂CH₂CH₃ and H substituents at C2] + H₂O ⇌ (H⁺) [open chain: HOCH₂-(CH₂)₃-CHO] + HOCH₂CH₃CH₃

14.50 a.

CH₃CH₂CH₂—OH —[O]→ CH₃CH₂CHO + H₂O

$$CH_3CH_2CHO + HOCH_2CH_3 \rightleftharpoons CH_3CH_2CH(OH)(OCH_2CH_3)$$

$$CH_3CH_2CH(OH)(OCH_2CH_3) + HOCH_2CH_3 \xrightleftharpoons{H^+} CH_3CH_2CH(OCH_2CH_3)_2 + H_2O$$

b.

$$\underset{\underset{CH_3}{|}}{CH_3CH}-\underset{\underset{O}{\parallel}}{C}-CH_3 + H_2 \xrightarrow{Pt} \underset{\underset{CH_3}{|}}{CH_3CH}-\underset{\underset{OH}{|}}{CH}-CH_3$$

$$\underset{\underset{CH_3}{|}}{CH_3CH}-\underset{\underset{OH}{|}}{CH}-CH_3 \xrightarrow[180°C]{H_2SO_4} \underset{\underset{CH_3}{|}}{CH_3C}=CH-CH_3 + H_2O$$

IMPORTANT ALDEHYDES AND KETONES (SECTION 14.4)

☑14.52 a. peppermint flavoring   menthone
        b. flavoring for margarines   biacetyl
        c. cinnamon flavoring   cinnamaldehyde
        d. vanilla flavoring   vanillin

CHEMISTRY AROUND US, CHEMISTRY AND YOUR HEALTH, HOW REACTIONS OCCUR, AND OVER THE COUNTER

14.54   Dihydroxyacetone is the active ingredient in some self-tanning lotions.

14.56   Norethynodrel mimics the action of another ketone, progesterone, a hormone that causes changes in the wall of the uterus to prepare it to accept a fertilized egg and maintain the resulting pregnancy.

14.58   Vanilloids have a hot and spicy flavor that might even be extremely irritating.

ADDITIONAL ACTIVITIES

Section 14.1 Review:
(1) Is the IUPAC *-al* ending for compounds that contain a carbonyl group adjacent to hydrogen atom meaningful?
(2) Is the IUPAC *-one* ending for compounds that contain a carbonyl group between two other carbon atoms meaningful?
(3) How is the longest carbon chain numbered for aldehydes?
(4) How is the longest carbon chain numbered for ketones?

Section 14.2 Review:
(1) Rank alcohols, ethers, and aldehydes/ketones in order of increasing boiling points.
(2) What intermolecular forces do aldehydes and ketones experience when surrounded by molecules of the same type?

Section 14.3 Review: Complete the following reaction diagrams.

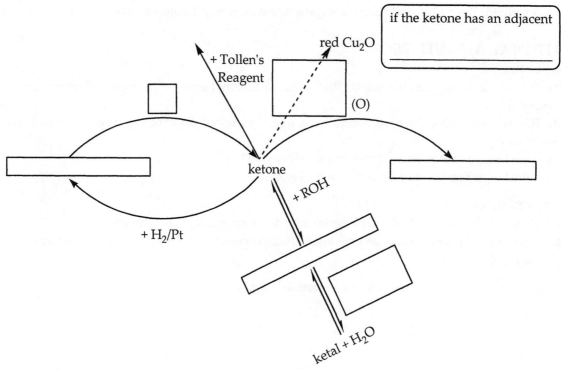

(1) What do hemiacetals and hemiketals have in common?
(2) What do hemiacetals and acetals have in common?
(3) What do hemiketals and ketals have in common?
(4) What do acetals and ketals have in common?
(5) How does hydrolysis differ from hydration?

Section 14.4 Review:
(1) What is the IUPAC name for formaldehyde?
(2) What is the IUPAC name for acetone?
(3) Do some aldehydes and ketones have an associated smell?

Tying It All Together with a Laboratory Application:
An instructor transfers approximately 50 mL of 2-propanone and propanal into two separate unlabeled containers and hands the bottles to a student with the challenge to determine what is in each bottle. One bottle has a piece of tape around it and the other does not.

The student knows that the two compounds have different (1) _____. The compound with the (2) _____ name 2-propanone has the common name (3) _____. 2-propanone is frequently used as (4) _____ remover. Propanal has an overwhelming fruity odor. Unfortunately, this student does not have the ability to smell the compounds.

The student also knows that the two compounds (5) _____ dissolve in water because they will experience (6) _____ intermolecular forces with water. These compounds are both (7) _____ molecules. Mixing these compounds with water (8) _____ differentiate these compounds.

The student also knows that the two compounds have different (9) _____ points; however, the student does not want to heat these compounds because they are (10) _____.

The student considers performing a reaction that could serve as a qualitative test. First the student draws and labels the structural formulas for the compounds (shown in (11)_____). Bromine, potassium dichromate, Tollen's, and Benedict's solutions are available. The student decides against using the

bromine solution because (12) _____ of the compounds react(s) with bromine. Both tests would be (13) _____. The student also decides against using the (14) _____ solution because both tests would be negative. The student decides not to use the (15) _____ solution because although it would make a silver mirror on the inside of a test tube containing (16) _____, the student does not want to ruin a test tube. Finally, the student decides to use the (17) _____ solution. This test should provide a
(18) _____ result for the 2-propanone and a (19) _____ result for the propanal. The student performs the test and the unknown from the bottle with the tape turns bright orange, while the unknown from the other bottle turns gray-green and feels extremely warm. The 2-propanone was in the bottle (20) _____ the tape.

(21) Suggest a reason the student did not add hydrogen to the compounds or react the compounds with alcohol.

SOLUTIONS FOR THE ADDITIONAL ACTIVITIES

Section 14.1 Review:
(1) Yes, this ending is meaningful. A carbonyl group adjacent to a hydrogen atom is the aldehyde key functional group. The *–al* ending indicates that the compound is an aldehyde.
(2) Yes, this ending is meaningful. A carbonyl group between two other carbon atoms is the ketone key functional group. The *–one* ending indicates that the compound is a ketone.
(3) The carbonyl carbon is always the first carbon in the chain.
(4) The chain is numbered so the carbonyl carbon has the lowest number possible. For open chain compounds, the carbonyl carbon will always have a number 2 or higher. For closed chain (cyclic) compounds, the carbonyl carbon will be given the number 1.

Section 14.2 Review:
(1) ethers < aldehydes/ketones < alcohols
(2) dipolar forces

Section 14.3 Review:

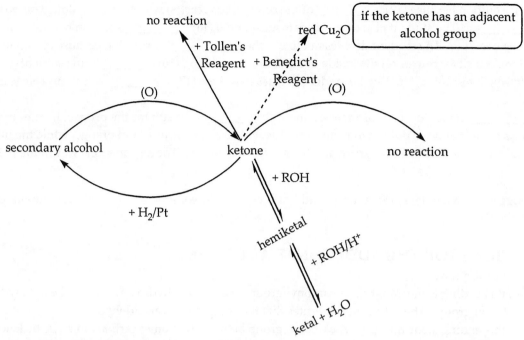

(1) Hemiacetal and hemiketals both have an –OH and an –OR bonded to the same carbon atom.
(2) Hemiacetals and acetals both have an –H and an –OR bonded to the same carbon atom.
(3) Hemiketals and ketals both have two –R groups and an –OR bonded to the same carbon atom.
(4) Acetals and ketals both have two –OR groups bonded to the same carbon atom.
(5) Hydrolysis involves breaking one of the bonds in a molecule using water. Hydration involves the addition of water to molecule (usually across a double bond).

Section 14.4 Review:
(1) Formaldehyde is called methanal in the IUPAC nomenclature system.
(2) Acetone is called 2-propanone in the IUPAC nomenclature system.
(3) Yes, some aldehydes and ketones have an associated smell. Formaldehyde smells like a dissecting lab, acetone is the smell of fingernail polish remover, and Table 14.3 lists other fragrant aldehydes and ketones.

Tying It All Together with a Laboratory Application:
(1) odors
(2) IUPAC
(3) acetone
(4) fingernail polish
(5) will
(6) hydrogen bonding
(7) polar
(8) will not
(9) boiling
(10) flammable

(12) neither
(13) negative
(14) Benedict's
(15) Tollen's
(16) propanal
(17) potassium dichromate
(18) negative
(19) positive
(20) with

(21) Reacting the compounds with hydrogen would produce a primary and a secondary alcohol; however, those compounds are colorless just like the reactants. Reacting the compounds with an alcohol would produce hemiacetal and/or acetal and hemiketal and/or ketal compounds; however, those compounds are also colorless like the reactants. Consequently, the student chose a colorful, qualitative test rather than either of these reactions.

SELF-TEST QUESTIONS
Multiple Choice

1. What is the correct IUPAC name for CH₃—CH₂—CH₂—CH₂—CH(=O) ?
   a. pentanal   b. 1-pentanol   c. 5-pentanal   d. 1-pentanone

2. What is the correct IUPAC name for [cyclopentanone with CH₂CH₃ substituent] ?
   a. 1-ethylcyclopentanone
   b. 2-ethylcyclopentanone
   c. 1-ethylcyclopentanal
   d. 2-ethylcyclopentanal

3. What is the correct IUPAC name for Br—CH₂—CH₂—CH(=O) ?
   a. 1-bromo-3-propanone
   b. 3-bromopropanal
   c. 1-bromopropanal
   d. 3-bromo-1-propanone

4. Which of the following pure compounds can exhibit hydrogen bonding with other molecules of the same compound?
   a. CH₃—CH₂—CH(=O)
   b. CH₃—CH₂—CH₂—OH
   c. CH₃—CH₂—C(=O)—CH₃
   d. more than one response is correct

5. Which of the chemical species below is a key component of a reagent mixture that may be used to test for the presence of an aldehyde?
   a. Ag⁺   b. Br₂   c. Cu₂O   d. HCl

6. The reaction of water and H⁺ with [CH₃—CH(O—CH₃)(O—CH₃)] produces
   a. CH₃—CH(=O) + 2 CH₃—OH
   b. CH₃—CH(=O) + 2 H—C(=O)—H
   c. H₃C—CH₂(O—CH₃) + CH₃—OH
   d. CH₃—CH(=O) + CH₃—O—CH₃

7. Which of the following compounds could be oxidized by Benedict's reagent?
   a. CH₃—C(=O)—CH₂—CH₃
   b. CH₃—C(=O)—CH(OH)—CH₃

   c. CH₃—CH₂—CH₂—C(=O)—H
   d. CH₃—C(=O)—CH₂—CH₂—OH

8. An important preservative for biological specimens is

a. CH₃—C(=O)—CH₃
b. H—C(=O)—H
c. CH₃—CH₂—CH₂—C(=O)—H
d. CH₃—C(=O)—H

9. A flavoring for margarine is
   a. biacetyl   b. benzaldehyde   c. 2-butanone   d. vanillin

10. Hydrogenation of an aldehyde produces a
    a. acetal   b. primary alcohol   c. secondary alcohol   d. ketone

Matching
For each structure on the left, select the correct class of compound from the responses on the right.

11. CH₃—CH₂—CH(OH)—O—CH₃
12. CH₃—CH₂—C(O—CH₃)(CH₃)—OH
13. CH₃—CH₂—C(O—CH₃)(CH₃)—O—CH₃
14. (tetrahydrofuran ring with CH₃ and OH substituents)

a. acetal
b. ketal
c. hemiacetal
d. hemiketal

Select the correct product for each of the reactions on the left.
15. aldehyde + [O] →
16. ketone + hydrogen →
17. aldehyde + alcohol ⇌
18. ketone + alcohol ⇌

a. hemiacetal
b. hemiketal
c. carboxylic acid
d. alcohol

True-False
19. Both aldehydes and ketones contain a carbonyl group.
20. Pure ketones can hydrogen bond.
21. Pure aldehydes can hydrogen bond.
22. Both aldehydes and ketones can form hydrogen bonds with water.
23. Acetone is an important organic solvent.
24. Camphor is used in foods as peppermint flavoring.
25. Cinnamon flavoring contains cinnamaldehyde.

Chapter 15: Carboxylic Acids and Esters

CHAPTER OUTLINE
15.1 The Nomenclature of Carboxylic Acids
15.2 Physical Properties of Carboxylic Acids
15.3 The Acidity of Carboxylic Acids
15.4 Salts of Carboxylic Acids
15.5 Carboxylic Esters
15.6 The Nomenclature of Esters
15.7 Reactions of Esters
15.8 Esters of Inorganic Acids

LEARNING OBJECTIVES/ASSESSMENT
When you have completed your study of this chapter, you should be able to:
1. Assign IUPAC names and draw structural formulas for carboxylic acids. (Section 15.1; Exercise 15.6)
2. Explain how hydrogen bonding affects the physical properties of carboxylic acids. (Section 15.2; Exercise 15.10)
3. Recognize and write key reactions of carboxylic acids. (Section 15.3; Exercise 15.26)
4. Assign common and IUPAC names to carboxylic acid salts. (Section 15.4; Exercise 15.28)
5. Describe uses for carboxylate salts. (Section 15.4; Exercise 15.32)
6. Recognize and write key reactions for ester formation. (Section 15.5; Exercise 15.36)
7. Assign common and IUPAC names to esters. (Section 15.6; Exercises 15.44 and 15.46)
8. Recognize and write key reactions of esters. (Section 15.7; Exercise 15.54)
9. Write reactions for the formation of phosphate esters. (Section 15.8; Exercise 15.56)

SOLUTIONS FOR THE END OF CHAPTER EXERCISES
THE NOMENCLATURE OF CARBOXYLIC ACIDS (SECTION 15.1)

15.2 R—C(=O)—OH carboxylic acid

The structural features of a fatty acid are the carboxylic acid functional group and a long aliphatic tail. They are called fatty acids because they were originally isolated from fats.

15.4 The carboxylic acid present in sour milk and sauerkraut is lactic acid.

☑15.6
a. $CH_3CH(CH_3)C(=O)OH$
2-methylpropanoic acid

b. $Br-CH_2CH_2CH_2-C(=O)-OH$
4-bromobutanoic acid

c. (benzoic acid with isopropyl at meta position)
m-isopropylbenzoic acid
3-isopropylbenzoic acid

d. $CH_3CH_2CH(OCH_3)CH_2-C(=O)-OH$
3-methoxypentanoic acid

e. (phenyl)—$CH_2CH_2CH_2-C(=O)-OH$
4-phenylbutanoic acid

213

15.8   a. pentanoic acid

$$CH_3CH_2CH_2CH_2-\underset{\underset{O}{\|}}{C}-OH$$

b. 2-bromo-3-methylhexanoic acid

$$CH_3CH_2CH_2\underset{\underset{CH_3}{|}}{CH}\underset{\underset{Br}{|}}{CH}-\underset{\underset{O}{\|}}{C}-OH$$

c. 4-propylbenzoic acid

[Structure: benzene ring with $-C(=O)-OH$ at top and $-CH_2CH_2CH_3$ at bottom]

PHYSICAL PROPERTIES OF CARBOXYLIC ACIDS (SECTION 15.2)

☑15.10 a. Acetic acid will have a higher boiling point than 1-propanol because acetic acid forms stronger hydrogen bonds than 1-propanol.

b. Propanoic acid will have a higher boiling point than butanone because propanoic acid forms stronger hydrogen bonds than butanone.

c. Butyric acid will have a higher boiling point than acetic acid because butyric acid has a higher molecular mass than acetic acid.

15.12

[Structure: cyclic dimer of propanoic acid showing two CH_3CH_2-C groups linked by O----H—O and O—H----O hydrogen bonds]

15.14 The carboxylic acid functional group allows caproic acid to be soluble in water. The solubility is limited by the aliphatic portion of the acid because it is hydrophobic.

15.16 pentane < butanal < 1-butanol < butyric acid

Alkanes are the least soluble in water. Aldehydes are more soluble than alkanes. Alcohols are more soluble than aldehydes. Carboxylic acids are more soluble than alcohols.

THE ACIDITY OF CARBOXYLIC ACIDS (SECTION 15.3)

15.18

$$CH_3CH_2-\underset{\underset{O}{\|}}{C}-O^-$$

15.20 Within the cell, lactic acid will dissociate to give the lactate ion and H⁺ because body fluids have a pH of 7.4 and lactic acid is a weak acid.

15.22

$$CH_3CH_2CH_2-\underset{\underset{O}{\|}}{C}-OH$$

15.24 a. $CH_3CH_2-C(=O)-OH + NaOH \longrightarrow CH_3CH_2-C(=O)-O^-Na^+ + H_2O$

b. $CH_3CH(OH)-C(=O)-OH + KOH \longrightarrow CH_3CH(OH)-C(=O)-O^-K^+ + H_2O$

☑15.26 a. $CH_3-C(=O)-OH + NaOH \longrightarrow CH_3-C(=O)-O^-Na^+ + H_2O$

b. $CH_3-C(=O)-OH + KOH \longrightarrow CH_3-C(=O)-O^-K^+ + H_2O$

c. $2\,CH_3-C(=O)-OH + Ca(OH)_2 \longrightarrow CH_3-C(=O)-O^-Ca^{2+}O^--C(=O)-CH_3 + 2\,H_2O$

This can also be written as: $Ca(CH_3COO)_2$

SALTS OF CARBOXYLIC ACIDS (SECTION 15.4)

☑15.28 a. $CH_3CH(Br)CH_2-C(=O)-O^-Na^+$
sodium 3-bromobutanoate

b. $(H-C(=O)-O^-)_2\,Ca^{2+}$
calcium methanoate

c. (phenyl)$-O-CH_2CH_2-C(=O)-O^-K^+$
potassium 3-phenoxypropanoate

15.30 a. potassium ethanoate
$CH_3-C(=O)-O^-K^+$

b. sodium m-methylbenzoate
(m-CH_3-phenyl)$-C(=O)-O^-Na^+$

c. sodium 2-methylbutanoate
$CH_3CH_2CH(CH_3)-C(=O)-O^-Na^+$

☑15.32
a. as a soap — sodium stearate
b. as a general food preservative used to pickle vegetables — acetic acid
c. as a preservative used in soft drinks — sodium benzoate
d. as a treatment for athlete's foot — zinc 10-undecylenate
e. as a mold inhibitor used in bread — calcium propanoate
   sodium propanoate
f. as a food additive noted for its pH buffering ability — sodium citrate
   citric acid

CARBOXYLIC ESTERS (SECTION 15.5)

15.34 b. CH₃—C(=O)—O—CH₂CH₃ (ester linkage circled, arrow points to C—O)

f. C₆H₅—C(=O)—O—CH₂CH₃ (ester linkage circled, arrow points to C—O)

d. C₆H₅—O—C(=O)—CH₃ (ester linkage, circled around —C(=O)—CH₃)

✓15.36 a. CH₃—C(=O)—O—C(=O)—CH₃ + CH₃—OH → CH₃—C(=O)—OCH₃ + HO—C(=O)—CH₃

b. (CH₃)₂CH—C(=O)—OH + C₆H₅—CH₂—OH ⇌ (H⁺, heat) H₂O + (CH₃)₂CH—C(=O)—O—CH₂—C₆H₅

c. (CH₃)₂CH—C(=O)—Cl + C₆H₅—OH → (CH₃)₂CH—C(=O)—O—C₆H₅ + HCl

15.38 a. propanoic acid and methyl alcohol

CH₃CH₂—C(=O)—O—CH₃

b. propanoic acid and phenol

CH₃CH₂—C(=O)—O—C₆H₅

c. propanoic acid and 2-methyl-1-propanol

$$CH_3CH_2-\overset{\overset{\displaystyle O}{\|}}{C}-O-CH_2-\overset{\overset{\displaystyle CH_3}{|}}{CH}-CH_3$$

15.40

[morphine structure with HO and OH groups] + 2 $CH_3-\overset{\overset{\displaystyle O}{\|}}{C}-Cl$ →

[diacetylated morphine (heroin) structure with $CH_3-\overset{\overset{\displaystyle O}{\|}}{C}-O-$ groups on both positions]

15.42

$$HO-\overset{\overset{\displaystyle O}{\|}}{C}-\overset{\overset{\displaystyle O}{\|}}{C}-OH + \overset{\overset{\displaystyle OH}{|}}{CH_2}-CH_2-\overset{\overset{\displaystyle OH}{|}}{CH_2} \rightleftharpoons$$

$$\left[-\overset{\overset{\displaystyle O}{\|}}{C}-\overset{\overset{\displaystyle O}{\|}}{C}-O-CH_2-CH_2-CH_2-O- \right]_n$$

THE NOMENCLATURE OF ESTERS (SECTION 15.6)

☑15.44   a.

$$CH_3\underset{\underset{\displaystyle OH}{|}}{CH}-\overset{\overset{\displaystyle O}{\|}}{C}-O-CH_2CH_3$$

ethyl lactate

b.

$$CH_3CH_2CH_2-\overset{\overset{\displaystyle O}{\|}}{C}-O-CH_2CH_2CH_3$$

propyl butyrate

15.46

a. methyl 2-methylpropanoate

$$CH_3CH(CH_3)-C(=O)-O-CH_3$$

b. ethyl 3,5-dichlorobenzoate

(3,5-dichlorophenyl)-C(=O)-O-CH$_2$CH$_3$

15.48
a. methyl propionate
b. methyl butyrate
c. methyl lactate

15.50

a. phenyl formate

C$_6$H$_5$-O-CH(=O)

c. ethyl 2-chloropropanoate

CH$_3$CH$_2$-O-C(=O)-CH(Cl)-CH$_3$

b. methyl 4-nitrobenzoate

O$_2$N-C$_6$H$_4$-C(=O)-O-CH$_3$

REACTIONS OF ESTERS (SECTION 15.7)

15.52 hydrolysis:

$$CH_3-C(=O)-OCH_2CH_3 + H_2O \xrightarrow{H^+} CH_3-C(=O)-OH + HOCH_2CH_3$$

saponification:

$$CH_3-C(=O)-OCH_2CH_3 + NaOH \longrightarrow CH_3-C(=O)-O^-Na^+ + HOCH_2CH_3$$

15.54

a. $$CH_3(CH_2)_{16}-C(=O)-OCH_2CH_3 + NaOH \longrightarrow CH_3(CH_2)_{16}-C(=O)-O^-Na^+ + HOCH_2CH_3$$

b. $$(CH_3)_2CH-C(=O)-O-C_6H_5 + H_2O \xrightleftharpoons{H^+} (CH_3)_2CH-C(=O)-OH + HO-C_6H_5$$

ESTERS OF INORGANIC ACIDS (SECTION 15.8)

☑15.56

$$\underset{\substack{|\\OH}}{\overset{\substack{OH\quad O\\|\quad\;\;||}}{CH_2-C-CH_2}} + HO-\overset{\substack{O\\||}}{\underset{\substack{|\\OH}}{P}}-OH \rightleftharpoons \underset{\substack{|\\OH}}{\overset{\substack{OH\quad O\\|\quad\;\;||}}{CH_2-C-CH_2}}-O-\overset{\substack{O\\||}}{\underset{\substack{|\\OH}}{P}}-OH \\ + H_2O$$

CHEMISTRY AROUND US, CHEMISTRY AND YOUR HEALTH, OVER THE COUNTER, AND HOW REACTIONS OCCUR

15.58

$$R-\underset{\substack{|\\OH}}{\overset{\substack{OH\quad O\\|\quad\;\;||}}{CH-C}}-OH$$

15.60   Aspirin can be used to treat arthritis and partially prevent first heart attack, first or recurrent strokes, and esophageal and colorectal cancer.

15.62   The carboxylate salt forms rather than a carboxylic acid because a strong base is used in saponification. In a high pH environment, the carboxylate salt will form rather than the carboxylic acid.

ADDITIONAL ACTIVITIES

Section 15.1 Review:
(1) What is importance of the *–oic acid* endings for compounds that contain hydroxy group bonded to a carbonyl carbon?
(2) How is the longest chain numbered for carboxylic acids?

Section 15.2 Review:
(1) Describe the type of smells associated with carboxylic acids.
(2) List alcohols, aldehydes, ethers, and carboxylic acids in order of decreasing boiling point.
(3) Can any of the other classes of compounds studied thus far for dimers? Explain.
(4) List alcohols, ketones, alkanes, and carboxylic acids in order of decreasing water solubility.

Section 15.3 Review:
(1) Classify carboxylic acids as strong or weak acids.
(2) What is the pH range for an aqueous carboxylic acid solution?
(3) When a titration is performed using sodium hydroxide and acetic acid, how does the pH at the equivalence point compare to pH 7?

Section 15.4 Review:
(1) What is the ending for an inorganic acid formed from a polyatomic ion with an *–ate* ending? (Hint: Refer back to Chapter 9.)
(2) All of the carboxylic acids form polyatomic ions when they dissociate in water. Carboxylic acids end with *–ic acid*. What is the ending for the polyatomic ions formed when a carboxylic acid dissociates in water or reacts with a base?
(3) Are the carboxylate salts edible?

Section 15.5 Review:
(1) What is(are) joined by an ester linkage?
(2) When a carboxylic acid reacts with an alcohol to produce an ester, does the odor associated with the carboxylic acid become one of the characteristics of the ester?
(3) How does condensation polymerization differ from addition polymerization?

220 Chapter 15

(4) What are three potential carbonyl containing starting materials for esters?
(5) What other reactant is needed to form an ester from the compounds in (4)?

Section 15.6 Review:
(1) Which carbon chain in an ester is named as a carboxylate group?
(2) Which carbon chain is named as an alkyl or aromatic group?

Section 15.7 Review:
(1) What happens when an ester is mixed with water in an acidic environment?
(2) What happens when an ester is mixed with an aqueous strong base?

Section 15.8 Review:
(1) How does the structure of phosphoric acid differ from the structure of a carboxylic acid?
(2) How do the differences from (1) affect the structures of the esters each can form?

Tying It All Together with a Laboratory Application:
A student is attempting to synthesize phenoxybenzoate. The structural formula for this compound is (1) _____. The carboxylic acid necessary for the synthesis is (2) _____. The other necessary components are (3) _____ and (4) _____ because the reaction is performed using a carboxylic acid. (5) Rank the reactants and the products of this reaction in order of increasing water solubility. The pH of the reactant mixture (assume water is present) is (6) _____ 7 because (7) _____. The pH of the ester in water is (8) _____ 7. This reaction system establishes an equilibrium. The forward reaction in called (9) _____ and the reverse reaction is called (10) _____.

Phenoxybenzoate can also be synthesized from (11) _____ and the acid chloride with the following structure, (12) _____. The product obtained in addition to phenoxybenzoate is (13) _____. A third technique for synthesizing phenoxybenzoate uses (14) _____ and the carboxylic anhydride with the following structure, (15) _____. The other product produced in addition to phenoxybenzoate is (16) _____.

Once the phenoxybenzoate is isolated, the student reacts the ester with potassium hydroxide in water with heat. The products produced by this reaction are (17) _____ and _____. The same carboxylate salt produced by this reaction could also have been synthesized by reacting potassium hydroxide with (18) _____. The pH of an aqueous solution of the carboxylate salt is (19) _____ 7.

(20) Identify the major interparticle forces in esters, carboxylate salts, and carboxylic acids.
(21) Rank esters, carboxylate salts, and carboxylic acids in order of increasing boiling point.
(22) Rank esters, carboxylate salts, and carboxylic acids in order of increasing water solubility.

SOLUTIONS FOR THE ADDITIONAL ACTIVITIES
Section 15.1 Review:
(1) The *–oic acid* endings for compounds that contain a hydroxy group bonded to a carbonyl carbon indicates that the compound is a carboxylic acid.
(2) The carbonyl carbon is always the first carbon in the longest chain of a carboxylic acid.

Section 15.2 Review:
(1) Unpleasant odors (like unwashed bodies and dirty laundry) are associated with carboxylic acids.
(2) carboxylic acids > alcohols > aldehydes > ethers
(3) No, carboxylic acids are the only compounds studied that can form dimers. They are the first compound class to have a hydroxy group and a carboxyl group attached to the same carbon atom.
(4) carboxylic acids > alcohols > ketones > alkanes

Section 15.3 Review:
(1) Carboxylic acids are usually weak acids.
(2) The pH of an aqueous carboxylic acid solution is below 7 (acidic).
(3) When a titration is performed using sodium hydroxide and acetic acid, the pH range of the mixture at the equivalence point is above 7 (basic).

Section 15.4 Review:
(1) –ic acid; (2) –ate; (3) Some of the carboxylate salts are edible.

Section 15.5 Review:
(1) A carbonyl carbon with an aliphatic chain and an oxygen atom bonded to an aliphatic chain are joined by an ester linkage.
(2) No, the odors associated with esters are usually much more pleasant than the odors associated with carboxylic acids.
(3) A condensation polymer generally forms from a monomer or a set of monomers that contains two key functional groups. In addition to producing a polymer, a small molecule is also produced during this reaction. An addition polymer generally forms from a monomer or a set of monomers that contains a carbon-carbon double bond. The polymer is the only product produced by this reaction.
(4) Three potential carbonyl containing starting materials for esters are carboxylic acids, carboxylic acid chlorides, and carboxylic acid anhyrides.
(5) All of the reactants from (4) can react with an alcohol or phenol to produce an ester. The carboxylic acid also requires a catalyst.

Section 15.6 Review:
(1) The carbon chain that includes the carboxyl group is named as a carboxylate.
(2) The carbon chain connected to an oxygen by a single bond is named as an alkyl or aromatic group.

Section 15.7 Review:
(1) When an ester is mixed with water in an acidic environment, a carboxylic acid and an alcohol form.
(2) When an ester is mixed with an aqueous strong base, a carboxylate salt and an alcohol form.

Section 15.8 Review:
(1) Phosphoric acid contains a phosphorus atom double bonded to an oxygen atom and single bonded to three other oxygen atoms. A carboxylic acid contains a carbon atom double bonded to an oxygen atom and single bonded to one other oxygen atom. Phosphoric acid is triprotic. Carboxylic acids are monoprotic, although diacids contain two carboxylic acid functional groups and are diprotic.
(2) The esters formed from phosphoric acid can be monoesters, diesters, or triesters depending on how many alkyl or aromatic groups attach to the 3 available oxygen atoms. Esters formed from carboxylic acids can only be monoesters because the carboxylic acid only has one available oxygen atom to bond with an alkyl or aromatic group. Diacids can form diesters if they react with an alcohol or polyesters if they react with a diol.

Tying It All Together with a Laboratory Application:
(1)
(2) benzoic acid
(3) phenol
(4) a catalyst
(5) phenoxyphenol < phenol < benzoic acid
(6) less than
(7) phenol and benzoic acid are acidic
(8) approximately
(9) esterification
(10) ester hydrolysis
(11) phenol
(12)

222 Chapter 15

(13) HCl
(14) phenol
(15)
(16) benzoic acid
(17) potassium benzoate, phenol
(18) benzoic acid
(19) greater than

(20) esters – dipolar forces;
    carboxylate salts – ionic bonding;
    carboxylic acids – hydrogen bonding
(21) esters < carboxylic acids
    < carboxylate salts
(22) esters < carboxylic acids
    < carboxylate salts

SELF-TEST QUESTIONS
Multiple Choice

1. What is the correct IUPAC name for [structure: 2-bromobenzoic acid]?
   a. bromobenzoic acid
   b. 2-bromobenzoic acid
   c. 1-bromobenzoic acid
   d. 2-bromo-1-benzoic acid

2. What is the correct IUPAC name for this compound? [CH$_3$CHCH$_2$—C(=O)—OH with CH$_3$ branch]
   a. 2-methylpentanoic acid
   b. 3-methylbutanoic acid
   c. 2-methylbutanoic acid
   d. 3-methylpentaoic acid

3. Which of the following pure substances would exhibit hydrogen bonding?
   a. aldehyde   b. ketone   c. ether   d. carboxylic acid

4. Which of the following substances would you expect to be the most soluble in water?
   a. CH$_3$CH$_2$—C(=O)—OH
   b. CH$_3$—C(=O)—OH
   c. CH$_3$—C(=O)—O—CH$_3$
   d. CH$_3$—C(=O)—O—CH$_2$CH$_3$

5. Which name is more appropriate for an organic acid under body conditions of pH 7.4?
   a. lactic acid   b. lactate acid   c. lactic   d. lactate

6. The reaction of butanoic acid and NaOH produces
   a. an ester
   b. a carboxylic acid and an alcohol
   c. a ketone
   d. a carboxylate salt

7. What reagent is needed to complete the following reaction?

    C₆H₅-C(=O)-OH + ? → C₆H₅-C(=O)-O⁻K⁺ + H₂O

    a. K   b. K⁺   c. KOH   d. KO₂

8. Which of the molecules a-d could be used as one of the reagents to prepare this compound?

    (cyclopentyl-O-C(=O)-CH₂CH₃)

    a. CH₃CH₂—OH   b. CH₃—C(=O)—OH   c. cyclopentanol   d. cyclopentyl-CHO

9. What is the organic product of the reaction CH₃—C(=O)—OH + CH₃—OH —H⁺→ ?

    a. CH₃—C(=O)—CH₂—OH
    b. CH₃—C(=O)—CH₃
    c. CH₃—C(=O)—O—CH₂—OH
    d. CH₃—C(=O)—O—CH₃

10. The ester formed by reacting propanoic acid and isopropyl alcohol is
    a. propyl propanoate
    b. isopropyl propanoate
    c. isopropyl propanoic acid
    d. 1-propyl propanoate

11. The IUPAC name for the ester formed in the reaction of isopropyl alcohol and benzoic acid is
    a. benzyl isopropyl ester
    b. isopropyl benzoate
    c. benzyl isopropanoate
    d. isopropyl benzoic acid

12. Which of the following materials might be obtained as one of the products from this reaction?

    CH₃—C(=O)—O—CH₂CH₂CH₃ + NaOH →

    a. CH₃CH₂—OH
    b. CH₃—C(=O)—O⁻Na⁺
    c. CH₃CH₂—C(=O)—O⁻Na⁺
    d. CH₃—C(=O)—OH

Matching
Select the best match for each of the following:

13. CH₃—C(=O)—OH

a. a preservative used in soda pop
b. a soap
c. present in vinegar
d. fingernail polish remover

14. C₆H₅—C(=O)—O⁻Na⁺

15. CH₃—(CH₂)₁₆—C(=O)—O⁻Na⁺

16. CH₃—C(=O)—O—CH₂CH₃

For each reaction on the left, choose the correct description from the responses on the right.

17. ester + H_2O $\xrightarrow{H^+}$
18. ester + NaOH →
19. carboxylic acid + H_2O →
20. carboxylic acid + alcohol $\xrightarrow[heat]{H^+}$

a. dissociation
b. esterification
c. hydrolysis
d. saponification

True-False

21. The boiling points of carboxylic acids are lower than those of the corresponding alcohols.
22. Salts of carboxylic acids are not usually soluble in water.
23. Hydrogen bonding increases both the boiling point and the water solubility of carboxylic acids.
24. CH_3COOH has a higher boiling point than $CH_3-C(=O)-H$.
25. Carboxylic acids are generally strong acids.
26. Both nitric acid and phosphoric acid can react with alcohols to form esters.
27. Certain phosphate esters are present in the body.
28. Esters are responsible for the pleasant fragrance of many flowers and fruits.
29. Carboxylate salts are fatty acids.
30. Esters form dimers more easily than carboxylic acids.

Chapter 16: Amines and Amides

CHAPTER OUTLINE
16.1 Classification of Amines
16.2 The Nomenclature of Amines
16.3 Physical Properties of Amines
16.4 Chemical Properties of Amines
16.5 Amines as Neurotransmitters
16.6 Other Biologically Important Amines
16.7 The Nomenclature of Amides
16.8 Physical Properties of Amides
16.9 Chemical Properties of Amides

LEARNING OBJECTIVES/ASSESSMENT
When you have completed your study of this chapter, you should be able to:
1. Given structural formulas, classify amines as primary, secondary, or tertiary. (Section 16.1; Exercise 16.4)
2. Assign common and IUPAC names to simple amines. (Section 16.2; Exercises 16.8 and 16.10)
3. Discuss how hydrogen bonding influences the physical properties of amines. (Section 16.3; Exercise 16.16)
4. Recognize and write key reactions for amines. (Section 16.4; Exercise 16.26)
5. Name amines used as neurotransmitters. (Section 16.5; Exercise 16.38)
6. Give uses for specific biological amines. (Section 16.6; Exercise 16.44)
7. Assign IUPAC names for amides. (Section 16.7; Exercise 16.46)
8. Show the formation of hydrogen bonds with amides. (Section 16.8; Exercise 16.50)
9. Give the products of acidic and basic hydrolysis of amides. (Section 16.9; Exercise 16.52)

SOLUTIONS FOR THE END OF CHAPTER EXERCISES
CLASSIFICATION OF AMINES (SECTION 16.1)

16.2 Primary amines have the formula: R—NH$_2$

Secondary amines have the formula: R—NH—R'

Tertiary amines have the formula: R—N(R')—R''

16.4 a. CH$_3$CH$_2$—NH—CH$_2$CH$_3$
    secondary

b. (phenyl)—N(CH$_3$)—CH$_2$CH$_3$
    tertiary

c. CH$_3$CH$_2$CH(CH$_3$)—NH$_2$
    primary

d.
    secondary

225

16.6   H—N—CH₂CH₂CH₃   CH₃   H—N—CH₂CH₃   H₃C—N—CH₃
         | | | |
         H   H—N—CHCH₃   CH₃   CH₃
       primary | secondary   tertiary
                            H
                         primary

THE NOMENCLATURE OF AMINES (SECTION 16.2)

16.8   a. CH₃—NH—CHCH₃   c. CH₃CH₂CH₂CH₂—NH₂
                  | butylamine
                  CH₃
           isopropylmethylamine

       b. [phenyl]—N—CH₂CH₃
                    |
                    CH₃
           ethylmethylphenylamine

16.10 a.   CH₃   c. CH₃CH═CHCHCH₃
                  | |
           CH₃CH₂CHCH₂—NH₂   NH₂
           1-amino-2-methylbutane   4-amino-2-pentene

       b.
            1-amino-2-methylcyclobutane

16.12 a.   NH—CH₂CH₂CH₃   b. [phenyl]—N—CH₃
            | |
            [phenyl]   CH₃
                                             N,N-dimethylaniline
            N-propylaniline

16.14 a.   2-amino-4-methylheptane   b. m-ethylaniline
                 NH₂   CH₃   NH₂
                  | | |
            CH₃—CH—CH₂—CHCH₂CH₂CH₃   [phenyl]—CH₂CH₃

c. *N,N*-diphenylaniline

PHYSICAL PROPERTIES OF AMINES (SECTION 16.3)

☑16.16 All low molecular weight amines are water soluble because all amines can hydrogen bond with water. Low molecular weight amines have small aliphatic portions, thus the hydrogen bonds to water are strong enough to allow the amines to dissolve.

16.18 The boiling points of tertiary amines are lower than the boiling points of primary and secondary amines because tertiary amines cannot hydrogen bond to each other, while primary and secondary amines can. The weaker intermolecular forces in tertiary amines allow them to boil at a lower temperature.

16.20 a.   [structure: CH_3CH_2—N(—CH_3)(—H) with H···O(H)H hydrogen bond above and H below on N]

b.   [structure: two pyrrolidine rings hydrogen-bonded N—H···N—H]

16.22 CH_3—N(—CH_3)—CH_2CH_3 < CH_3CH_2—NH—CH_2CH_3 < H_2N—$CH_2CH_2CH_2CH_3$

Tertiary amines have the lowest boiling point, secondary amines have an intermediate boiling point, and primary amines have the highest boiling point.

CHEMICAL PROPERTIES OF AMINES (SECTION 16.4)

16.24 CH_3CH_2—NH—CH_2CH_3 + H_2O ⇌ CH_3CH_2—$^+NH(H)$—CH_2CH_3 + OH^-

☑16.26 a. $CH_3CH_2CH_2CH_2$—NH_2 + HCl → $CH_3CH_2CH_2CH_2$—$NH_3^+Cl^-$

b. [cyclopentane—NH_2] + H_2O ⇌ [cyclopentane—NH_3^+] + OH^-

c.

cyclopentyl-NH₂ + CH₃C(=O)—OH ⟶ cyclopentyl-NH₃⁺ ⁻O—C(=O)—CH₃

d.
$$\underset{CH_3CHCH_3}{NH_3^+Cl^-} + NaOH \longrightarrow \underset{CH_3CHCH_3}{NH_2} + H_2O + NaCl$$

e.
$$\underset{CH_3}{\underset{|}{CH_3CH}}-\overset{O}{\overset{\|}{C}}-Cl + CH_3-NH_2 \longrightarrow \underset{CH_3}{\underset{|}{CH_3CH}}-\overset{O}{\overset{\|}{C}}-\underset{CH_3}{\underset{|}{NH}} + HCl$$

f.

Ph—C(=O)—O—C(=O)—Ph + NH₃ ⟶ Ph—C(=O)—OH + H₂N—C(=O)—Ph

16.28 Amine salts are more soluble in water, and therefore in blood, than their corresponding amines; consequently, the amine salts are the preferred form for drugs.

16.30

$$H_3C-\overset{O}{\overset{\|}{C}}-Cl + CH_3-\underset{CH_3}{\underset{|}{NH}} \longrightarrow CH_3-\overset{O}{\overset{\|}{C}}-\underset{CH_3}{\underset{|}{N}}-CH_3 + HCl$$

$$CH_3-\overset{O}{\overset{\|}{C}}-O-\overset{O}{\overset{\|}{C}}-CH_3 + CH_3-\underset{CH_3}{\underset{|}{NH}} \longrightarrow CH_3-\overset{O}{\overset{\|}{C}}-\underset{CH_3}{\underset{|}{N}}-CH_3 + CH_3-\overset{O}{\overset{\|}{C}}-OH$$

16.32 a.

Ph—NH₃⁺Cl⁻ ⟶ Ph—NH₂   NaOH

b.

$$CH_3CH_2-\overset{O}{\overset{\|}{C}}-Cl \longrightarrow CH_3CH_2-\overset{O}{\overset{\|}{C}}-NH_2 \quad NH_3$$

c.
$$CH_3-NH-CH_3 \longrightarrow CH_3-NH_2^+Cl^- \text{ with } CH_3$$

HCl

AMINES AS NEUROTRANSMITTERS (SECTION 16.5)

16.34 The space between neurons is called a synapse.

16.36 The two amines often associated with the biochemical theory of mental illness are norepinephrine and serotonin.

☑16.38 Four neurotransmitters important in the central nervous system are acetylcholine, norepinephrine, dopamine, and serotonin.

OTHER BIOLOGICALLY IMPORTANT AMINES (SECTION 16.6)

16.40 The clinical uses of epinephrine are as a component of injectable local anesthetics, to reduce hemorrhage, to treat asthma attacks, and to combat anaphylactic shock.

16.42 Alkaloids are derived from plants.

☑16.44
a. found in cola drinks — caffeine
b. used to reduce saliva flow during surgery — atropine
c. present in tobacco — nicotine
d. a cough suppressant — codeine
e. used to treat malaria — quinine
f. an effective painkiller — morphine

THE NOMENCLATURE OF AMIDES (SECTION 16.7)

☑16.46
a. CH₃CH₂CH₂—C(=O)—NH—CH₃
   N-methylbutanamide

b. CH₃—C(=O)—N(CH₃)—CH₃
   N,N-dimethylethanamide

c. (benzene ring)—C(=O)—NH₂ with CH₃ at ortho position
   o-methylbenzamide or 2-methylbenzamide

d. CH₃CH₂CH(CH₃)—C(=O)—NH₂
   2-methylbutanamide

16.48
a. propanamide
   CH₃CH₂—C(=O)—NH₂

b. N-ethylpentanamide
   CH₃CH₂CH₂CH₂—C(=O)—NH—CH₂CH₃

c. N,N-dimethylmethanamide

$$\text{H}-\overset{\overset{\text{O}}{\|}}{\text{C}}-\underset{\underset{\text{CH}_3}{|}}{\text{N}}-\text{CH}_3$$

PHYSICAL PROPERTIES OF AMIDES (SECTION 16.8)
☑16.50 a. [Structure showing cyclopentyl-C(=O)-N with hydrogen bonding to multiple water molecules]

b. [Structure showing two isobutyramide molecules hydrogen-bonded to each other]

CHEMICAL PROPERTIES OF AMIDES (SECTION 16.9)
☑16.52 a.

$$\text{CH}_3-\overset{\overset{\text{O}}{\|}}{\text{C}}-\text{NH}-\text{CH}_2\text{CH}_3 + \text{NaOH} \xrightarrow{\text{heat}} \text{CH}_3-\overset{\overset{\text{O}}{\|}}{\text{C}}-\text{O}^-\text{Na}^+ + \text{NH}_2-\text{CH}_2\text{CH}_3$$

b.

$$\text{CH}_3\text{CH}_2-\overset{\overset{\text{O}}{\|}}{\text{C}}-\text{NH}_2 + \text{H}_2\text{O} + \text{HCl} \xrightarrow{\text{heat}} \text{CH}_3\text{CH}_2-\overset{\overset{\text{O}}{\|}}{\text{C}}-\text{OH} + \text{NH}_4^+\text{Cl}^-$$

16.54 [Reaction of m-methylbenzamide N,N-diethyl derivative with NaOH/heat yielding m-methylbenzoate sodium salt and diethylamine]

CHEMISTRY AROUND US, CHEMISTRY AND YOUR HEALTH, AND OVER THE COUNTER

16.56   The chemical substance responsible for watery eyes and other symptoms of hay fever is histamine.

16.58   The three categories of products available to combat the symptoms of hay fever are decongestant, antihistamine, and allergy preventive.

16.60   Acetaminophen does not irritate the intestinal tract like aspirin.
Acetaminophen has similar analgesic and antipyretic effects to aspirin.
Acetaminophen does not have the anti-inflammatory effect of aspirin.
Overuse of acetaminophen has been linked to liver and kidney damage.

16.62 Four substances that normally contain caffeine are coffee, tea, chocolate, and cola beverages.

ADDITIONAL ACTIVITIES

Section 16.1 Review:
How does the classification of amines differ from the classification of alcohols?

Section 16.2 Review:
What prefixes, stems, and endings indicate that a compound is an amine?

Section 16.3 Review:
(1) Which intermolecular force is responsible for the solubility of low molecular weight amines in water?
(2) Which intermolecular force is responsible for the difference in the melting points and boiling points of primary, secondary, and tertiary amines?
(3) Classify the types of odors associated with low molecular weight amines.

Section 16.4 Review:
(1) Classify amines as weak or strong bases.
(2) Why are amines classified as bases?
(3) If a titration were performed with hydrochloric acid and an amine, how would the pH of the reaction mixture be at the equivalence point compare to pH 7?
(4) What is the Lewis structure for the inorganic polyatomic ammonium ion?
(5) What do the organic and inorganic ammonium ions have in common?
(6) Which have higher melting points: ammonium compounds or their corresponding amines?
(7) Which have higher water solubility: ammonium compounds or their corresponding amines?
(8) How can an amine salt be returned to an amine?
(9) Why is it possible to have a quaternary ammonium salt, but not a quaternary amine?
(10) Acid chlorides and acid anhydrides react with amines to form amides. Which other class of previously-studied molecules were synthesized from acid chlorides and acid anhydrides? Which class of molecules reacted with the acid chlorides or acid anhydrides?
(11) Why can polymerization occur when diacid chlorides react with diamines?
(12) Which class of biomolecules includes natural polyamides?

Section 16.5 Review:
(1) Where are neurotransmitters stored?
(2) How far do neurotransmitters travel?
(3) What are beta blockers?
(4) How do serotonin levels affect the body?

Section 16.6 Review:
(1) What are other names for epinephrine?
(2) What are some other names for amphetamines?
(3) Why are drugs from plant sources known as alkaloids?
(4) Why do alkaloids affect the central nervous system?

Section 16.7 Review:
(1) Draw the structures for *N*-methylpropanamide and methyl propanoate.
(2) Label the linkages in these structures.
(3) Circle the portion of the structure related to a carboxylic acid.
(4) Look at the structure. To which class of compounds does the uncircled portion of the structure belong?

232 Chapter 16

(5) Is a reaction between a carboxylic acid and the compound class identified in (4) the most efficient way to make an amide or an ester? Explain.

Section 16.8 Review:
(1) Rank unsubstituted, monosubstituted, and disubstituted amides of similar molecular weight in order of increasing melting point.
(2) Which intermolecular force is responsible for the solubility of some amides in water?

Section 16.9 Review:
(1) Amides can be thought of as a combination of two compound classes as shown in the Section 16.7 Review. Based on their components, suggest a reason amides are neutral.
(2) When amides react with strong acids, what are the resulting products?
(3) When amides react with strong bases, what are the resulting products?

Tying It All Together with a Laboratory Application:
A student is attempting to identify three unknown amines labeled A, B, and C. Help the student complete the following chart (1).

IUPAC Name	N,N-dimethylaniline	(Hint: Use *amino-*.)	
Common Name(s)		phenethylamine	ethylphenylamine
Structure		H_2N-CH_2 attached to CH_2-phenyl	
Molecular Formula			
Melting Point	2.45°C	-60°C	-64°C
Boiling Point	194°C	200°C	204.7°C
Solubility in water	insoluble	soluble	insoluble
Amine Classification (1°, 2°, 3°)			

(2) Which of the compounds isomers? (3) Which type of isomers? If these compounds were derived from plant sources they would be called (4) _____. If these compounds were chemical messengers of the nervous system, they would be called (5) _____.

Working in a fume hood, the student places a small amount of each of the compounds into labeled, individual test tubes, and then places the test tubes into an ice bath. The student observes that compound B crystallizes after a few minutes. The identity of unknown B is (6) _____. Unknown B freezes at a higher temperature than the other two unknowns because (7) _____.

The student decides to mix unknowns A and C with water. Compound C dissolves in water, but compound A does not. The identity of unknown A is (8) _____ and the identity of unknown C is (9) _____. The difference in water solubility is caused by (10) _____. When the student checks the pH of the solution of compound C in water, the pH value is (11) _____ 7.

Unknown C reacts with hydrochloric acid to produce the compound (12) _____. The structure for this compound is (13) _____. In order to produce an amide, unknown C would have to react with (14) _____ or _____. The pH in water of the amide produced in the reaction would be (15) _____ 7. In order to regenerate unknown C from the amide, (16) _____ hydrolysis could be performed. The other product from that reaction is the (17) _____. (18) _____ hydrolysis of the amide would produce a(n)

Amines and Amides 233

(19) _____ and an amine salt. The amine salt could be converted back to unknown C by a reaction with (20) _____.

SOLUTIONS FOR THE ADDITIONAL ACTIVITIES

Section 16.1 Review:
Amines are classified based on the number of carbon atoms bonded to the nitrogen atom. Alcohols are classified based on the number of carbon atoms bonded to the carbon atom with the attached hydroxy group.

Section 16.2 Review:
An amine can be identified by the prefix *amino-*, the stem *aniline*, and the ending *–amine*.

Section 16.3 Review:
(1) Hydrogen bonding with water is the intermolecular force responsible for the solubility of low molecular weight amines.
(2) Hydrogen bonding (or lack thereof) is the intermolecular force is responsible for the difference in the melting points and boiling points of primary, secondary, and tertiary amines. Primary and secondary amines can hydrogen bond with molecules of the same type; however, tertiary amines cannot. Consequently, primary and secondary amines have higher melting points and boiling points than tertiary amines.
(3) Low molecular weight amines are foul smelling compounds (i.e. fishy or meat-like).

Section 16.4 Review:
(1) Amines are weak bases.
(2) Amines are classified as bases because they will produce hydroxide ions in an aqueous solution because of their reversible reactions with water.
(3) If a titration were performed with hydrochloric acid and an amine, the pH of the reaction mixture would be less than 7 at the equivalence point because the titration involves a strong acid and a weak base.

(4) $\left[\begin{array}{c} H \\ | \\ H-N-H \\ | \\ H \end{array} \right]^{+}$

(5) Organic and inorganic ammonium ions both contain a nitrogen atom with four covalent bonds and a charge of +1.
(6) Ammonium compounds have higher melting points than their corresponding amines.
(7) Ammonium compounds have higher water solubility than their corresponding amines.
(8) Reacting an amine salt with a strong base will produce an amine.
(9) Nitrogen atoms typically form three bonds as seen in amines. When a nitrogen makes four bonds, it has a positive charge and forms an ammonium ion as seen in ammonium salts.
(10) Acid chlorides and acid anhydrides react with alcohols to form esters.
(11) Diacid chlorides and diamines both have two functional groups. Consequently, a polymer can form if both of the reactive ends of these molecules react.
(12) Proteins are natural polyamides.

Section 16.5 Review:
(1) Neurotransmitters are stored in small pockets in the axon near the synapse.
(2) Neurotransmitters do not travel very far. They are released from the axon, travel across the synapse, and bind to receptors on the dendrites of the next neuron.

(3) Beta blockers are drugs that reduce the stimulant action of epinephrine and norepinephrine on cells.
(4) Serotonin levels influence sleeping, the regulation of body temperature, and sensory perception.

Section 16.6 Review:
(1) Epinephrine is also known as adrenaline and the fight-or-flight hormone.
(2) Some amphetamines are also known as Benzedrine®, Methedrine®, speed, STP, and mescaline.
(3) Alkaloids are nitrogen containing compounds derived from plants. They are weak bases. The word alkaloid is derived from alkali, which means basic, and the *–oid* ending, which means like or similar to. These alkaloids are similar to bases.
(4) Alkaloids have similar chemical structures to neurotransmitters. These compounds also affect the central nervous system.

Section 16.7 Review:

(1), (2), (3)
(4) The uncircled portion of the amide belongs to the amine class. The uncircled portion of the ester belongs to the alcohol class.
(5) The reaction between a carboxylic acid and an amine results in an ammonium carboxylate salt. The reaction between a carboxylic acid and an alcohol requires a catalyst and heat. The most efficient way to make an amide is to react an amine and an acid chloride or an amine and an acid anhydride. The most efficient way to make an ester is to react an alcohol and an acid chloride or an alcohol and an acid anhydride.

Section 16.8 Review:
(1) melting point: disubstituted < monosubstituted < unsubstituted
(2) Hydrogen bonding is the intermolecular force responsible for the solubility of some amides in water.

Section 16.9 Review:
(1) Carboxylic acids are weak acids and amines are weak bases. Consequently, an amide which has a carboxylic acid portion and an amine portion are neutral. It is important to note that amide formation does not occur when carboxylic acids react with amines. This is not a neutralization reaction.
(2) When amides react with strong acids, the products are carboxylic acids and amine (ammonium) salts.
(3) When amides react with strong bases, the products are carboxylate salts and amines.

Tying It All Together with a Laboratory Application:

IUPAC Name	*N,N-dimethylaniline*	1-amino-2-phenylethane	*N-ethylaniline*
Common Name(s)	**dimethylphenylamine**	*phenethylamine*	*ethylphenylamine*
Structure	(benzene ring)–N(CH$_3$)–CH$_3$	(benzene ring)–CH$_2$–CH$_2$–NH$_2$	(benzene ring)–NH–CH$_2$–CH$_3$
Molecular Formula	C$_8$H$_{11}$N	C$_8$H$_{11}$N	C$_8$H$_{11}$N
Amine Classification	3°	1°	2°

(2) All of the compounds are isomers.
(3) They are structural isomers.
(4) alkaloids
(5) neurotransmitters
(6) N,N-dimethylaniline
(7) it is a tertiary amine; consequently, it has weaker intermolecular forces than the other compounds.
(8) ethylphenylamine
(9) phenethylamine
(10) the greater hydrogen bonding that occurs with the primary amine with water.
(11) greater than
(12) phenethylammonium chloride

(13)
(14) an acid chloride or an acid anhydride
(15) approximately
(16) basic
(17) salt of a carboxylic acid
(18) Acidic
(19) carboxylic acid
(20) a strong base (like NaOH)

SELF-TEST QUESTIONS
Multiple Choice

1. The amine, $CH_3-CH(CH_3)-NH_2$, is classified as
   a. primary
   b. secondary
   c. tertiary
   d. quaternary

2. Which of the following is a secondary amine?
   a. $CH_3-CH_2-CH(NH_2)-CH_3$
   b. $CH_3-CH_2-NH_2$
   c. $CH_3-N(CH_3)-CH_2-CH_3$
   d. pyrrolidine (cyclic NH)

3. A common name for CH₃CH₂NH—CH₂CH₃ is
   a. ethylaminoethane
   b. diethylamine
   c. aminoethylethane
   d. diethylammonia

4. What is the IUPAC name for $CH_3-CH(NH_2)-CH_2-CH(CH_3)-CH_3$?
   a. 4-amino-2-methylpentane
   b. 2-amino-4-methylpentane
   c. 4-amino-2-isopropylpropane
   d. 2-amino-1-isopropylpropane

5. The reaction of CH$_3$—CH(NH$_2$)—CH$_3$ with HCl produces
   a. CH$_3$—CH(NH$_3^+$Cl$^-$)—CH$_3$
   b. CH$_3$—CH(NH—C(=O)—CH$_3$)—CH$_3$ (shown as CH$_3$—CH—CH$_3$ with NH—C(=O)—CH$_3$ substituent)
   c. CH$_3$—CH(HN—Cl)—CH$_3$
   d. CH$_3$—CH(H$_2$N—Cl)—CH$_3$ — CH$_3$—CH—CH$_3$

6. The reaction of a primary amine with a carboxylic acid chloride produces a(n)
   a. secondary amine   b. tertiary amine   c. amide   d. carboxylate salt

7. What is the IUPAC name for CH$_3$—CH$_2$—C(=O)—NH$_2$?
   a. 1-aminopropanamide
   b. 1-aminobutanamide
   c. propanamide
   d. butanamide

8. What is the IUPAC name for CH$_3$—CH$_2$—CH$_2$—C(=O)—NH—CH$_3$?
   a. pentanamide
   b. 1-methylbutanmide
   c. 1-methylpentanamide
   d. N-methylbutanamide

9. One of the products produced when CH$_3$CH$_2$CH$_2$—C(=O)—NHCH$_3$ is treated with NaOH is
   a. CH$_3$—NH$_2$
   b. CH$_3$—CH$_2$—CH$_2$—NH$_2$
   c. CH$_3$—NH$_3^+$Cl$^-$
   d. CH$_3$CH$_2$—C(=O)—NH$_2$

10. One of the products produced when CH$_3$—C(=O)—NH—CH$_2$—CH$_3$ is treated with HCl and H$_2$O is
    a. CH$_3$—C(=O)—O$^-$
    b. CH$_3$—CH$_2$—NH$_3^+$Cl$^-$
    c. CH$_3$—C(=O)—NH$_2$
    d. CH$_3$—CH$_2$—NH$_2$

Matching
For each description on the left, select an amide from the list on the right.
11. a tranquilizer
12. an aspirin substitute
13. an antibiotic

a. penicillin
b. benzamide
c. acetaminophen
d. valium

For each description on the left, select the correct alkaloid.
14. present in coffee
15. present in tobacco
16. used as a cough suppressant
17. used to dilate the pupil of the eye

a. atropine
b. nicotine
c. codeine
d. caffeine

True-False
18. Triethylamine is a tertiary amine.
19. The structure of 1-methylaniline is [benzene ring with NH—CH₃ substituent]
20. Tertiary amines have higher boiling points than primary and secondary amines.
21. Low molecular weight amines have a characteristic pleasant odor.
22. Both primary and tertiary amines can hydrogen bond with water molecules.
23. Amines under basic conditions exist as amine salts.
24. Disubstituted amides have lower boiling points than unsubstituted amides.
25. Amides with fewer than six carbons are water soluble.
26. Amide molecules are neither basic nor acidic.
27. Serotonin is an important neurotransmitter.
28. Amphetamine is a powerful nervous system stimulant.
29. Amine salts are frequently used in medication.
30. Alkaloids are acidic.

Chapter 17: Carbohydrates

CHAPTER OUTLINE

17.1 Classes of Carbohydrates
17.2 The Stereochemistry of Carbohydrates
17.3 Fischer Projections
17.4 Monosaccharides
17.5 Properties of Monosaccharides
17.6 Important Monosaccharides
17.7 Disaccharides
17.8 Polysaccharides

LEARNING OBJECTIVES/ASSESSMENT

When you have completed your study of this chapter, you should be able to:
1. Describe the four major functions of carbohydrates in living organisms. (Section 17.1; Exercise 17.2)
2. Classify carbohydrates as monosaccharides, disaccharides, or polysaccharides. (Section 17.1; Exercise 17.4)
3. Identify molecules possessing chiral carbon atoms. (Section 17.2; Exercise 17.8)
4. Use Fischer projections to represent D and L compounds. (Section 17.3; Exercise 17.12)
5. Classify monosaccharides as aldoses or ketoses, and classify them according to the number of carbon atoms they contain. (Section 17.4; Exercise 17.22)
6. Write reactions for monosaccharide oxidation and glycoside formation. (Section 17.5; Exercise 17.34)
7. Describe uses for important monosaccharides. (Section 17.6; Exercise 17.42)
8. Draw the structures and list sources and uses for important disaccharides. (Section 17.7; Exercise 17.44)
9. Write reactions for the hydrolysis of disaccharides. (Section 17.7; Exercise 17.46)
10. Describe the structures and list sources and uses for important polysaccharides. (Section 17.8; Exercise 17.54)

SOLUTIONS FOR THE END OF CHAPTER EXERCISES

CLASSES OF CARBOHYDRATES (SECTION 17.1)

☑17.2
a. cellulose – structural material in plants
b. sucrose, table sugar – an energy source in our diet
c. glycogen – form of stored energy in animals
d. starch – form of stored energy in plants

☑17.4
a. table sugar – carbohydrate, disaccharide
b. – carbohydrate, monosaccharide
c. starch – carbohydrate, polysaccharide
d. fructose – carbohydrate, monosaccharide
e. cellulose – carbohydrate, polysaccharide
f. – carbohydrate, disaccharide
g. glycogen – carbohydrate, polysaccharide
h. amylose – carbohydrate, polysaccharide

THE STEREOCHEMISTRY OF CARBOHYDRATES (SECTION 17.2)

17.6   Carbon atom 1 is not chiral because it is only bonded to three other atoms.
Carbon atom 3 is not chiral because it is bonded to two hydrogen atoms.
Chiral carbon atoms are bonded to four unique groups.

240 Chapter 17

☑17.8 a. CH₃CH₂—CH—CH₂CH₃ with OH below — no chiral carbon atoms; no enantiomers

c. cyclopentane with C attached bearing OH (chiral), CH₂CH₃, and CH₃ — 1 chiral carbon atom; 2 stereoisomers; 1 enantiomer

b. CH₃CH₂—CH(OH)—C(=O)—CH₃ (chiral at CH) — 1 chiral carbon atom; 2 stereoisomers; 1 enantiomer

FISCHER PROJECTIONS (SECTION 17.3)

17.10
Fischer projection: CHO on top, H—C—OH, CH₂CH₃ on bottom.
The intersection of the lines represents a chiral carbon atom. The H and the OH groups stick out in front of the plane of the paper and the CHO and CH₂CH₃ groups stick out behind the plane of the paper.

☑17.12 a. D-form b. L-form

(Fischer projections shown for aldohexose with CHO top, CH₂OH bottom; and ketohexose with CH₂OH top, C=O, CH₂OH bottom — D and L forms)

17.14 a.

COOH / H—C—NH₂ / CH₂ / OH   (D-form)
COOH / H₂N—C—H / CH₂ / OH   (L-form)

b.

COOH / H—C—NH₂ / CH(CH₃)₂   D-form
COOH / H₂N—C—H / CH(CH₃)₂   L-form

17.16 a. CH₂(OH)—CH(OH)—C(=O)—CH₂—OH
1 chiral carbon atoms; 2 stereoisomers

b. CH₂(OH)—CH(OH)—CH(OH)—CH(OH)—CHO
3 chiral carbon atoms; 8 stereoisomers

17.18 Aldopentoses contain 3 chiral carbon atoms; therefore, eight possible stereoisomers exist. Four of these stereoisomers are D form and four of these stereoisomers are L form.

17.20 Optically active molecules contain at least one chiral carbon atom and rotate the plane of polarized light.

MONOSACCHARIDES (SECTION 17.4)

☑17.22

PROPERTIES OF MONOSACCHARIDES (SECTION 17.5)

17.24 Certain carbohydrates are called sugars because they taste sweet.

17.26

The arrows indicate the sites on the molecule where water can hydrogen bond to glucose.

17.28

17.30

17.32 When a sugar fails to react with Cu^{2+}, it is a nonreducing sugar.

☑17.34

c. [chemical reaction scheme showing furanose sugar + 2 CH₃—OH with H⁺ catalyst producing methylated products + 2 H₂O]

IMPORTANT MONOSACCHARIDES (SECTION 17.6)

17.36   Ribose and deoxyribose are monosaccharides used in the synthesis of nucleic acids.

17.38   Dextrose and blood sugar are two other names for glucose.

17.40   Glucose and galactose have similar structures. The only difference is the orientation of the hydroxy group attached to carbon 4.

☑17.42 Fructose can be a low-calorie sweetener because it is sweeter, gram for gram, than the other common sugars; consequently, less fructose is needed to obtain the same degree of sweetness.

DISACCHARIDES (SECTION 17.7)

☑17.44 a. The most common household sugar   sucrose
        b. Converted to lactic acid when milk sours   lactose
        c. An ingredient of synthetic human milk   lactose
        d. Formed in germinating grain   maltose
        e. Hydrolyzes when cooked with acidic foods to give invert sugar   sucrose
        f. Found in high concentrations in sugar cane   sucrose

☑17.46 a. sucrose + water $\xrightarrow{H^+}$ glucose + fructose   c. lactose + water $\xrightarrow{H^+}$ glucose + galactose
        b. maltose + water $\xrightarrow{H^+}$ glucose + glucose   d. cellobiose + water $\xrightarrow{H^+}$ glucose + glucose

17.48   Adding Benedict's solution to a sample of the sweet-tasting water solution would provide a qualitative test for the presence of sucrose or honey. Honey contains invert sugar that will react with Benedict's solution. Sucrose will not react with Benedict's solution.

17.50   a. [structure of lactose disaccharide]

        Lactose is a reducing sugar because it contains a hemiacetal group.

        b. [structure of sucrose disaccharide]

        Sucrose is not a reducing sugar because it does not contain a hemiacetal or a hemiketal group.

17.52   a. When melibiose is hydrolyzed, glucose and galactose are the monosaccharides produced.
        b. Melibiose is a reducing sugar because the glucose ring contains a hemiacetal group.
        c. The glycosidic linkage in melibiose is a α (1→6) linkage.

POLYSACCHARIDES (SECTION 17.8)

☑17.54
    a. The unbranched polysaccharide in starch   amylose
    b. A polysaccharide widely used as a textile fiber   cellulose
    c. The most abundant polysaccharide in starch   amylopectin
    d. The primary constituent of paper   cellulose
    e. A storage form of carbohydrates in animals   glycogen

CHEMISTRY AROUND US, CHEMISTRY AND YOUR HEALTH, AND OVER THE COUNTER

17.56   Three chemical substances used as artificial sweeteners in the United States are saccharine, aspartame, and acesulfame-K.

17.58   Fat is not used as a source of energy until the level of physical activity reaches at least 70% of a person's aerobic exercise capacity. Regular exercise can create a caloric deficit that stimulates the body to use fat as a source of energy.

17.60   Insoluble fiber is indigestible and passes through the body without providing many calories; it causes people to feel full and therefore eat less, thereby helping them avoid weight gain. Insoluble fiber may also contribute to weight control by inhibiting the absorption of some calorie-dense dietary fats from the digestive system.

ADDITIONAL ACTIVITIES

Section 17.1 Review:
(1) Carbohydrates are polyfunctional compounds. Which key functional groups do they contain?
(2) Complete the following table.

Prefix	Meaning	Classes of Carbohydrates	Definition
mono-		monosaccharide	
di-		disaccharide	
oligo-		oligosaccharide	
poly-		polysaccharide	

Section 17.2 Review:
(1) What are the major types of isomers?
(2) Which class of isomers from (1) includes enantiomers?
(3) Which other subclass of isomers belongs to the same class as enantiomers?
(4) Are all mirror images stereoisomers? Explain.
(5) What is the requirement for a chiral carbon atom?
(6) Are all stereoisomers enantiomers? Explain.

Section 17.3 Review:
(1) Draw the Fischer Projection of glyceraldehyde or look at the drawing on page 538. Imagine that this is a stick figure of a person facing you.
(2) What is the "head" of the glyceraldehyde?
(3) What is the "body" of the glyceraldehyde?
(4) What are the "hands" of the glyceraldehyde?
(5) What is the "feet" of the glyceraldehyde?
(6) What are the relative positions of the "head" and the "body"?
(7) What are the relative positions of the "body" and the "hands"?
(8) What are the relative positions of the "body" and the "feet"?
(9) What are the relative positions of the "head" and the "feet"?
(10) What are the relative positions of the "head" and the "hands"?

(11) What are the relative positions of the "hands" and the "feet"?
(12) Could the glyceraldehyde be described as "doing a back bend"? Explain.
(13) Could the glyceraldehyde be described as "giving a hug"? Explain.
(14) What is the difference between D-glyceraldehyde and L-glyceraldehyde?
(15) Does D-glyceraldehyde rotate light to the right?
(16) Which isomers of monosaccharides are used by humans?
(17) Which isomers of amino acids are used by humans?

Section 17.4 Review:
(1) How many chiral carbon atoms are found in an L-ketohexose?
(2) How many stereoisomers are classified as L-ketohexoses?
(3) Are L-ketohexoses abundant in nature? Explain.

Section 17.5 Review:
(1) How do the many hydroxy groups on monosaccharides influence the state of matter of monosaccharides at room temperature?
(2) When monosaccharides are in cyclic form, do they contain aldehyde or ketone groups? Explain.
(3) How might the expression, "Birds fly. Fish swim.", help you to remember the difference between α- and β-anomers?
(4) What is the difference between a Fischer projection and a Haworth structure?
(5) What are the rules for drawing Haworth structures?
(6) Are all reducing sugars monosaccharides? Explain.
(7) What is a glycoside?

Section 17.6 Review:
(1) Do the cyclic monosaccharides ribose and deoxyribose contain aldehyde, ketone, hemiacetal, hemiketal, acetal, or ketal key functional groups?
(2) Which internal organ converts other sugars into glucose?
(3) Where is lactose synthesized in the body?
(4) What is the major difference between fructose and the other monosaccharides in this section?

Section 17.7 Review:
(1) In glycosidic linkages of disaccharides, which ring determines whether the linkage is an α or β?
(2) In glycosidic linkages of disaccharides, which carbon atom is identified first?
(3) In glycosidic linkages of disaccharides, which carbon atom is identified second?
(4) Are all disaccharides reducing sugars? Explain.

Section 17.8 Review:
(1) Polysaccharides are condensation polymers. What is the small molecule produced in addition to the polysaccharide during polymerization?
(2) When polysaccharides are mixed with water and a colloidal dispersion forms, what is the dispersing medium?
(3) When polysaccharides are mixed with water and a colloidal dispersion forms, what is the dispersed phase?
(4) Are polysaccharide-water colloidal dispersions transparent? Explain.
(5) Which disaccharide has the same type of glycosidic linkage as amylose?
(6) What is the major difference between amylose and the disaccharide from (5)?
(7) When peeled potatoes are boiled, the water become cloudy. How could you check to see if the "potato water" contains amylose?
(8) Will "potato water" contain glycogen? Explain.

(9) How does the molecular formula of an amylose molecule with 1500 monosaccharide units compare to the molecular formula of a cellulose molecule with 1500 monosaccharide units?

(10) How does the structural formula for an amylose molecule with 1500 monosaccharide units compare to the structural formula of a cellulose molecule with 1500 monosaccharide units?

Tying It All Together with a Laboratory Application:

While cooking sweet-and-sour sauce at home, a chemistry student realizes the ingredients are similar to the compounds studied in class. Pineapple juice, lemon juice, granulated sugar, and corn starch all contain (1) _____. The pineapple and lemon juices as well as the red wine vinegar contain (2) _____ in the form of acetic acid and citric acid.

The recipe calls for mixing the pineapple juice and water in a sauce pan over low heat and slowly adding the table sugar while stirring. The solubility of sugar (3) _____ when heat is applied and the (4) _____ of dissolving increases as a result of the stirring. After the table sugar is dissolved, the vinegar and lemon juice are added to the sauce pan. When table sugar reacts with water in an (5) _____ environment, it undergoes (6) _____ to produce (7) _____ and _____. The resulting mixture is (8) _____ than sugar-water.

The corn starch is mixed in enough cold water to make a slurry. The student remembers that (9) _____ and _____ are the molecules contained in starch. They are both (10) _____ that contain many hydroxy groups. No matter how much water is added the mixture will remain cloudy because starch and water form a(n) (11) _____. Starch is the (12) _____ and water is the (13) _____. The corn starch slurry is then added to the sauce pan and stirred constantly. The corn starch slurry was added to the mixture to (14) _____.

The student decides to take the ingredients and final product to chemistry class and obtain permission to test the materials for the presence of reducing sugars. The student selects (15) _____ (Tollen's or Benedict's) reagent and makes the following predictions:

The reagent has a (16) _____ color. The product of a negative test will have a (17) _____ color. The product of a positive test will be (18) _____.
Pineapple juice should test positive for reducing sugars because it contains (19) _____, fruit sugar.
Lemon juice will be similar to pineapple juice.
Table sugar and water should test (20) _____ for reducing sugars because it is a(n) (21) _____ with locked (22) _____ carbon atoms and no acid catalyst is present.
After heating, table sugar, acetic acid, and water should test (23) _____ for reducing sugars because (24) _____ will occur.
Corn starch should test (25) _____ for reducing sugars because it contains (26) _____ with (27) _____ linkages.
The red wine vinegar should test (28) _____ for reducing sugars because the sugar in the wine was converted to ethanol which was then (29) _____ to form acetic acid.
The sweet-and-sour sauce should test (30) _____ for reducing sugars.

The student also asks for permission to use a brown iodine test solution. Corn starch should be the only ingredient tested that has a (31) _____ test with iodine. The corn starch-iodine mixture should have a (32) _____ color because (33) _____. The student does not know whether the iodine will have a (34) _____ test with the sweet-and-sour sauce. Although the sauce contains starch, it also contains acids that can (35) _____ starch and form (36) _____, _____, and _____.

246 Chapter 17

SOLUTIONS FOR THE ADDITIONAL ACTIVITIES

Section 17.1 Review:

(1) Carbohydrates are polyfunctional compounds that contain aldehyde or ketone key functional groups (or can form aldehyde or ketone key functional groups) as well as several hydroxy groups.

(2)

Prefix	Meaning	Classes of Carbohydrates	Definition
mono-	1	monosaccharide	a simple carbohydrate
di-	2	disaccharide	a carbohydrate formed by combining two monosaccharides
oligo-	small	oligosaccharide	a carbohydrate formed by combining 3-10 monosaccharides
poly-	many	polysaccharide	a carbohydrate formed by combining many monosaccharides

Section 17.2 Review:

(1) structural isomers, stereoisomers
(2) stereoisomers
(3) geometric isomers (*cis-trans*)
(4) Only nonsuperimposable mirror images are stereoisomers. Superimposable mirror images are the same molecule.
(5) A chiral carbon atom must be bonded to four unique groups.
(6) No, not all stereoisomers are enantiomers. If a molecule contains 2 chiral carbon atoms, it is one of four stereoisomers of that structure. It has a mirror image (enantiomer) and two other stereoisomers that are enantiomers of each other, but not of the original molecule.

Section 17.3 Review:

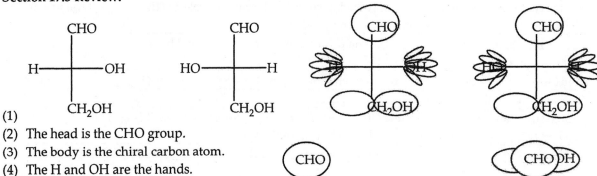

Side view - back bend
Question 12

Top view - hug
Question 13

(2) The head is the CHO group.
(3) The body is the chiral carbon atom.
(4) The H and OH are the hands.
(5) The CH₂OH is the feet.
(6) The head is behind the body.
(7) The body is behind the hands.
(8) The body is in front of the feet.
(9) The head is above the feet, but both are in the same plane.
(10) The head is behind the hands.
(11) The hands are in front of the feet.
(12) Yes, the glyceraldehyde could be described as "doing a back bend" because the head and feet are behind the body.
(13) Yes, the glyceraldehyde could be described as "giving a hug" because the hands are outstretched in front of the body.
(14) The hydroxy group is on the right side of the chiral carbon in D-glyceraldehyde and the hydroxy group is on the left side of the chiral carbon in L-glyceraldehyde. (Note: This is from the perspective of the observer, not the glyceraldehyde stick person.)

(15) D-glyceraldehyde may rotate light to the right or to the left because it is optically active. The name and structure alone are not enough to determine which direction it will rotate the light.
(16) Humans use D-isomers of monosaccharides.
(17) Humans use L- isomers of amino acids.

Section 17.4 Review:
(1) An L-ketohexose contains 3 chiral carbon atoms.
(2) Four stereoisomers are classified as L-ketohexoses.
(3) L-ketohexoses are not abundant in nature because naturally occurring monosaccharides are primarily D-form aldoses.

Section 17.5 Review:
(1) The hydroxy groups on monosaccharides allow them to hydrogen bond to other monosaccharide molecules. These strong intermolecular forces between monosaccharide molecules cause them to be solids at room temperature.
(2) When monsaccharides are in cyclic form, they do not contain aldehyde or ketone groups, but rather hemiacetal or hemiketal groups.
(3) The β-anomers have an –OH group on the anomeric carbon that points up. Birds fly up in the sky like the –OH group on the anomeric carbon of a β-anomer. The α-anomers have an –OH group on the anomeric carbon that points down. The α symbol looks like a fish. Fish swim down in the water like the –OH group on the anomeric carbon of a α-anomer.
(4) A Fischer projection is used to draw open chain structure of molecules, including carbohydrates. A Haworth structure is used to draw a closed ring structure of carbohydrates.
(5) The rules for drawing Haworth structures are:
    a. the ring is drawn with the oxygen in the back, in the upper right;
    b. the anomeric carbon is on the right side of the ring;
    c. the terminal –CH_2OH group is always shown above the ring for D-monosaccharides.
(6) No, not all reducing sugars are monosaccharides. Reducing sugars must contain an aldehyde with an adjacent hydroxy group, ketone with an adjacent carbonyl group or adjacent hydroxy group, cyclic hemiacetal with additional hydroxy groups, or cyclic hemiketal group with additional hydroxy groups. Monosaccharides fit these criteria, but so do other molecules including some disaccharides.
(7) A glycoside is a molecule that contains a cyclic monosaccharide bonded at the anomeric carbon atom through an oxygen atom to another carbon containing group.

Section 17.6 Review:
(1) Ribose and deoxyribose contain hemiacetal key functional groups.
(2) The liver converts other sugars into glucose.
(3) Lactose is synthesized in the mammary glands.
(4) Fructose is the only ketose in this section. All of the other important monosaccharides are aldoses.

Section 17.7 Review:
(1) In glycosidic linkages of disaccharides, the ring containing an anomeric carbon in the glycosidic linkage determines whether the linkage is classified as α or β.
(2) In glycosidic linkages of disaccharides, the anomeric carbon atom involved in the glycosidic linkage is identified first. If two anomeric carbon atoms are involved in the glycosidic linkage, then the anomeric carbon atom with the smaller number is identified first.
(3) In glycosidic linkages of disaccharides, the carbon atom that is not anomeric is identified second. If two anomeric carbon atoms are involved in the glycosidic linkage, then the anomeric carbon atom with the larger number is identified second.

(4) No, not all disaccharides reducing sugars because a reducing sugar must be able to form an open-chain. Disaccharides that have two anomeric carbon atoms in the glycosidic linkage cannot form an open-chain.

Section 17.8 Review:
(1) Water is the small molecule formed during the condensation reaction to form a polysaccharide.
(2) When polysaccharides are mixed with water and a colloidal dispersion forms, water is the dispersing medium.
(3) When polysaccharides are mixed with water and a colloidal dispersion forms, the polysaccharide is the dispersed phase.
(4) Polysaccharide-water colloidal dispersions are not transparent because they are colloids. Colloids appear cloudy because the dispersed phase scatters light.
(5) Maltose has the same type of glycosidic linkage as amylose.
(6) Amylose contains 1000-2000 glucose units. Maltose contains 2 glucose units.
(7) Add iodine test solution to a sample of the "potato water." If the brown iodine turns dark bluish-purple, amylose is present.
(8) No, "potato water" will not contain glycogen because glycogen is the storage carbohydrate for animals, not plants.
(9) The molecular formula of an amylose molecule with 1500 monosaccharide units is the same as the molecular formula of a cellulose molecule with 1500 monosaccharide units.
(10) An amylose molecule with 1500 monosaccharide units is joined by α(1→4) linkages that create a helical structure. A cellulose molecule with 1500 monosaccharide units is joined by β(1→4) linkages that create a extended straight chain.

Tying It All Together with a Laboratory Application:
(1) carbohydrates
(2) carboxylic acids
(3) increases
(4) rate
(5) acidic
(6) hydrolysis
(7) D-glucose, D-fructose
(8) sweeter
(9) amylose, amylopectin
(10) polymers
(11) colloidal dispersion
(12) dispersed phase
(13) dispersing medium
(14) thicken it
(15) Benedict's
(16) blue
(17) blue
(18) a brick red solid
(19) fructose
(20) negative
(21) disaccharide
(22) anomeric
(23) positive
(24) hydrolysis
(25) negative
(26) polysaccharides
(27) glycosidic
(28) negative
(29) oxidized
(30) positive
(31) positive
(32) deep purple
(33) starch contains amylose
(34) positive
(35) hydrolyze
(36) dextrins, maltose, glucose

SELF-TEST QUESTIONS
Multiple Choice
1. Which of the following is a monosaccharide?
   a. amylose
   b. ribose
   c. cellulose
   d. lactose

2. Which of the following is a polysaccharide?
   a. amylose
   b. lactose
   c. maltose
   d. galactose

3. How many chiral carbon atoms are in $CH_3-CH(CH_3)-CH_2-CH(Br)-CH_2-CH_2(OH)$?
   a. 0
   b. 1
   c. 2
   d. 3

4. How many chiral carbon atoms are in this molecule?

   a. 0   b. 1   c. 2   d. 3

5. How many stereoisomers are possible for CH₂—CH—CH—CHO with OH OH OH?
   a. 0   b. 2   c. 4   d. 8

6. Glucose is a(n)
   a. ketopentose   b. ketohexose   c. aldopentose   d. aldohexose

7. Fructose is a(n)
   a. ketopentose   b. ketohexose   c. aldopentose   d. aldohexose

8. A positive Benedict's test is indicated by the formation of
   a. Ag   b. CuOH   c. Cu$_2$O   d. Cu^{2+}

Matching

For each monosaccharide described on the left, select the best response from the right.
9. given intravenously
10. present with glucose in invert sugar
11. combines with glucose to form lactose
12. found in genetic material

a. fructose
b. galactose
c. glucose
d. ribose

For each disaccharide described on the left, select the best match from the responses on the right.
13. used as household sugar
14. found in milk
15. formed in germinating grain

a. glycogen
b. sucrose
c. maltose
d. lactose

For each disaccharide on the left, select the correct hydrolysis products from the right.
16. sucrose
17. maltose
18. lactose

a. glucose and galactose
b. glucose and fructose
c. only glucose
d. hydrolysis does not occur

Select the correct polysaccharide for each description on the left.
19. a storage form of carbohydrates in animals
20. most abundant polysaccharide in starch
21. primary constituent of paper

a. amylopectin
b. amylose
c. glycogen
d. cellulose

True-False

22. A D enantiomer is the mirror image of an L enantiomer.
23. In a D carbohydrate, the hydroxy group on the chiral carbon farthest from the carbonyl group points to the left.
24. The L carbohydrates are preferred by the human body.
25. Sugars that contain a hemiacetal group are reducing sugars.
26. In β-galactose, the hydroxy group at carbon 1 points up.
27. Maltose contains a glycosidic linkage.
28. The glucose ring of lactose can exist in an open-chain form.
29. Sucrose contains a hemiacetal group.
30. Linen is prepared from cellulose.

Chapter 18: Lipids

CHAPTER OUTLINE
18.1 Classification of Lipids
18.2 Fatty Acids
18.3 The Structure of Fats and Oils
18.4 Chemical Properties of Fats and Oils
18.5 Waxes
18.6 Phosphoglycerides
18.7 Sphingolipids
18.8 Biological Membranes
18.9 Steroids
18.10 Steroid Hormones
18.11 Prostaglandins

LEARNING OBJECTIVES/ASSESSMENT
When you have completed your study of this chapter, you should be able to:
1. Classify lipids as saponifiable or nonsaponifiable and list five major functions of lipids. (Section 18.1; Exercises 18.2 and 18.4)
2. Describe four general characteristics of fatty acids. (Section 18.2; Exercise 18.6)
3. Draw structural formulas of triglycerides given the formulas of the component parts. (Section 18.3; Exercise 18.14)
4. Describe the structural similarities and differences of fats and oils. (Section 18.3; Exercise 18.12)
5. Write key reactions for fats and oils. (Section 18.4; Exercise 18.18)
6. Compare the structures of fats and waxes. (Section 18.5; Exercise 18.24)
7. Draw structural formulas and describe uses for phosphoglycerides. (Section 18.6; Exercises 18.28 and 18.30)
8. Draw structural formulas and describe uses for sphingolipids. (Section 18.7; Exercise 18.34)
9. Describe the major features of cell membrane structure. (Section 18.8; Exercise 18.42)
10. Identify the structural characteristic typical of steroids and list important groups of steroids in the body. (Section 18.9; Exercises 18.44 and 18.46)
11. Name the major categories of steroid hormones. (Section 18.10; Exercise 18.50)
12. Describe the biological importance and therapeutic uses of the prostaglandins. (Section 18.11; Exercise 18.58)

SOLUTIONS FOR THE END OF CHAPTER EXERCISES
INTRODUCTION AND CLASSIFICATION OF LIPIDS (SECTION 18.1)
☑18.2 Lipids are a storage form of energy, a structural component, and some hormones in the human body.

☑18.4
a. A steroid — nonsaponifiable
b. A wax — saponifiable
c. A triglyceride — saponifiable
d. A phosphoglyceride — saponifiable
e. A glycolipid — saponifiable
f. A prostaglandin — nonsaponifiable

FATTY ACIDS (SECTION 18.2)
☑18.6 Four structural characteristics exhibited by most fatty acids are:
(1) they are usually straight-chain carboxylic acids;
(2) the sizes of most common fatty acids range from 10 to 20 carbon atoms;
(3) fatty acids usually have an even number of carbon atoms;
(4) fatty acids can be saturated or unsaturated.

18.8 Two essential fatty acids are linolenic acid and linoleic acid. The human body cannot synthesize these acids; therefore, they must be obtained through diet.

252 Chapter 18

18.10 Unsaturated fatty acids have lower melting points than saturated fatty acids because they contain double bonds which form kinks and prevent the unsaturated fatty acids from packing as well as the saturated fatty acids. The greater separation between the molecules causes the intermolecular forces in the unsaturated fatty acids to be weaker than the intermolecular forces in the saturated fatty acids. Weaker intermolecular forces result in lower melting points.

THE STRUCTURE OF FATS AND OILS (SECTION 18.3)

☑18.12 Fats and oils are both triglycerides that contain a glycerol backbone and three fatty acid chains. Fats contain more saturated than unsaturated fatty acids and oils contain more unsaturated than saturated fatty acids.

☑18.14

$$\begin{array}{l} CH_2-O-\overset{O}{\underset{\Vert}{C}}-(CH_2)_{14}CH_3 \\ CH-O-\overset{O}{\underset{\Vert}{C}}-(CH_2)_7CH=CH(CH_2)_7CH_3 \\ CH_2-O-\overset{O}{\underset{\Vert}{C}}-(CH_2)_{16}CH_3 \end{array}$$

(arrows point to the three ester C=O carbonyl groups)

18.16 Triglyceride B should have the lower melting point because it contains a higher percentage of unsaturated fatty acids than Triglyceride A.

CHEMICAL PROPERTIES OF FATS AND OILS (SECTION 18.4)

☑18.18
   a. Acid hydrolysis of a fat   glycerol, saturated fatty acids
   b. Acid hydrolysis of an oil   glycerol, unsaturated fatty acids
   c. Saponification of a fat   glycerol, salts of saturated fatty acids
   d. Saponification of an oil   glycerol, salts of unsaturated fatty acids

18.20 Hydrogenation of vegetable oils is of great commercial importance because it increases the melting point of the fat or oil and allows the product to be used as margarine and cooking shortenings.

18.22 a. Glycerol from beef fat:
     Acid-catalyzed hydrolysis

$$\begin{array}{l} CH_2-O-\overset{O}{\underset{\Vert}{C}}-(CH_2)_{14}CH_3 \\ CH-O-\overset{O}{\underset{\Vert}{C}}-(CH_2)_7CH=CH(CH_2)_7CH_3 \\ CH_2-O-\overset{O}{\underset{\Vert}{C}}-(CH_2)_{16}CH_3 \end{array} + 3\,H_2O \xrightarrow{H^+}$$

$$\begin{array}{l} \boxed{CH_2-OH} \\ \boxed{CH-OH} \\ \boxed{CH_2-OH} \end{array} + HO-\overset{O}{\underset{\Vert}{C}}-(CH_2)_{14}CH_3 + HO-\overset{O}{\underset{\Vert}{C}}-(CH_2)_{16}CH_3 + HO-\overset{O}{\underset{\Vert}{C}}-(CH_2)_7CH=CH(CH_2)_7CH_3$$

(glycerol is circled)

Saponification

[Reaction: Triglyceride with (CH₂)₁₄CH₃, (CH₂)₇CH=CH(CH₂)₇CH₃, and (CH₂)₁₆CH₃ chains + 3 NaOH → glycerol (circled: CH₂—OH, CH—OH, CH₂—OH) + Na⁺⁻O—CO—(CH₂)₁₄CH₃ + Na⁺⁻O—CO—(CH₂)₁₆CH₃ + Na⁺⁻O—CO—(CH₂)₇CH=CH(CH₂)₇CH₃]

b. Stearic acid from beef fat:
Acid-catalyzed hydrolysis

[Reaction: Same triglyceride + 3 H₂O, H⁺ → glycerol + HO—CO—(CH₂)₁₄CH₃ + HO—CO—(CH₂)₁₆CH₃ (circled) + HO—CO—(CH₂)₇CH=CH(CH₂)₇CH₃]

c. A margarine from corn oil:
Hydrogenation

[Reaction: Triglyceride with (CH₂)₆(CH₂CH=CH)₂(CH₂)₄CH₃, (CH₂)₆(CH₂CH=CH)₃CH₂CH₃, and (CH₂)₂(CH₂CH=CH)₄(CH₂)₄CH₃ chains + 9 H₂, Pt → Triglyceride with (CH₂)₁₆CH₃, (CH₂)₁₆CH₃, and (CH₂)₁₈CH₃ chains]

d. Soaps from lard:
Saponification

[Reaction: Triglyceride with (CH₂)₁₄CH₃, (CH₂)₇CH=CH(CH₂)₇CH₃, and (CH₂)₁₆CH₃ chains + 3 NaOH → glycerol + Na⁺⁻O—CO—(CH₂)₁₄CH₃ + Na⁺⁻O—CO—(CH₂)₁₆CH₃ + Na⁺⁻O—CO—(CH₂)₇CH=CH(CH₂)₇CH₃ (boxed)]

WAXES (SECTION 18.5)

☑18.24 Waxes and fats are esters of long-chain fatty acids; however, waxes do not contain glycerol backbones and fats do. Waxes contain one fatty acid chain, while fats contain three fatty acid chains.

18.26 Waxes are protective coatings on feathers, fur, skin, leaves, and fruits. They also occur in secretions of the sebaceous glands to keep skin soft and prevent dehydration.

PHOSPHOGLYCERIDES (SECTION 18.6)

☑18.28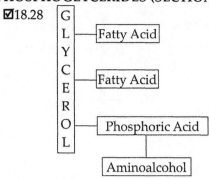

☑18.30 Lecithins are important structural components in cell membranes as well as emulsifying and micelle-forming agents.

18.32 Lecithins are phosphoglycerides that contain the aminoalcohol choline. Cephalins are phosphoglycerides that contain ethanolamine or serine.

SPHINGOLIPIDS (SECTION 18.7)

☑18.34 Sphingolipids include sphingomyelins and glycolipids. Both of these subclasses contain the sphingosine backbone and a fatty acid.

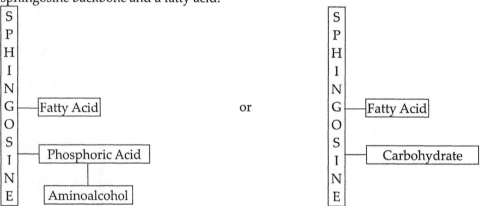

18.36 Three diseases caused by abnormal metabolism and accumulation of sphingolipids are Tay-Sachs, Gaucher's, and Niemann-Pick disease.

18.38 Another name for cerebrosides is glycolipids. These compounds are abundant in brain tissue.

BIOLOGICAL MEMBRANES (SECTION 18.8)

18.40 The three classes of lipids found in membranes are phosphoglycerides, sphingomyelin, and steroids.

☑18.42 The fluid-mosaic model contains lipids organized in a bilayer in such a way that the hydrophilic heads are pointed toward the outside of the bilayer and the hydrophobic tails are pointed toward the inside of the bilayer. Some proteins float in the lipid bilayer and other proteins extend completely through the bilayer. The lipid molecules move freely laterally within the bilayer.

Lipids 255

STEROIDS (SECTION 18.9)
☑18.44 All steroids contain a 4 fused ring system composed of 3 six membered rings and 1 five membered ring.

☑18.46 Important groups of compounds the body synthesizes from cholesterol are bile salts, male and female sex horomones, vitamin D, and the adrenocorticoid hormones.

18.48 The major component in most gallstones is cholesterol.

STEROID HORMONES (SECTION 18.10)
☑18.50 The two groups of adrenocorticoid hormones are mineralcorticoids and glucocorticoids. Mineralcorticoids regulate the concentration of ions in body fluids. An example of a mineralcorticoid is aldosterone which regulates the retention of sodium and chloride ions in urine formation. Glucocorticoids enhance carbohydrate metabolism. An example of a glucocorticoid is cortisol which helps increase glucose and glycogen concentrations in the body.

18.52 The primary male sex hormone is testosterone. The three principle female sex hormones are estradiol, estrone, and progesterone.

18.54 The estrogens are involved in egg development in the ovaries. Progesterone causes changes in the wall of the uterus to prepare it to accept a fertilized egg and maintain the resulting pregnancy.

PROSTAGLANDINS (SECTION 18.11)
18.56 The starting material for prostaglandin synthesis is arachidonic acid.

☑18.58 Therapeutic uses of prostaglandins include inducing labor, therapeutic abortion, treating asthma, inhibiting gastric sectretions, and treating peptic ulcers.

CHEMISTRY AROUND US, CHEMISTRY AND YOUR HEALTH, AND OVER THE COUNTER
18.60 Testosterone and estrogen are hormones formed from DHEA.

18.62 Some foods that may contain trans fatty acids are hard and soft margarines.

18.64 Certain fish oil supplements are not recommended because of the possibility of negative side effects from overconsumption.

18.66 Synthetic dipalmitoyl phosphatidylcholine is used to treat respiratory distress syndrome.

256 Chapter 18

ADDITIONAL ACTIVITIES

Section 18.1 Review:

Complete the following diagram by adding the subclasses of lipids (waxes, fats, oils, phosphoglycerides, sphingolipids, steroids, and prostaglandins) into the appropriate locations.

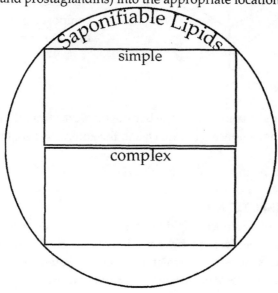

Section 18.2 Review:

(1) Draw a micelle of fatty acid anions in an oil based environment with small amounts of aqueous materials.
(2) What is the angle between the *cis*-carbon groups attached to the carbon atoms in a double bond? The following drawing may help along with the reminder that a straight line has an angle of 180° and the sum of the angles of a triangle is 180°.

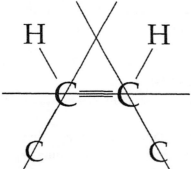

(3) What is the angle of the kink in a *cis*-unsaturated fatty acid?
(4) What is the angle between *trans*-carbon groups attached to the carbon atoms in a double bond?

(5) Would *trans*-unsaturated fatty acids have a kink? Explain.
(6) Are the two essential fatty acids saturated, monounsaturated, or polyunsaturated? Explain.

Section 18.3 Review:
(1) How many possible isomers exist for a fat that contains three unique fatty acids?
(2) How many possible isomers exist for an oil that contains two unique fatty acids?
(3) How does the percentage of saturated fatty acids affect the phase of matter of the compound?

Section 18.4 Review:
(1) What are the products for the acid hydrolysis of a fat or oil?
(2) What are the products for the basic hydrolysis of a fat or oil?
(3) What is the product for the complete hydrogenation of a polyunsaturated fatty acid?
(4) How does the hydrogenation of a polyunsaturated fatty acid change its physical properties?

Section 18.5 Review:
(1) What are the products of acid hydrolysis of a wax?
(2) What are the products of basic hydrolysis of a wax?
(3) Suggest a reason that waxes are less easily hydrolyzed than fats and oils.

Section 18.6 Review:
(1) What does the name phosphoglycerides imply about the structure of these compounds?
(2) What parts of the phosphoglyceride structure is not implied by the name of this group?
(3) What are the two categories of phosphoglycerides?

Section 18.7 Review:
(1) What does the name sphingolipid imply about the structures of these compounds?
(2) What are the structural differences between a sphingomyelin and a glycolipid?

Section 18.8 Review:
(1) What role do lipids play in cell structure?
(2) How does a lipid bilayer differ from a micelle?

Section 18.9 Review:
(1) What is the common physical property of all lipids?
(2) What are the major steroid subgroups discussed in this section?
(3) Which of the major steroid subgroups from (2) has negative health effects when the levels in the body are high?
(4) Which of the major steroid subgroups from (2) has negative health effects when the levels in the body are low?

Section 18.10 Review:
(1) What is the name of the steroid subgroup discussed in this section?
(2) What are the major classes of molecules that belong to this subgroup?

Section 18.11 Review:
(1) What is the common feature of prostaglandins?
(2) What is a potential application of prostaglandin research?

Tying It All Together with a Laboratory Application:
A student in the biochemistry laboratory was testing the properties of lipids. All lipids are (1) _____ in water because lipids are (2) _____ and water is (3) _____. All lipids are (4) _____ in cyclohexane because cyclohexane is (5) _____, (6) _____ lipids. The solubility concept of "(7) _____ dissolves like" applies even in biochemistry.

The student determined that a hydrolysis reaction under (8) _____ conditions would separate the class of lipids into two subclasses, (9) _____ and _____. The (10) _____ lipids contain two groups: (11) _____ and prostaglandins. Of these two groups only prostaglandins contain a(n) (12) _____ key functional group. This might cause a difference in their (13) _____ properties because of the reactivity of this key functional group.

The other major lipid subclass is divided into two smaller groups, (14) _____ and _____. One similarity between all of these compounds is they contain a (15) _____ chain, which can be saturated or (16) _____. A bromine test would be (17) _____ for a saturated (18) _____ chain because it (19) _____ carbon-carbon double bonds.

When these compounds undergo saponification, (20) _____, _____, and _____ will produce glycerol. These compounds are classified in (21) _____ subclass(es) because an alcohol and (22) _____ are the only components of (23) _____ lipids. (24) _____ contain a phosphate group in addition to the other components. These compounds contain an (25) _____ linkage as do (26) _____, which are (27) _____ lipids. The saponification of (28) _____ will produce sphingosine as one of the products. These compounds are joined by an (29) _____ linkage to the part of the molecule that can become a soap. These compounds are (30) _____ lipids.

SOLUTIONS FOR THE ADDITIONAL ACTIVITIES
Section 18.1 Review:

Section 18.2 Review:

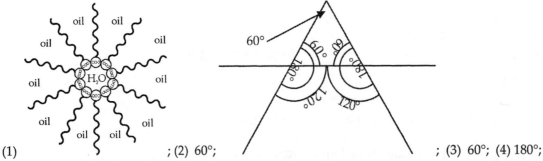

(1) _____ ; (2) 60°; ; (3) 60°; (4) 180°;

(5) The *trans*-unsaturated fatty acids do not have a kink because the angle between the carbon groups attached to the double bonded carbon atoms is 180°, which is a straight line.

(6) The two essential fatty acids are polyunsaturated because they contain more than one double bond.

Section 18.3 Review:
(1) The glycerol backbone of fats is symmetrical; therefore, only three isomers exist.
(2) The glycerol backbone of oils is symmetrical; therefore, only two isomers exist.

(3) The higher the percentage of saturated fatty acids in a compound, the more likely the compound is to be a solid. The lower the percentage of saturated fatty acids in a compound, the more likely the compound is to be a liquid.

Section 18.4 Review:
(1) The products for the acid hydrolysis of a fat or oil are glycerol and three fatty acid molecules.
(2) The products for the basic hydrolysis of a fat or oil are glycerol and three salts of fatty acids.
(3) The product for the complete hydrogenation of a polyunsaturated fatty acid is a saturated fatty acid.
(4) The hydrogenation of a polyunsaturated fatty acid will increase the melting and boiling points.

Section 18.5 Review:
(1) The products of acid hydrolysis of a wax are an alcohol and a fatty acid.
(2) The products of basic hydrolysis of a wax are an alcohol and a salt of a fatty acid.
(3) Waxes are less easily hydrolyzed than fats and oils because their ester linkages are less accessible than the ester linkages in fats and oils. Waxes have hydrophobic groups on either side of the ester linkage, while fats and oils have hydrophobic groups on only one side of the ester linkage.

Section 18.6 Review:
(1) The name phosphoglycerides implies these compounds contain a glycerol backbone and a phosphate group.
(2) The name phosphoglyceride does not imply that these compounds contain two fatty acids or that an aminoalcohol is attached to the phosphate group.
(3) Lecithins and cephalins are the two categories of phosphoglycerides.

Section 18.7 Review:
(1) The name sphingolipid implies these compounds contain a sphingosine backbone.
(2) Sphingomyelins contain a fatty acid and an aminoalcohol connected to a phosphate group in addition to the sphingosine backbone. Glycolipids contain a fatty acid and a carbohydrate in addition to the sphingosine backbone.

Section 18.8 Review:
(1) Lipids help form membranes both in and around cells.
(2) A lipid bilayer contains two sets of lipids interacting tail to tail in order to form an extended structure. A micelle contains several lipids interacting tail to tail in order to form a sphere.

Section 18.9 Review:
(1) The common physical property of all lipids is their ability to dissolve in nonpolar solvents.
(2) The major steroid subgroups discussed in this section are cholesterol and bile salts.
(3) High levels of cholesterol are associated with negative health effects.
(4) Low levels of bile salts are associate with negative health effects.

Section 18.10 Review:
(1) Steroid hormones are the steroid subgroup discussed in this section.
(2) Adrenocorticoid hormones and sex hormones belong to the steroid hormone subgroup.

Section 18.11 Review:
(1) Prostaglandins are synthesized from arachidonic acid, a 20-carbon unsaturated fatty acid.
(2) A potential application of prostaglandin research is new treatments for a variety of ailments.

Tying It All Together with a Laboratory Application:
(1) insoluble
(2) nonpolar
(3) polar
(4) soluble
(5) nonpolar
(6) like

(7) like
(8) basic
(9) saponifiable, nonsaponifiable
(10) nonsaponifiable
(11) steroids
(12) carboxylic acid
(13) chemical
(14) simple, complex
(15) fatty acid
(16) unsaturated
(17) negative
(18) fatty acid
(19) lacks (does not contain)
(20) fats, oils, phosphoglycerides
(21) different
(22) fatty acid chain
(23) simple
(24) Phosphoglycerides
(25) ester
(26) waxes
(27) simple
(28) sphingolipids
(29) amide
(30) complex

SELF-TEST QUESTIONS

Multiple Choice

1. All simple lipids are
   a. salts of fatty acids
   b. esters of fatty acids with various alcohols
   c. esters of fatty acids with glycerol
   d. esters of fatty acids with alcohol and other additional components

2. The esters of fatty acids and long chain alcohols are known as
   a. waxes   b. phospholipids   c. compound lipids   d. fats

3. Which of the following is a glycerol-containing lipid?
   a. sphingolipid   b. glycolipid   c. phosphoglyceride   d. prostaglandin

4. Which of the following is a complex lipid?
   a. steroid   b. sphingomyelin   c. prostaglandin   d. triacylglycerol

5. Which fatty acid is most likely to be found in an oil?
   a. $CH_3(CH_2)_7CH=CH(CH_2)_7COOH$
   b. $CH_3(CH_2)_{16}COOH$
   c. $CH_3(CH_2)_{14}COOH$
   d. $CH_3(CH_2)_{18}COOH$

6. Which of the following food sources would most likely be highest in saturated fatty acids?
   a. cottonseed   b. corn   c. beef   d. sunflower seeds

7. Generally, the structural difference between a fat and an oil is the
   a. alcohol
   b. chain length of fatty acid
   c. degree of fatty-acid unsaturation
   d. degree of fatty-acid chain-branching

8. In triglycerides, fatty acids are joined to glycerol by
   a. ester linkages   b. ether linkages   c. amide linkages   d. hydrogen bonds

9. Which of the following is an essential fatty acid?
   a. stearic acid   b. myristic acid   c. linoleic acid   d. palmitic acid

Matching
Match the following formulas to the correct lipid classification given as a response

a. steroid
b. phosphoglyceride
c. fat or oil
d. wax

10. [steroid structure with OH, CH₃, R groups]

11. [triglyceride structure: H₂C-O-CO-R, HC-O-CO-R', H₂C-O-CO-R'']

12. $CH_3-(CH_2)_6-CO-O-CH_2-(CH_2)_8-CH_3$

13. [phosphoglyceride structure: H₂C-O-CO-R, HC-O-CO-R', H₂C-O-P(O)(O⁻)-O-R'']

Materials can be obtained from lipids, and the lipids changed by chemical processes. Choose the correct process to accomplish each change described below.

14. obtain a high molecular weight alcohol from a plant wax
15. obtain glycerol from an oil
16. obtain soaps from an oil
17. raise the melting point of an oil

a. hydrogenation
b. acid-catalyzed hydrolysis
c. saponification
d. more than one listed process would work

Match the lipids on the right with their possible hydrolysis products, W, X, Y, and Z.

18. W = fatty acids, X = glycerol, Y = phosphoric acid, Z = choline
19. W = fatty acids, X= glycerol, no other products
20. W = fatty acids, X = sphingosine, Y = phosphoric acid, Z = choline
21. W = fatty acids, X = sphingosine, Y = carbohydrate, no other product forms

a. simple lipid
b. glycolipid
c. phospholipid
d. sphingolipid

True-False

22. Cell membranes contain about 60% lipid and 40% carbohydrate.
23. Cell membranes are thought to be relatively flexible.
24. A compound containing nine carbon atoms could not be a steroid.
25. In their physiological action, the prostaglandins resemble hormones.
26. All of the male and female sex hormones are steroids.
27. The hormone aldosterone exerts its influence at the pancreas.
28. Lipids are polar molecules.
29. Lipids are soaps.
30. Bile salts are steroids.

Chapter 19: Proteins

CHAPTER OUTLINE
19.1 The Amino Acids
19.2 Zwitterions
19.3 Reactions of Amino Acids
19.4 Important Peptides
19.5 Characteristics of Proteins
19.6 The Primary Structure of Proteins
19.7 The Secondary Structure of Proteins
19.8 The Tertiary Structure of Proteins
19.9 The Quaternary Structure of Proteins
19.10 Protein Hydrolysis and Denaturation

LEARNING OBJECTIVES/ASSESSMENT
When you have completed your study of this chapter, you should be able to:
1. Identify the characteristic parts of alpha-amino acids. (Section 19.1; Exercise 19.2)
2. Draw structural formulas to illustrate the various ionic forms assumed by amino acids. (Section 19.2; Exercise 19.12)
3. Write reactions to represent the formation of peptides and proteins. (Section 19.3; Exercise 19.16)
4. Describe uses for important peptides. (Section 19.4; Exercise 19.22)
5. Describe proteins in terms of the following characteristics: size, function, classification as fibrous or globular, and classification as simple or conjugated. (Section 19.5; Exercises 19.30 and 19.32)
6. Explain what is meant by the primary structure of proteins. (Section 19.6; Exercise 19.34)
7. Describe the role of hydrogen bonding in the secondary structure of proteins. (Section 19.7; Exercise 19.38)
8. Describe the role of side-chain interactions in the tertiary structure of proteins. (Section 19.8; Exercise 19.42)
9. Explain what is meant by the quaternary structure of proteins. (Section 19.9; Exercise 19.46)
10. Describe the conditions that can cause proteins to hydrolyze or become denatured. (Section 19.10; Exercise 19.50)

SOLUTIONS FOR THE END OF CHAPTER EXERCISES
THE AMINO ACIDS (SECTION 19.1)
☑19.2 All amino acids contain an amine and a carboxylic acid key functional group.

19.4 a. threonine

c. serine

b. aspartate

d. phenylalanine

263

264 Chapter 19

19.6

19.8 a. D-form (COO⁻, H—C—NH₃⁺, CH₂SH); L-form (COO⁻, H₃N⁺—C—H, CH₂SH)

b. D-form (COO⁻, H—C—NH₃⁺, CH₂CH₂COO⁻); L-form (COO⁻, H₃N⁺—C—H, CH₂CH₂COO⁻)

ZWITTERIONS (SECTION 19.2)

19.10 The characteristics that indicate amino acids exist as zwitterions are amino acids are white crystalline solids with relatively high melting points and high water solubilities.

☑19.12 a. H₂N—CH—COO⁻ with side chain CH₂—CH₂—C(=O)—NH₂

b. HN—CH—COO⁻ in a ring with CH₂—CH₂—CH₂ (proline)

19.14 a. H₃N⁺—CH(CH₂OH)—COO⁻ + OH⁻ → H₂N—CH(CH₂OH)—COO⁻ + H₂O

b. H₃N⁺—CH(CH₂OH)—COO⁻ + H₃O⁺ → H₃N⁺—CH(CH₂OH)—COOH + H₂O

REACTIONS OF AMINO ACIDS (SECTION 19.3)

☑19.16 H₃N⁺—CH(CH(CH₃)₂)—C(=O)—NH—CH(CH₂COO⁻)—C(=O)—NH—CH(CH₂SH)—C(=O)—O⁻

Val-Asp-Cys

19.18 Ala-Phe-Arg, Ala-Arg-Phe, Phe-Ala-Arg, Phe-Arg-Ala, Arg-Phe-Ala, Arg-Ala-Phe

19.20 All tripeptides that contain three unique amino acids can have 6 possible isomers.

IMPORTANT PEPTIDES (SECTION 19.4)

☑19.22 Cysteine forms disulfide bridges in the peptides vasopressin and oxytocin.

CHARACTERISTICS OF PROTEINS (SECTION 19.5)

19.24 The presence of certain proteins in urine or blood can indicate that cellular damage has occurred in the body because most proteins are too large to leave the cell in which they were synthesized. The presence of these proteins outside of their original cells indicates that the cell walls have leaked the proteins into body fluids.

19.26 A protein is least soluble in an aqueous medium that has a pH equal to the isoelectric point of the protein because the protein does not have a net charge at its isoelectric point. The protein molecules will clump together at the isoelectric point because the individual molecules do not contain like charges that repel other molecules.

19.28 The eight principle functions of proteins are: (1) catalysis, (2) structure, (3) storage, (4) protection, (5) regulation, (6) nerve impulse transmission, (7) movement, and (8) transport.

☑19.30
a. α-keratin — This protein is more likely to be fibrous than globular because it is a structural protein found in hair, feather, and nails.
b. DNA polymerase — This protein is more likely to be globular than fibrous because it is a catalyst that must work to replicate and repair segments of DNA.

☑19.32 Simple proteins only contain only amino acid residues. Conjugated proteins contain amino acid residues and other organic or inorganic components called prosthetic groups.

THE PRIMARY STRUCTURE OF PROTEINS (SECTION 19.6)

☑19.34 The primary structure of proteins is the order of the amino acid residues in a protein.

19.36

$$\text{---NH---CH(R)---C(=O)---NH---CH(R')---C(=O)---NH---CH(R'')---C(=O)---NH---CH(R''')---C(=O)---}$$

THE SECONDARY STRUCTURE OF PROTEINS (SECTION 19.7)

☑19.38 Hydrogen bonding between amide hydrogen atoms and carbonyl oxygen atoms of amino acid residues in the backbone of the protein or polypeptide chain is the most important type of bonding for maintaining a specific secondary configuration.

THE TERTIARY STRUCTURE OF PROTEINS (SECTION 19.8)

19.40 The hydrogen bonding in the tertiary structure of a protein is between the side chains on the amino acid residues. The hydrogen bonding in the secondary structure of a protein in between the atoms in the backbone of the protein.

☑19.42
a. threonine and asparagine — hydrogen bonds
b. glutamate and arginine — salt bridge
c. phenylalanine and leucine — hydrophobic interactions
d. cysteine and cysteine — disulfide bridge

19.44   The alanine, phenylalanine, and methionine would be directed toward the inside of the protein structure. The lysine and glutamate would be directed toward the aqueous surroundings.

THE QUATERNARY STRUCTURE OF PROTEINS (SECTION 19.9)
☑19.46   Hydrophobic forces, hydrogen bonds, disulfide bridges, and salt bridges between subunits give rise to quaternary structure.

19.48   The term subunit describes a polypeptide chain that has its own primary, secondary, and tertiary structure that combines with other polypeptide chains to form the quaternary structure of a protein.

PROTEIN HYDROLYSIS AND DENATURATION (SECTION 19.10)
☑19.50   A protein that is completely hydrolyzed will result in amino acids. A protein that is denatured will maintain its primary structure, but will lose its secondary, tertiary, and quaternary structure.

19.52   The primary structure of the protein in a raw egg and a cooked egg are the same.

19.54   Cooking an egg is a permanent change to its three dimensional structure. This process is irreversible protein denaturation. Beating an egg white is not a completely permanent change to its three dimensional structure. This process is reversible protein denaturation.

19.56   Egg white can serve as an emergency antidote for heavy-metal poisoning because the proteins in the egg bind to the metal ions and form a precipitate that resists further digestion in the stomach. The metal-containing precipitate is then removed from the stomach by pumping or is ejected by induced vomiting.

OVER THE COUNTER, CHEMISTRY AND YOUR HEALTH, AND CHEMISTRY AROUND US
19.58   Some drugs that should never be taken by a nursing mother are bromocriptine, many cancer chemotherapy drugs, ergotamine, lithium, methotrexate, nicotine, and illegal drugs.

19.60   Memory loss is one of the aspects of Alzheimer's disease that can be treated with some success.

19.62   Sickle-cell disease is caused by one amino acid residue difference in each of the two β-chains of hemoglobin.

ADDITIONAL ACTIVITIES
Section 19.1 Review:
(1) Which are larger biomolecules: lipids or proteins?
(2) Number the carbon chain of the unionized amino acid on the right.
(3) Name the amino acid on the right as a carboxylic acid.
(4) Label the carbon atoms in the amino acid with Greek letters (α, β, γ, δ, ε, etc.).
(5) What is the Greek letter for the carbon atom connected to the –NH₂ group?
(6) Circle the side chain.

(7) What is the Greek letter for the carbon atom to which the side chain is connected?
(8) Is the side chain polar or nonpolar?
(9) Is the side chain acidic, basic, or neutral?
(10) What is the common name for this amino acid?
(11) What are the abbreviations for this amino acid?
(12) Draw this amino acid as a Fischer Projection of the L-form.

Section 19.2 Review:
(1) Label the acidic and basic areas of the amino acid from the Section 19.1 Review.
(2) Use an arrow to show the internal acid-base reaction within the amino acid from (1).
(3) What is a general name of structures like the one produced by the reaction in (2)?
(4) What is the net charge on the amino acid in (2)?
(5) Write an equation that shows what happens when hydronium ions are added to a solution of (2).
(6) What is the net charge on the amino acid after the addition of hydronium ions?
(7) Write an equation that shows what happens when hydroxide ions are added to a solution of (2).
(8) What is the net charge on the amino acid after the addition of hydroxide ions?
(9) What is the name of the pH at which the amino acid exists in the form produced in (2)?
(10) Is this amino acid able to act as a buffer? Explain.

Section 19.3 Review:
(1) Can the amino acid from the Section 19.1 Review produce a disulfide bond with another amino acid? Explain.
(2) Show the two possible amide-forming reactions between glycine and the amino acid from the Section 19.1 Review.
(3) What is the general term used to describe the larger products from (2)?
(4) Name the products from (2).
(5) Write the abbreviations for the products from (2).
(6) Label the peptide linkages in the structures from (2).
(7) What are the other names for a peptide linkage?
(9) Circle the N-terminal end of the products from (2).
(10) Box the C-terminal end of the products from (2).

Section 19.4 Review:
(1) What happens to the polypeptide structure when two cysteine residues form a disulfide bridge in a polypeptide?
(2) Does the order of the amino acid residues influence biological function? Explain.

Section 19.5 Review:
(1) Which class of carbohydrates has similar molecular weights to proteins?
(2) What else is similar between proteins and the class of carbohydrates identified in (1)?
(3) Which of the following expressions best help to explain the solubility of proteins: "Like dissolves like." or "Opposites attract."? Explain.
(4) Differentiate between the following protein functions:
   a. catalytic and regulatory
   b. structural and protective
   c. storage and nerve impulse transmission
   d. movement and transport
(5) Can a protein be both fibrous and simple? Explain.

268 Chapter 19

(6) Can a protein be both globular and conjugated? Explain.
(7) Which class of proteins contains prosthetic groups?

Section 19.6 Review:
(1) What do all proteins have in common with regard to primary structure?
(2) What is unique about the primary structure of different proteins?

Section 19.7 Review:
(1) Which intermolecular force is responsible for secondary structure?
(2) How is primary structure related to secondary structure?

Section 19.8 Review:
(1) Distinguish between disulfide bridges and salt bridges.
(2) Can the amino acid residues with neutral side chains be involved in tertiary structure hydrogen bonds? Explain.
(3) Which groups of amino acid residues are involved in hydrophobic interactions?

Section 19.9 Review:
(1) What is the difference between a subunit and a prosthetic group?
(2) Do all conjugated proteins have quaternary structure?

Section 19.10 Review:
(1) What was the result of the hydrolysis reactions involving simple lipids?
(2) What is the result of the hydrolysis reactions involving proteins?
(3) Which level(s) of structure is(are) changed for a protein that undergoes hydrolysis?
(4) What factors will denature a protein?
(5) Which level(s) of structure is(are) changed for a protein that undergoes denaturation?
(6) Which level(s) of structure is(are) important for a protein to perform its original biological function?

Tying It All Together with a Laboratory Application:
A pentapeptide contains the amino acid residues of valine, cysteine, asparagine, lysine, and glutamate. These residues are the (1) _____-form stereoisomers found in most living organisms. At its isoelectric point, the Val-Cys-Asn-Lys-Glu pentapeptide has the structure (2) _____. The N-terminal amino acid residue in this pentapeptide is (3) _____. The C-terminal amino acid residue in this pentapeptide is (4) _____. This is one of (5) _____ possible pentapeptides. The major difference between the pentapeptides is the (6) _____ of the amino acid residues, which is called the (7) _____ level of structure.

Hydrogen bonding between the carbonyl group of the glutamate amino acid residue and the amide group of the cysteine amino acid residue of the Val-Cys-Asn-Lys-Glu pentapeptide would be considered the (8) _____ level of structure. If this pattern of hydrogen bonding occurred again and again in a longer polypeptide or protein it would result in a(n) (9) _____.

If the side chains of these amino acids were to interact, the (10) _____ level of structure would result.
(11) The following table lists the classification of side chains of each amino acid as well as the type of interactions that side chain is likely to experience:

Amino Acid Residue	Classification of Side Chain	Probable Side Chain Interactions
valine		
cysteine		
asparagine		
lysine		
glutamate		

Side chain interaction is (12) _____ (likely or unlikely) because this peptide chain is much (13) _____ than the polypeptides with this level of structure. The (14) _____ level of structure is unlikely to occur with this pentapeptide because (15) _____.

This pentapeptide will likely have its isoelectric point (16) _____ 7 because it contains (17) _____ neutral side chain(s), (18) _____ acidic side chain(s), and (19) _____ basic side chain(s). Increasing the pH of a solution containing this pentapeptide would initially (20) _____ the pentapeptide by changing its (21) _____ state; however, adding a strong base to this pentapeptide could result in (22) _____ which destroys the (23) _____ level of structure.

The amino acids that result from (24) _____ under basic conditions could be isolated. The techniques that could be employed include gel electrophoresis, electrodialysis, or ion-exchange chromatography. All of these techniques rely on the (25) _____ of the amino acids at different pH values. The two amino acids that will be most easily separated by these techniques are (26) _____ and _____. The amino acids with (27) _____ side chains will be more difficult to separate; however, their side chains differ in (28) _____ and reactivity.

(29) _____ could be isolated by mixing the aqueous solution with a nonpolar solvent like benzene in a separatory funnel because it is more soluble in benzene than water. The remaining residues could be mixed with an oxidizing agent or heavy metal ions. Only the (30) _____ will react with those reagents. The (31) _____ is the better choice because this reaction can be reversed by adding (32) _____; however, the (33) _____ will form insoluble salts that can be filtered from the aqueous solution, leaving (34) _____ as the only amino acid in the aqueous solution.

SOLUTIONS FOR THE ADDITIONAL ACTIVITIES

Section 19.1 Review:
(1) Proteins are larger biomolecules than lipids. Proteins are polymers, but lipids are not.
(3) 2-amino-3-methylpentanoic acid
(5) The amino group is attached to the α-carbon atom.
(7) The side chain is attached to the α-carbon atom.
(8) The side chain is nonpolar.
(9) The side chain is neutral.
(10) The common name for this amino acid is isoleucine.
(11) The three letter abbreviation for this amino acid is Ile. The one letter abbreviation for this amino acid is I.

Section 19.2 Review:

(3) zwitterion; (4) net charge = 0;

270 Chapter 19

(5)
$$H_3N^+-CH(CH(CH_3)CH_2CH_3)-C(=O)-O^- + H_3O^+ \longrightarrow H_3N^+-CH(CH(CH_3)CH_2CH_3)-C(=O)-OH + H_2O$$

(6) net charge = +1;

(7)
$$H_3N^+-CH(CH(CH_3)CH_2CH_3)-C(=O)-O^- + OH^- \longrightarrow H_2N-CH(CH(CH_3)CH_2CH_3)-C(=O)-O^- + H_2O$$

(8) net charge = -1
(9) isoelectric point
(10) Yes, this amino acid is able to act as a buffer because it has the ability to react with both H_3O^+ and OH^-.

Section 19.3 Review:

(1) No, the amino acid from the Section 19.1 Review cannot produce a disulfide bond with another amino acid because it is not cysteine. Cysteine is the only amino acid that forms disulfide bonds.

(2)
$$H_2N-CH_2-C(=O)-OH + H_2N-CH(CH(CH_3)CH_2CH_3)-C(=O)-OH \longrightarrow H_2N-CH_2-C(=O)-NH-CH(CH(CH_3)CH_2CH_3)-C(=O)-OH + H_2O$$

$$H_2N-CH(CH(CH_3)CH_2CH_3)-C(=O)-OH + H_2N-CH_2-C(=O)-OH \longrightarrow H_2N-CH(CH(CH_3)CH_2CH_3)-C(=O)-NH-CH_2-C(=O)-OH + H_2O$$

(3) dipeptides; (4) glycylisoleucine, isoleucylglycine; (5) Gly-Ile, Ile-Gly;
(7) Peptide linkages are also called peptide bonds or amide linkages.

Section 19.4 Review:
(1) A loop forms when two cysteine residues form a disulfide bridge in a polypeptide.
(2) Yes, the order of the amino acids influences biological function. The order of the side chains effects the three dimensional structure of the polypeptide.

Section 19.5 Review:
(1) Polysaccharides have similar molecular weights to proteins.
(2) Both polysaccharides and proteins are polymers.
(3) "Like dissolves like" best explains protein solubility. Proteins will dissolve in water when they are above or below their isoelectric point because the protein will be polar (like water) at those pH values.

"Opposites attract" best explains protein insolubility. When proteins are at their isoelectric point, they do not have a net charge; however, they do have positive and negative charges within the molecule. These positive and negative charges are attracted to each other and cause the proteins to be more insoluble.
(4) a. Catalytic proteins lower the activation energy for reactions. Regulatory proteins affect the rate of body processes which may involve more than one chemical reaction.
   b. Structural proteins are responsible for the mechanical strength of skin, bone, hair, and fingernails. Protective proteins are antibodies that counteract viruses, bacteria, and other foreign substances.
   c. Storage proteins provide a way to store small molecules or ions for nourishment. Nerve impulse transmission proteins receive small molecules that pass between the synapses of nerve cells.
   d. Movement proteins are found in muscles. Transport proteins are responsible for moving small molecules and ions through the body.
(5) Yes, a protein can be both fibrous and simple because fibrous proteins are rod-shaped or stringlike molecules that intertwine to form strong fibers. These proteins are also simple if they only contain amino acid residues.
(6) Yes, a protein can be both globular and conjugated. Globular proteins are more spherical in shape than fibrous proteins. These proteins are also conjugated if they contain other organic or inorganic components in addition to the amino acid residues.
(7) Conjugated proteins contain prosthetic groups.

Section 19.6 Review:
(1) All proteins have a primary structure that contains a backbone of carbon and nitrogen atoms held together by peptide bonds.
(2) Different proteins have different lengths of backbones and different sequences of side chains.

Section 19.7 Review:
(1) Hydrogen bonding is responsible for secondary structure.
(2) The carbonyl oxygen atoms and the amide hydrogen atoms in the backbone of a protein (primary structure) hydrogen bond to each other in order to form the secondary structure.

Section 19.8 Review:
(1) Disulfide bridges form when two cysteine amino acid residues undergo oxidation and form a covalent bond. Salt bridges form when ionized acidic amino acid residue side chains and ionized basic amino acid residue side chains form an ionic bond.
(2) Neutral side chains can be involved in hydrogen bonding, if they are polar and contain oxygen or nitrogen atoms.
(3) Amino acid residues with neutral, nonpolar side chains can be involved in hydrophobic interactions.

272 Chapter 19

Section 19.9 Review:
(1) A subunit is a polypeptide chain. A prosthetic group is a non-amino acid part of a conjugated protein. It can be organic or inorganic.
(2) Quaternary structure is defined as the arrangement of subunits to form a larger protein. If a conjugated protein contains two or more subunits, then it will have quaternary structure. If the conjugated protein only contains one subunit, then it will not have quaternary structure.

Section 19.10 Review:
(1) When simple lipids are hydrolyzed, glycerol or an alcohol is produced as well as fatty acids or salts of fatty acids.
(2) When proteins are hydrolyzed, smaller peptides or amino acids are produced.
(3) Primary structure is changed for a protein that undergoes hydrolysis; consequently, all other levels of structure are affected.
(4) A protein can be denatured by extreme temperatures, extreme pH values, organic solvents, detergents, and heavy-metal ions.
(5) The higher levels of structure (secondary, tertiary, and quaternary) are changed for a protein that undergoes denaturation. Primary structure is not affected.
(6) All four levels of structure are important for a protein to perform its original biological function.

Tying It All Together with a Laboratory Application:
(1) L-
(2)

$$^+H_3N-CH(CH(CH_3)_2)-C(=O)-NH-CH(CH_2SH)-C(=O)-NH-CH(CH_2C(=O)NH_2)-C(=O)-NH-CH((CH_2)_4NH_3^+)-C(=O)-NH-CH((CH_2)_2COO^-)-C(=O)-O^-$$

(3) valine
(4) glutamate
(5) 120
(6) order
(7) primary
(8) secondary
(9) α-helix
(10) tertiary

(11)

Residue	Classification	Interactions
valine	neutral, nonpolar	hydrophobic interactions
cysteine	netural, polar	disulfide bridges
asparagine	neutral, polar	hydrogen bonding
lysine	basic, polar	salt bridges
glutamate	acidic, polar	salt bridges

(12) unlikely
(13) shorter
(14) quaternary
(15) it only contains 1 peptide chain
(16) near
(17) 3
(18) 1
(19) 1
(20) denature
(21) native
(22) hydrolysis (or digestion)
(23) primary
(24) hydrolysis
(25) charges
(26) lysine, glutamate
(27) neutral
(28) polarity
(29) Valine
(30) cysteine
(31) oxidizing agent
(32) a source of hydrogen
(33) heavy metal ions
(34) asparagine

SELF-TEST QUESTIONS

Multiple Choice

1. The main distinguishing feature between various amino acids is
   a. the length of the carbon chain
   b. the number of amino groups
   c. the composition of the side chains
   d. the number of acid groups

2. The amino acid valine is represented below. Which of the lettered carbon atoms is the alpha carbon atom?

   $$H_3N^+-C_bH-C_a-O^-$$ with C_cH-CH_3 and C_dH_3 (with C=O double bond on C_a)

   a. a
   b. b
   c. c
   d. d

3. What kind of amino acid is this compound?

   a. acidic
   b. basic
   c. neutral
   d. more than one response is correct

4. Which of the following best represents alanine at its isoelectric point?
   a. $H_3N^+-CH(CH_3)-C(=O)-O^-$
   b. $H_2N-CH(CH_3)-C(=O)-O^-$
   c. $H_2N-CH(CH_3)-C(=O)-OH$
   d. $H_3N^+-CH(CH_3)-C(=O)-OH$

5. A linkage present in all peptides is
   a. NH_2-CH_2-
   b.
   c. $NH_2-CH(R)-$
   d.

6. Prosthetic groups are found in _____ proteins.
   a. all
   b. simple
   c. conjugated
   d. no

7. A protein that is relatively spherical in shape and fairly soluble in water is a _____ protein.
   a. simple
   b. conjugated
   c. fibrous
   d. globular

8. Which of the following characteristics of a protein would be classified as a primary structural feature?
   a. amino acid sequence
   b. pleated-sheet configuration
   c. α-helix configuration
   d. the shape of the protein molecule

9. Which protein serves as antibodies?
   a. myoglobin
   b. hemoglobin
   c. collagen
   d. immunoglobulin

10. Which of the following side-group interactions involves nonpolar groups?
    a. salt bridges
    b. hydrogen bonds
    c. disulfide bonds
    d. hydrophobic interactions

11. Which of the following does **NOT** contribute to the tertiary structure of proteins?
    a. salt bridges
    b. hydrogen bonds
    c. disulfide bonds
    d. peptide bonds

12. The quaternary structure of hemoglobin involves _____ subunits.
    a. two
    b. four
    c. six
    d. eight

13. Denaturation of a protein involves the breakdown of
    a. primary structure
    b. secondary and tertiary structures
    c. primary and secondary structures
    d. primary, secondary, and tertiary structures

14. Ions of heavy metals (Hg^{2+} or Pb^{2+}) denature proteins by combining with _____ groups.
    a. —NH_2
    b. —OH
    c. —SH
    d. —C(=O)—NH—

Matching
Match the following definitions to the correct words given as responses.
15. a dipolar amino acid molecule containing both a + and a − charge
16. a substance composed of 25 amino acids linked together
17. the pH at which amino acids exist in the form that has no net charge

   a. polypeptide
   b. ninhydrin
   c. isoelectric point
   d. zwitterion

Select the correct polypeptide for each description on the left.
18. stimulates milk production
19. controls carbohydrate metabolism
20. decreases urine production

   a. prolactin
   b. insulin
   c. oxytocin
   d. vasopressin

For each description on the left, select the correct protein class.
21. hemoglobin belongs to this class
22. proteins that function as enzymes
23. insulin belongs to this class
24. keratin belongs to this class
25. collagen belongs to this class

   a. regulatory proteins
   b. transport proteins
   c. structural proteins
   d. catalytic proteins

For each of the bonds given on the left, choose a response that indicates the level of protein structure associated with each type of bond.

26. hydrogen bonds
27. amide bonds
28. hydrophobic bonds
29. salt bonds

a. primary structure
b. secondary structure
c. tertiary structure
d. more than one response is correct

True-False
30. Amino acids found in living systems are generally in the D form.
31. Amino acids are white crystalline solids.
32. Cysteine is the only amino acid that can form disulfide bridges.
33. Six structural isomers could be formed if tripeptides were made from serine, valine, and glycine.
34. Proteins are the building blocks of amino acids.
35. Acids cannot hydrolyze proteins.

Chapter 20: Enzymes

CHAPTER OUTLINE
20.1 General Characteristics of Enzymes
20.2 Enzyme Nomenclature and Classification
20.3 Enzyme Cofactors
20.4 The Mechanism of Enzyme Action
20.5 Enzyme Activity
20.6 Factors Affecting Enzyme Activity
20.7 Enzyme Inhibition
20.8 The Regulation of Enzyme Activity
20.9 Medical Application of Enzymes

LEARNING OBJECTIVES/ASSESSMENT
When you have completed your study of this chapter, you should be able to:
1. Describe the general characteristic of enzymes and explain why enzymes are vital in body chemistry. (Section 20.1; Exercise 20.2)
2. Determine the function and/or substrate of an enzyme on the basis of its name. (Section 20.2; Exercise 20.12)
3. Identify the general function of cofactors. (Section 20.3; Exercise 20.14)
4. Use the lock-and-key theory to explain specificity in enzyme action. (Section 20.4; Exercise 20.20)
5. List two ways of describing enzyme activity. (Section 20.5; Exercise 20.26)
6. Identify the factors that affect enzyme activity. (Section 20.6; Exercise 20.28)
7. Compare the mechanisms of competitive and noncompetitive enzyme inhibition. (Section 20.7; Exercise 20.34)
8. Describe the three methods of cellular control over enzyme activity. (Section 20.8; Exercise 20.38)
9. Discuss the importance of measuring enzyme levels in the diagnosis of disease. (Section 20.9; Exercise 20.46)

SOLUTIONS FOR THE END OF CHAPTER EXERCISES
GENERAL CHARACTERISTICS OF ENZYMES (SECTION 20.1)
☑20.2 Enzyme catalysis of a reaction is superior to normal laboratory conditions because they are specific in the type of reaction they catalyze and enzyme activity can be regulated by the cell.

20.4 Many types of enzymes are needed because enzymes can only catalyze specific reactions and a body requires many reactions to occur in order to maintain life.

20.6 Absolute specificity means that an enzyme can only catalyze the reaction of one substance.

ENZYME NOMENCLATURE AND CLASSIFICATION (SECTION 20.2)
20.8 The IEC names and most common names of enzymes end with –ase.

20.10

	Enzyme	Substrate		Enzyme	Substrate
a.	urease	urea	d.	arginase	arginine
b.	fumarase	fumarate	e.	lactase	lactose
c.	fructase	fructose			

☑20.12

	Enzyme	Substrate	Type of Reaction
a.	cholinesterase	choline	formation or hydrolysis of an ester
b.	cytochrome oxidase	cytochrome	oxidation

	Enzyme	Substrate	Type of Reaction
c.	glucose 6-phosphate isomerase	glucose 6-phosphate	isomerization
d.	lactate dehydrogenase	lactate	dehydrogenation

ENZYME COFACTORS (SECTION 20.3)

☑20.14 When an apoenzyme and a cofactor combine they form an active enzyme.

20.16 Some minerals (inorganic ions) from our diets are cofactors for apoenzymes.

THE MECHANISM OF ENZYME ACTION (SECTION 20.4)

20.18 The equation, $E + S \rightleftarrows ES \rightarrow E + P$, shows that the enzyme (E) and substrate (S) establish an equilibrium with the enzyme-substrate complex (ES). This is a reversible reaction. The enzyme-substrate complex can break apart into the enzyme and substrate or react to produce the enzyme and the product (P). The reaction to produce the product is not a reversible reaction.

☑20.20 The lock-and-key theory explains enzyme specificity because only one substrate will fit into the active site of the enzyme. The active site of the enzyme specifies which substrates can bind to the enzyme.

20.22 The induced-fit theory best explains the ability of one enzyme to catalyze the reactions of propanoic acid, butanoic acid, and pentanoic acid because the active site would need to be able to accommodate several similar, but different, substrates.

ENZYME ACTIVITY (SECTION 20.5)

20.24 Enzyme activity in an experiment is observed by any method that allows the rate of product formation or reactant usage to be determined. The disappearance or appearance of a characteristic color is an example.

☑20.26 An enzyme international unit is the quantity of enzyme that catalyzes the conversion of 1 micromole of substrate per minute. This is useful in medical diagnosis because a specific enzyme activity of a patient can be measured and compared to normal activity.

FACTORS AFFECTING ENZYME ACTIVITY (SECTION 20.6)

☑20.28 As substrate concentration increases, the enzyme activity increases until a maximum velocity is reached. As enzyme concentration increases, the enzyme activity increases linearly. As pH increases, the enzyme activity increases until the optimum pH value is attained, then the enzyme activity decreases. As temperature increases, the enzyme activity increases until the optimum temperature is attained, then the enzyme activity decreases.

20.30 V_{max} for an enzyme can be determined by measuring the turnover rate for a specific enzyme concentration with increasing concentrations of substrate. The turnover rates will approach a steady value (V_{max}) when the substrate concentration is high enough for all of the enzymes to be saturated.

20.32 The pH of enzymes is maintained near 7.0 to prevent the enzymes from denaturing. Enzymes are proteins that can lose their three dimensional structure if the pH is drastically changed.

ENZYME INHIBITION (SECTION 20.7)

☑20.34 In competitive enzyme inhibition, an inhibitor binds to the active site of an enzyme and prevents the substrate from binding to the active site. In noncompetitive enzyme inhibition, an inhibitor binds to the enzyme at a location other than the active site and changes the shape of the active site, thus preventing the substrate from binding to the active site.

20.36 One antidote for cyanide poisoning is sodium thiosulfate. This antidote works because the cyanide ion reacts with the thiosulfate ion to produce the thiocyanate ion and the sulfite ion. Unlike cyanide, thiocyanate does not bind to the iron of cytochrome oxidase.

One antidote for heavy-metal poisoning is a chelating agent, like ethylenediaminetetraacetic acid. This antidote works because the heavy-metal ions combine with the EDTA and are tightly bound. Consequently, the heavy-metal ions cannot react with proteins in the body.

THE REGULATION OF ENZYME ACTIVITY (SECTION 20.8)

☑20.38 Enzyme activity is controlled by the activation of zymogens, allosteric regulation, and genetic control.

20.40 Enzymes that catalyze blood clotting are zymogens because under normal circumstances, blood needs to flow freely through the body. Only when bleeding occurs is clotting action necessary. Consequently, prothrombin flows through the body in the blood stream, but it does not cause the blood to clot until it is converted into thrombin which catalyzes clotting.

20.42 Activators are modulators that increase the activity of an enzyme. Inhibitors are modulators that decrease the activity of an enzyme.

20.44 Genetic control of enzyme activity is the cellular production of additional enzyme molecules in order to increase the production from an enzyme-catalyzed reaction.

Enzyme induction is the synthesis of enzymes in response to a temporary need of the cell.

MEDICAL APPLICATION OF ENZYMES (SECTION 20.9)

☑20.46
  a. acid phosphatase   This enzyme assay is used to diagnose prostate cancer.
  b. ALT   This enzyme assay is used to diagnose hepatitis.
  c. AST   This enzyme assay is used to diagnose heart attack and hepatitis.

20.48 LDH is a good initial diagnostic test because LDH occurs in multiple forms (isoenzymes) and can be used to diagnose a wide range of diseases (anemias, acute liver diseases, congestive heart failure, and muscular diseases).

CHEMISTRY AROUND US, CHEMISTRY AND YOUR HEALTH, AND OVER THE COUNTER

20.50 Inborn errors of metabolism is an expression sometimes used to refer to hereditary diseases caused by the absence of enzymes.

20.52 Extremozymes are enzymes that exist in extreme environments, such as temperatures greater than 100°C, temperatures lower than 0°C, extreme pH values, and high salt concentrations in solution.

20.54 Herbal supplements are not under regulation by the FDA because they are classified as dietary supplements, not food or drugs.

20.56 Ethanol is an antidote for ethylene glycol poisoning because it is a competitive inhibitor for the metabolism of ethylene glycol. Consequently, the ethylene glycol is excreted without undergoing dehydrogenation.

ADDITIONAL ACTIVITIES

Section 20.1 Review:
(1) What is catalytic power?
(2) Imagine you have a box of pens and pencils of all different descriptions. If you were to place the pen or pencil you are currently using in the box, how could you describe it to someone else and have them successfully find your current writing implement?
(3) If your friend finds a copy of your writing implement, but not your exact writing implement, is that absolute specificity, relative specificity, or stereochemical specificity?
(4) If your friend finds a writing implement that produces the same color of ink or graphite on the paper as your writing implement, but it is not your writing implement or a copy of your writing implement, is that absolute specificity, relative specificity, or stereochemical specificity?
(5) Are enzymes self-regulating? Explain.

Section 20.2 Review:
(1) Which two endings might indicate a substance is an enzyme?
(2) How are enzymes classified?
(3) What is a substrate?

Section 20.3 Review:
(1) What is the difference between a prosthetic group and a cofactor?
(2) What is the difference between a cofactor and a coenzyme?
(3) What is an apoenzyme?
(4) How do vitamins relate to enzymes?

Section 20.4 Review:
(1) What structural feature on an enzyme is identified in both the lock-and-key theory and the induced-fit theory?
(2) How does the structural feature in (1) differ in the lock-and-key theory and the induced-fit theory?

Section 20.5 Review:
(1) What is the turnover number for an enzyme?
(2) What type of experiments could be used to determine the turnover number of an enzyme?
(3) Who is more likely to determine enzyme activity as a turnover number: a research biochemist or a hospital lab technician?
(4) Who is more likely to determine enzyme activity levels in enzyme international units: a research biochemist or a hospital lab technician?

Section 20.6 Review:
(1) What is typically the limiting factor in an enzyme-catalyzed reaction?
(2) What does a maximum rate imply about the enzyme molecules?
(3) Temperature and pH have similar trends with respect to enzyme activity. Explain.

Section 20.7 Review:
(1) Inhibitors are divided into two categories. One of the categories is shown on the diagram on the next page. Label the other category.
(2) What types of inhibitors fall into the category you labeled in (1)?

(3) Reversible inhibitors are divided into two categories. One of the categories is shown on the diagram below. Label the other category.

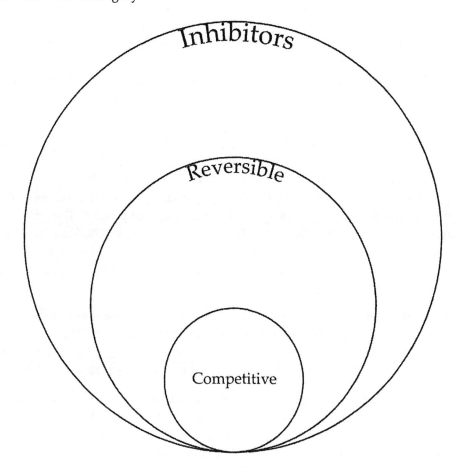

(4) What is the competition for competitive inhibitors?

Section 20.8 Review:
(1) What is another name for zymogens?
(2) How do zymogens differ from apoenzymes?
(3) What are the two categories of modulators?
(4) How does the name "allosteric enzymes" indicate that these enzymes have distinctive binding sites for modulators?
(5) What biomolecules are responsible for genetic control of enzymes?

Section 20.9 Review:
(1) Name a body fluid checked for changes in enzyme concentration as the result of injury or disease.
(2) What remains constant in all of the LDH isoenzymes?

Tying It All Together with a Laboratory Application:
Our stomachs contain gastric juices with a low pH that help to digest the foods we eat, including proteins. Food that has been cooked will probably contain proteins that have been (1) _____ and no longer maintain their native state. The (2) _____ environment of the stomach favors digestion; however, this can be a time consuming process. The digestive enzyme pepsin (3) _____ the rate of digestion of proteins by (4) _____ the activation energy of the reaction. This enzyme does not effect the digestion of lipids because enzymes have (5) _____. The two theories that help explain this phenomenon both refer

to the (6) _____ of the enzyme. The (7) _____ says the (6) is rigid, while (8) _____ says the (6) adapts to the substrate.

Pepsin is stored in the inactive form pepsinogen. Pepsinogen is an example of a (9) _____ or _____ because when pepsin is needed, the gastric glands of the stomach secret pepsinogen into the stomach where the change in pH activates the enzyme by cleaving a bond. If pepsin had required a cofactor, which could have been a(n) (10) _____ or a(n) _____, then the inactive form of pepsin would be called a(n) (11) _____.

The enzyme (12) _____ of pepsin can be measured by adding 0.1 g of lean hamburger meat to two test tubes (A & B) and 0.1 g of lard to two additional test tubes (C & D), then adding 3 mL of 0.01 M HCl to test tubes A and C and 3 mL of a 0.01 M HCl solution containing 10 µg pepsin/mL solution to test tubes B and D. The test tubes are then covered and allowed to react for 1 hour at 25°C. The amount of protein digestion can be compared between test tubes (13) _____ and _____ by checking the (14) _____ of the resulting peptides. Test tube (15) _____ should have more digested protein. The amount of lipid digestion can be compared between test tubes (16) _____ and _____ by checking the amounts of (17) _____ and _____ that were produced. The contents of (18) _____ test tube should have undergone more digestion than the contents of the other test tube. The test tube with the highest enzyme activity should be (19) _____.

To measure the effect of temperature on the activity of this enzyme, several test tubes like test tube (20) _____ should be prepared. They should be placed in water baths with (21) _____ temperatures for (22) _____ amount(s) of time. The activity of this enzyme should (23) _____ until the optimum temperature is reached and then should (24) _____ as the temperature is raised further.

To measure the effect of substrate concentration on the activity of this enzyme, several test tubes like test tube (25) _____ should be prepared; however, the amount of (26) _____ should be varied. The reaction time(s) and temperature(s) should be (27) _____ for the test tubes. The activity of the enzyme should (28) _____ until (29) _____ is reached. This occurs because the enzyme (30) _____.

To measure the effect of enzyme concentration on the activity of this enzyme, several test tubes like test tube (31) _____ should be prepared; however the amount of (32) _____ should be varied. The reaction time(s) and temperature(s) should be (33) _____ for the test tubes. The activity of the enzyme should increase with (34) _____ enzyme concentration.

To measure the effect of pH on the activity of this enzyme, several test tubes like test tube (35) _____ should be prepared; however, the concentration of (36) _____ should be varied. The reaction time(s) and temperature(s) should be (37) _____ for the test tubes. The activity of the enzyme should increase with (38) _____ pH until the (39) _____ pH is reached. A student performs this reaction and observes that at pH values of 6 or higher the enzyme activity appears to be zero. The student hypothesizes the pepsin is inactivated at higher pH values because pepsin is a (40) _____ that has multiple levels of (41) _____ which are necessary for biological function. At high pH values, pepsin is (42) _____.

The addition of a heavy metal ion to a test tube containing hydrochloric acid, pepsin, and lean hamburger meat would (43) _____ the enzyme activity because heavy metal ions are (44) _____. They are not classified as competitive or noncompetitive because they are (45) _____ (44), not (46) _____ (44).

SOLUTIONS FOR THE ADDITIONAL ACTIVITIES

Section 20.1 Review:
(1) Catalytic power is the ability of a catalyst to speed up a reaction by lowering the activation energy.
(2) You might describe it as a pen or a pencil, the color of the outside, the color of the ink or graphite, the status of the eraser, the appearance of the tip, whether or not it has a cap, whether or not it has a clip, the brand, the length, the width, and any other distinguishing characteristic of the writing implement.
(3) Finding a copy of your writing implement is absolute specificity. The writing implement fits all of the criteria of the original.
(4) Finding a writing implement with the same color of ink or graphite is an example of relative specificity. This new writing implement shares common characteristics with the original; however, it is not the original, nor is it a copy of the original.
(5) Enzymes are not self-regulating. The cell controls the regulating action of enzymes.

Section 20.2 Review:
(1) The endings –*in* and –*ase* indicate a substance is an enzyme.
(2) Enzymes are classified according to the types of reactions they catalyze.
(3) A substrate is a the substance upon which an enzyme acts.

Section 20.3 Review:
(1) A prosthetic group is tightly bound to and forms an integral part of the enzyme structure. A cofactor is weakly bound to the enzyme and is easily separated from the protein structure.
(2) A cofactor is any nonprotein component of an enzyme that is weakly bound to the enzyme and is easily separated from the protein structure. A coenzyme is an organic cofactor. All coenzymes are cofactors, not all cofactors are coenzymes.
(3) An apoenzyme is the protein component of an enzyme that requires a cofactor.
(4) Vitamins serve as the precursors for coenzymes which are needed for some enzymes to function properly.

Section 20.4 Review:
(1) The active site of an enzyme is the structural feature on an enzyme is identified in both the lock-and-key theory and the induced-fit theory.
(2) The active site in the lock-and-key theory is fixed or rigid and the active site in the induced-fit theory is flexible and adapts to the incoming substrates.

Section 20.5 Review:
(1) The turnover number for an enzyme is the rate at which one molecule of enzyme converts its substrate into product.
(2) An enzyme assay could be used to determine the turnover number of an enzyme.
(3) A research biochemist is more likely to determine enzyme activity as a turnover number.
(4) A hospital lab technician is more likely to determine enzyme activity levels in enzyme international units.

Section 20.6 Review:
(1) The limiting factor in the rate of an enzyme-catalyzed reaction is the enzyme.
(2) A maximum rate implies the enzyme molecules are saturated with substrate and cannot work any faster under the conditions imposed.
(3) Extreme temperatures and pH values can denature enzymes. Enzymes have optimum temperatures and pH values.

284 Chapter 20

Section 20.7 Review:

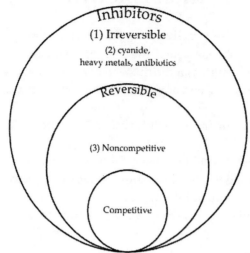

(4) Competitive inhibitors compete with the substrate for the active site of the enzyme.

Section 20.8 Review:
(1) Another name for zymogens is proenzymes.
(2) Zymogens must be released from storage and activated at the location of the reaction. Activation involves the cleavage of one or more peptide bonds of the zymogen. Apoenzymes require a cofactor in order to form an active enzyme.
(3) Modulators are either activators or inhibitors.
(4) The –*steric* part of the name "allosteric enzymes" might indicate they have distinctive binding sites for modulators because it reflects the same root word found in stereochemistry and stereoisomers that indicates a specific three dimensional structure.
(5) Nucleic acids are the biomolecules responsible for genetic control of enzymes.

Section 20.9 Review:
(1) Blood is a body fluid checked for changes in enzyme concentration as the result of injury or disease.
(2) All of the LDH isoenzymes contain 4 subunits.

Tying It All Together with a Laboratory Application:
(1) denatured
(2) acidic
(3) increases
(4) decreasing
(5) specificity
(6) active site
(7) lock-and-key theory
(8) induced-fit theory
(9) zymogen, proenzyme
(10) coenzyme, inorganic ion
(11) apoenzyme
(12) activity
(13) A, B
(14) size (concentration, amount, molecular weight)
(15) B
(16) C, D
(17) glycerol, fatty acids
(18) neither
(19) B
(20) B
(21) different
(22) the same
(23) increase
(24) decrease
(25) B
(26) lean hamburger
(27) constant
(28) increase
(29) V_{max}
(30) saturates
(31) B
(32) enzyme
(33) constant
(34) increasing
(35) B
(36) hydrochloric acid
(37) constant
(38) increasing
(39) optimum
(40) protein
(41) structure
(42) denatured
(43) decrease
(44) inhibitors
(45) irreversible
(46) reversible

SELF-TEST QUESTIONS
Multiple Choice
1. Enzymes which act upon only one substrate exhibit
   a. catalytic specificity
   b. relative specificity
   c. binding specificity
   d. absolute specificity

2. Which of the following enzyme properties is explained by the lock-and-key model?
   a. specificity
   b. high molecular weight
   c. high turnover rate
   d. susceptibility to denatuartion

3. The induced-fit theory of enzyme action extends the lock-and-key theory in which of the following ways?
   a. assumes enzymes and substrates are rigid
   b. assumes the shape of the substrates changes (conforms) to fit the enzyme
   c. assumes enzymes have no active site
   d. assumes the enzyme shape changes to accommodate the substrate

4. Enzyme turnover numbers are expressed in
   a. activity/mg
   b. units/minute
   c. units/mg
   d. molecules/minute

5. Under saturating conditions, an enzyme-catalyzed reaction has a velocity, v. Which of the following would increase the rate of the reaction?
   a. a decrease in the substrate concentration
   b. an increase in the substrate concentration
   c. a decrease in the enzyme concentration
   d. an increase in the enzyme concentration

6. Which of the following has no effect on the rate of an enzyme-catalyzed reaction?
   a. the volume of the reaction mixture
   b. the pH of the reaction mixture
   c. the temperature of the reaction mixture
   d. the enzyme concentration in the reaction mixture

7. In noncompetitive inhibition,
   a. substrate and inhibitor bind at separate locations on the enzyme.
   b. the inhibitor forms a covalent bond with the enzyme.
   c. substrate and inhibitor bind at the same location on the enzyme.
   d. the enzyme becomes permanently inactivated.

8. The use of ethanol as a treatment for ethylene glycol poisoning is an excellent example of
   a. feedback inhibition
   b. noncompetitive inhibition
   c. competitive inhibition
   d. enzyme denaturation

9. An enzyme is sometimes generated initially in an inactive form called
   a. a coenzyme
   b. an activator
   c. a cofactor
   d. a zymogen

10. Enzymes which exist in more than one form are called
    a. proenzymes
    b. apoenzymes
    c. isoenzymes
    d. allosteric enzymes

11. Which of the following is an enzyme whose activity is changed by the binding of modulators?
    a. isoenzymes
    b. allosteric enzymes
    c. proenzymes
    d. apoenzymes

Matching
Select an enzyme that matches each description on the left.
12. name is based only on nature of reaction catalyzed
13. name is based on enzyme substrate
14. name is based on both the substrate and the nature of the reaction catalyzed
15. name is an early nonsystematic type

a. pepsin
b. oxidase
c. succinate dehydrogenase
d. sucrase

Match the enzyme components below to the correct term given as a response.
16. the protein portion of an enzyme
17. an organic, but nonprotein portion of an enzyme
18. a nonprotein molecule or ion required by an enzyme

a. cofactor
b. apoenzyme
c. coenzyme
d. proenzyme

Match each description on the left with a response from the right.
19. reacts with cytochrome oxidase
20. stops bacterial synthesis of folic acid
21. inhibits succinate dehydrogenase
22. inhibits bacterial cell-wall construction

a. penicillin
b. cyanide
c. sulfanilamide
d. malonate

Match the correct disease or condition given as a response to the enzyme useful in diagnosing the disease or condition.
23. amylase
24. lysozyme
25. acid phosphatase

a. infectious hepatitis
b. pancreatic diseases
c. prostate cancer
d. monocytic leukemia

True-False
26. The presence of enzymes enables reactions to proceed at lower temperatures.
27. Enzymes are destroyed during chemical reactions in which they participate.
28. Some enzymes are carbohydrates.
29. All enzymes act only on a single substance.
30. Coenzymes are always organic compounds.
31. The optimum pH of most of the enzymes in the human body is near pH 7.
32. A plot of reaction rate (vertical axis) versus temperature of enzyme-catalyzed reactions most frequently yields a straight line.
33. Enzyme induction is an example of genetic control of enzyme activity.
34. Lactate dehydrogenase exists in five different isomeric forms.
35. The enzyme-substrate complex always produces an enzyme and a product.

Chapter 21: Nucleic Acids and Protein Synthesis

CHAPTER OUTLINE
21.1 Components of Nucleic Acids
21.2 The Structure of DNA
21.3 DNA Replication
21.4 Ribonucleic Acid (RNA)
21.5 The Flow of Genetic Information
21.6 Transcription: RNA Synthesis
21.7 The Genetic Code
21.8 Translation and Protein Synthesis
21.9 Mutations
21.10 Recombinant DNA

LEARNING OBJECTIVES/ASSESSMENT
When you have completed your study of this chapter, you should be able to:
1. Identify the components of nucleotides and correctly classify the sugars and bases. (Section 21.1; Exercises 21.2 and 21.4)
2. Describe the structure of DNA. (Section 21.2; Exercise 21.10)
3. Outline the process of DNA replication. (Section 21.3; Exercise 21.20)
4. Contrast the structures of DNA and RNA and list the function of the three types of cellular RNA. (Section 21.4; Exercises 21.26 and 21.28)
5. Describe what is meant by the terms *transcription* and *translation*. (Section 21.5; Exercise 21.32)
6. Describe the process by which RNA is synthesized in cells. (Section 21.6; Exercise 21.34)
7. Explain how the genetic code functions in the flow of genetic information. (Section 21.7; Exercise 21.38)
8. Outline the process by which proteins are synthesized in cells. (Section 21.8; Exercise 21.44)
9. Describe how genetic mutations occur and how they influence organisms. (Section 21.9; Exercise 21.48)
10. Describe the technology used to produce recombinant DNA. (Section 21.10; Exercise 21.52)

SOLUTIONS FOR THE END OF CHAPTER EXERCISES
COMPONENTS OF NUCLEIC ACIDS (SECTION 21.1)
☑21.2 DNA contains 2-deoxyribose. RNA contains ribose.

☑21.4
a. guanine — purine
b. thymine — pyrimidine
c. uracil — pyrimidine
d. cytosine — pyrimidine
e. adenine — purine

21.6 cytidine 5'-monophosphate

THE STRUCTURE OF DNA (SECTION 21.2)
21.8 The 3' end of the DNA segment TAGCC is C.
The 5' end of the DNA segment TAGCC is T.

288 Chapter 21

☑ 21.10 The secondary structure of DNA proposed by Watson and Crick is two complementary polynucleotide chains held together by hydrogen bonds between the A/T and C/G base pairs. The resulting structure is frequently called the double helix.

21.12   a. GCATGC   16 hydrogen bonds   b. TATGGC   15 hydrogen bonds

21.14   5′ ATGCATC 3′ original strand
3′ TACGTAG 5′ complementary strand   5′ GATGCAT 3′ complementary strand read 5′→3′

DNA REPLICATION (SECTION 21.3)

21.16 A chromosome is a structure within a cell that contains one molecule of DNA coiled about small, basic proteins called histones. Each human cell contains 46 chromosomes.

21.18 A replication fork is the point along the DNA helix at which the DNA helix unwinds for replication to begin.

☑ 21.20 The steps for DNA replication are: (1) unwinding the double helix, (2) synthesizing DNA segments, and (3) closing the nicks.

21.22 The new DNA strand is formed from the 5′ end to the 3′ end.

21.24

RIBONUCLEIC ACID (RNA) (SECTION 21.4)

☑ 21.26 The sugar in the sugar phosphate backbone of RNA contains a hydroxy group on the 2′ carbon, while in DNA it does not. RNA contains ribose and DNA contains deoxyribose.

☑ 21.28 Messenger RNA functions as a carrier of genetic information from the DNA of the cell nucleus directly to the cytoplasm. It contains 75-3000 nucleotides.
Ribosomal RNA is located in the cytoplasm of the cell in organelles called ribosomes which

serve as the sites of protein synthesis. It contains 120, 1700, or 3700 nucleotides.
Transfer RNA delivers amino acids to the site of protein synthesis. They contain 73-93 nucleotides.

21.30 The two important regions of the tRNA molecule are the anticodon that binds to the mRNA and the 3' end that binds to an amino acid.

THE FLOW OF GENETIC INFORMATION (SECTION 21.5)
☑21.32 Transcription is the transferring of necessary information from a DNA molecule onto a molecule of messenger RNA.
Translation is the decoding of the messenger RNA into an amino acid sequence.

TRANSCRIPTION: RNA SYNTHESIS (SECTION 21.6)
☑21.34 5' AGGTAGGTA 3' DNA template
3' UCCAUCCAU 5' mRNA

> 5' UACCUACCU 3' mRNA read 5'→3'

21.36 Introns are segments of DNA that do not carry amino acid code. Exons are segments of DNA that carry an amino acid code. hn-RNA is heterogeneous nuclear RNA. It is formed when both introns and exons are transcribed.

THE GENETIC CODE (SECTION 21.7)
☑21.38 a. CAU   histidine   b. UCA   serine   c. UCU   serine

21.40 a. False Each codon is composed of ~~four~~ bases. (Correction: not four bases, three bases)
b. True   Some amino acids are represented by more than one codon.
c. False ~~All~~ codons represent an amino acid. (Correction: Most, not all; some codons represent stop signals)
d. False Each living species is thought to ~~have its own unique genetic code~~. (Correction: share the same genetic code, not have its own unique genetic code)
e. True   The codon AUG at the beginning of a sequence is a signal for protein synthesis to begin at that codon.
f. False ~~It is not known whether or not~~ the code contains stop signals for protein synthesis. (Correction: UAG, UAA, and UGA are the known stop signals for protein synthesis.)

TRANSLATION AND PROTEIN SYNTHESIS (SECTION 21.8)
21.42 A codon is a three-letter sequence of the bases on a mRNA strand that represents a code for an amino acid or a stop signal.

☑21.44 The stages of protein synthesis are: (1) initiation of the polypeptide chain, (2) elongation of the chain, and (3) termination of polypeptide synthesis.

21.46 The A site is the aminoacyl site located on the mRNA-ribosome complex next to the P site where an incoming tRNA carrying the next amino acid will bond. The P site is the peptidyl site and it is the location on the ribosome that the initiating codon must occupy for the initiation process to begin.

MUTATIONS (SECTION 21.9)

☑21.48 A genetic mutation is a change in the genetic code that results from an incorrect sequence of bases on a DNA molecule.

21.50 The result of a genetic mutation that causes the mRNA sequence GCC to be replaced by CCC is that instead of the amino acid alanine, the polypeptide formed will contain the amino acid proline.

RECOMBINANT DNA (SECTION 21.10)

☑21.52 Recombinant DNA contains segments of DNA from two different organisms.

21.54 Substances likely to be produced on a large scale by genetic engineering are: human insulin (needed by diabetics allergic to insulin from pigs or cattle), human growth hormone (treatment of dwarfism), interferon (needed to fight viral infections), hepatitis B vaccine (protection against hepatitis), malaria vaccine (protection against malaria), and tissue plasminogen activator (needed to promote the dissolving of blood clots).

CHEMISTRY AROUND US, CHEMISTRY AND YOUR HEALTH, AND OVER THE COUNTER

21.56 The two categories of drugs approved by the FDA to combat HIV are reverse transcriptase inhibitors and protease inhibitors.

21.58 The AIDS virus is lethal because the HIV virus has an affinity for the helper T cell, white blood cells that play a critical role in the body's immune system. When these cells cannot perform their immune function, the body becomes susceptible to rare infections and unusual cancers.

21.60 Coupling PCR techniques with DNA fingerprinting allows for the ability to obtain DNA fingerprints from exceedingly small samples of genetic material.

21.62 Nuclei from adult mammary gland cells were the cellular source used to provide the DNA that led to the cloned sheep Dolly.

ADDITIONAL ACTIVITIES

Section 21.1 Review:
(1) Insert the following words into the appropriate area of the diagram on the right: only found in the cell nucleus, in cytoplasm, contains nucleotides, polymer, uracil, thymine, cytosine, adenine, guanine, D-ribose, and D-deoxyribose.
(2) Insert the bases uracil, thymine, cytosine, adenine, and guanine into the appropriate area of the pyrimidine/purine diagram to the right.

(3) Number the carbon atoms in the structure below.

(4) Which nucleic acid could contain the nucleotide in (3)?

Section 21.2 Review:
(1) Why are the linkages between nucleotides called *phosphodiester bonds*?
(2) Why is the sugar-phosphate chain called the *nucleic acid backbone*?
(3) How can the 5′ end of a nucleotide be identified?
(4) How can the 3′ end of a nucleotide be identified?
(5) Why are the percentages of guanine and cytosine in DNA always the same?

Section 21.3 Review:
(1) Rank these components in order increasing size: cell, DNA, chromosome, gene.
(2) What is semiconservative replication?
(3) Why does one DNA daughter strand not have any nicks, but the other one does?
(4) PCR requires heating and cooling DNA repeatedly. How can the human body replicate DNA without these extreme temperatures?

Section 21.4 Review:
Complete the diagram below by adding the following phrases in the appropriate location:

80-85% of RNA in the cell	contains loops	in nucleus	smallest RNA
carries genetic information	at least 20 types	single-stranded	lasts < 1 hr.
present during protein synthesis	has codon	in cytoplasm	has anticodon
component of ribosomes	polymer	deliver amino acids	

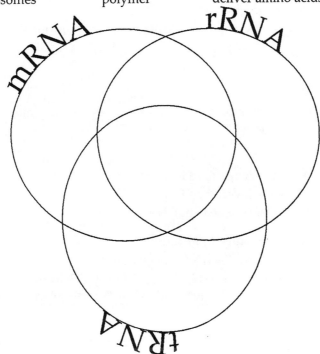

Section 21.5 Review:
(1) Which of the following only involve the "language" of nucleotides: replication, translation, transcription?
(2) DNA contains the "blueprint" for proteins, but it does not directly participate in protein synthesis. Describe how this occurs using the words replicate, translate, and transcribe.

Section 21.6 Review:
(1) RNA is synthesized using the "language" of nucleotides. How does the "vocabulary" differ between the template and the RNA produced?
(2) What is the direction of mRNA synthesis?
(3) How does (2) compare to the direction of daughter DNA strand synthesis?
(4) What is the direction for "reading" a strand of DNA during replication and transcription?
(5) Distinguish between introns and exons.
(6) When hnRNA undergoes enzyme-catalyzed reactions to cut and splice the hnRNA, what type of RNA is produced?

Section 21.7 Review:
Identify each of the descriptions below as degenerate, precise, universal, chain initiation, or chain termination.
(1) The code is almost the same for all living organisms.
(2) Each codon represents only one amino acid or stop.
(3) The codon required for protein synthesis to begin.
(4) One amino acid may be represented by several codons.
(5) One of the codons required for protein synthesis to end.

Section 21.8 Review:
(1) Protein synthesis occurs in three major steps: start, growth, stop. What are the scientific names for these steps?
(2) What is the direction of protein growth?
(3) Which end of the mRNA is closest to the P site of the ribosome during the "start" step of protein synthesis?
(4) What is *translocation*?

Section 21.9 Review:
(1) Will all mutations cause a difference in the synthesized protein? Explain.
(2) What are the causes of mutations?

Section 21.10 Review:
(1) What are the major types of enzymes used in genetic engineering?
(2) What is the difference between a vector and a plasmid?
(3) What are the steps for forming recombinant DNA?

Tying It All Together with a Laboratory Application:
The formation of (1) _____ DNA uses the DNA from two different organisms. The formation of (1) DNA is possible because the genetic code is (2) _____. When (1) DNA undergoes replication, (3) _____ are formed. This is a (4) _____ process because each of the "new" molecules contains (5) _____ and _____.

When (1) DNA undergoes transcription, (6) _____ is formed. (6) can then be used for (7) _____ synthesis. (8) _____ binds to the amino acids and delivers them to (6) which is "complexed" with a(n) (9) _____. The (10) _____ on the (6) only allows the (8) with the correct (11) _____ to leave an amino acid.

Nucleic Acids and Protein Synthesis 293

In order to synthesize (1) DNA, a(n) (12) _____ must be isolated from one organism using a(n) (13) _____. The (13) is also used on a(n) (14) _____, like a plasmid. Then, (15) _____ is used to join the (16) _____ of the fragments. A technique, like (17) _____, could be used to copy the (1) DNA. In addition to the (1) DNA, DNA polymerase, the cofactor MgCl$_2$, the four (18) _____ building blocks, and primers will be needed. The (18)s contain a phosphate group, (19) _____ (a sugar), and (20) _____, _____, _____, or _____ (nitrogen-containing bases).

SOLUTIONS FOR THE ADDITIONAL ACTIVITIES
Section 21.1 Review:

(1)   (2)

(3)   (4) DNA

Section 21.2 Review:
(1) The linkages between nucleotides are called *phosphodiester bonds* because a phosphate is bonded to two sugars through oxygen atoms; therefore, two phosphate ester bonds are present.

(2) The sugar-phosphate chain is called the *nucleic acid backbone* because it is the repeating unit of the polymer. The bases are the side chains.
(3) The 5' end of a nucleotide be identified as the end with a phosphate group.
(4) The 3' end of a nucleotide be identified as the end with a sugar group.
(5) DNA contains two complementary strands hydrogen bonded to form a double helix. The hydrogen bonds are formed between complementary base pairs. Guanine and cytosine are complementary base pairs; therefore, DNA always contains the same percentages of these bases.

Section 21.3 Review:

(1) cell ⊃ chromosome ⊃ DNA ⊃ gene

(2) Semiconservative replication is the process by which one molecule of DNA is used to produce two "new" molecules of DNA. Each of the "new" DNA molecules contains 1 new strand (called "daughter") and 1 old strand (called "parent"). The process is considered semiconservative because each new molecule contains half of the old molecule.

(3) As the DNA unwinds at a replication fork, both of the parent strands are read as templates from the 3' end to the 5' end. The daughter strands grow from the 5' end to the 3' end. The template strands are antiparallel to each other; therefore, one of the daughter strands grows continuously as the DNA parent strands unwind while the other grows in Okazaki fragments separated by nicks as the parent strands unwind.

(4) The human body replicates DNA without these extreme temperatures by using enzymes to aid in the replication process.

Section 21.4 Review:

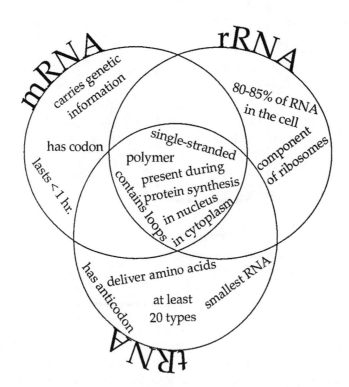

Section 21.5 Review:

(1) Replication and transcription only involve the "language" of nucleotides. Replication is the formation of new DNA molecules from old DNA molecules. Transcription is the process of copying segments of DNA into mRNA that can leave the nucleus.

(2) DNA is replicated within the nucleus. When a protein is needed, the DNA unwinds and is transcribed into mRNA. The mRNA leaves the nucleus and enters the cytoplasm in order to align with a ribosome. The tRNA delivers amino acids to the mRNA-ribosome complex and the mRNA is translated into a protein.

Section 21.6 Review:
(1) The DNA template has a sugar-phosphate backbone that contains D-deoxyribose rather than the D-ribose of the synthesized RNA. Also, the DNA contains thymine and does not contain uracil, while the synthesized RNA contains uracil and does not contain thymine.
(2) A molecule of mRNA is synthesized from 5' end to 3' end.
(3) DNA daughter strand synthesis also occurs from the 5' end to the 3' end.
(4) A strand of DNA is read from the 3' end to the 5' end during replication and transcription.
(5) Introns do not carry genetic code and exons carry genetic code.
(6) mRNA is produced when hnRNA undergoes enzyme-catalyzed reactions to cut and splice the hnRNA.

Section 21.7 Review:
(1) universal; (2) precise; (3) chain initiation; (4) degenerate; (5) chain termination

Section 21.8 Review:
(1) Start = initiation of the polypeptide chain; growth = elongation of the chain; stop = termination of polypeptide synthesis
(2) Proteins are synthesized from the N-terminal end to the C-terminal end.
(3) The 5' end of the mRNA is closest to the P site of the ribosome during the "start" step of protein synthesis.
(4) *Translocation* is the movement of the ribosome along the mRNA during protein synthesis.

Section 21.9 Review:
(1) No, not all mutations will cause a difference the synthesized protein. The genetic code is degenerate, so it is possible that a mutation in the code could still "call" for the same amino acid. Some mutations will "call" for a different amino acid, may lose the "start" codon, or may lose the "stop" codon.
(2) The causes of mutations are naturally occurring errors in DNA replication, ionizing radiation, and chemicals called mutagens.

Section 21.10 Review:
(1) Restriction enzymes and DNA ligases are used in genetic engineering.
(2) A vector carries foreign DNA into a cell. One example of a vector is a plasmid which is a circular piece of double-stranded DNA.
(3) The steps for forming recombinant DNA are isolating plasmids from bacterium, using restriction enzymes on the plasmid and the human chromosome to produce sticky ends, splicing the plasmid and the human chromosome using DNA ligase.

Tying It All Together with a Laboratory Application:
(1) recombinant
(2) universal
(3) 2 molecules of DNA
(4) semiconservative
(5) 1 strand of parent DNA, 1 strand of daughter DNA
(6) mRNA
(7) protein
(8) tRNA
(9) ribosome
(10) codon
(11) anticodon
(12) gene
(13) restriction enzyme
(14) vector
(15) DNA ligase
(16) "sticky ends"
(17) PCR (polymerase chain reaction)
(18) nucleotide
(19) D-deoxyribose
(20) adenine, cytosine, guanine, thymine

SELF-TEST QUESTIONS

Multiple Choice

1. One of the primary differences between DNA and RNA involves
   a. the phosphate
   b. the pentose sugar
   c. the bonding between sugar and base
   d. the bonding between sugar and phosphate

2. Which of the following bases is a purine?
   a. uracil
   b. adenine
   c. thymine
   d. cytosine

3. A sample of nucleic acid is to be analyzed. For which of the following would an analysis be useful in deciding whether the sample was DNA or RNA?
   a. guanine
   b. cytosine
   c. adenine
   d. thymine

4. The structural backbone of all nucleic acids consists of alternating molecules of
   a. a sugar and a base
   b. a sugar and a phosphate
   c. a base and a phosphate
   d. purine and pyrimidine bases

5. If the nucleotide sequence in one strand of DNA were T-C-G, the complementary strand written in 5'-3' sequence would be
   a. A-C-G
   b. T-G-C
   c. C-G-A
   d. A-G-C

6. A sample of double helical DNA contained 26% of the base thymine (T). The amount of adenine (A) should be
   a. 26%
   b. 24%
   c. 22%
   d. 20%

7. The number of hydrogen bonds between C and G in double-stranded helical DNA is
   a. 0
   b. 1
   c. 2
   d. 3

8. mRNA synthesis is called
   a. replication
   b. translocation
   c. translation
   d. transcription

9. Histidine is represented by the codons CAC and CAU. This is an example of
   a. degeneracy of the genetic code
   b. punctuation in the genetic code
   c. a nonsense codon
   d. the universal nature of the genetic code

10. The genetic code consists of _____ three-letter codons.
    a. 16
    b. 32
    c. 64
    d. 86

11. Segments of eukaryotic DNA which code for amino acids are called
    a. exons
    b. introns
    c. hnRNA
    d. codons

12. Recombinant DNA is DNA
    a. formed by combining portions of DNA from two different sources
    b. found within plasmids
    c. with an unusual ability to recombine
    d. which contains two template strands

Matching
Characteristics are given below for various nucleic acids. Match each description to the correct type of nucleic acid as a response.

13. represents a large percentage of cellular RNA
14. serves as master template for the formation of all nucleic acids in the body
15. contains base sequences called codons
16. has lowest molecular weight of all nucleic acids
17. replicated during cell division
18. delivers amino acids to site of protein synthesis
19. contains anti-codon
20. helps to serve as a site for protein synthesis
21. carries the directions for protein synthesis to the site of protein synthesis

a. DNA
b. mRNA
c. tRNA
d. rRNA

True-False
22. Purine bases contain both a five- and a six-member ring.
23. DNA and RNA contain identical nucleic acid backbones.
24. During DNA replication, only one DNA strand serves as a template for the formation of a complementary strand.
25. A complementary DNA strand formed during replication is identical to the strand serving as a template.
26. RNA is usually double-stranded.
27. Some amino acids are represented by more than one codon.
28. The genetic code has some code words that don't code for any amino acids.
29. The genetic code is thought to be different for humans and bacteria.
30. Incoming tRNA, carrying an amino acid, attaches at the peptidyl site.
31. N-formylmethionine is always the first amino acid in a growing peptide chain in prokaryotic cells.
32. Protein synthesis generally takes place within the cell nucleus.
33. Some genetic mutations might aid an organism rather than hinder it.
34. Viruses contain both nucleic acids and protein.
35. Restriction enzymes act at sites on DNA called palindromes.

Chapter 22: Nutrition and Energy for Life

CHAPTER OUTLINE
22.1 Nutritional Requirements
22.2 The Macronutrients
22.3 Micronutrients I: Vitamins
22.4 Micronutrients II: Minerals
22.5 The Flow of Energy in the Biosphere
22.6 Metabolism and an Overview of Energy Production
22.7 ATP: The Primary Energy Carrier
22.8 Important Coenzymes in the Common Catabolic Pathway

LEARNING OBJECTIVES/ASSESSMENT
When you have completed your study of this chapter, you should be able to:
1. Describe the difference between macronutrients and micronutrients in terms of amounts required and their functions in the body. (Section 22.1; Exercise 22.2)
2. Describe the primary functions in the body of each macronutrient. (Section 22.2; Exercise 22.4)
3. Distinguish between and classify vitamins as water-soluble or fat-soluble on the basis of name and behavior in the body. (Section 22.3; Exercise 22.10)
4. List a primary function in the body for each major mineral. (Section 22.4; Exercise 22.16)
5. Describe the major steps in the flow of energy in the biosphere. (Section 22.5; Exercise 22.24)
6. Differentiate among metabolism, anabolism, and catabolism. (Section 22.6; Exercise 22.26)
7. Outline the three stages in the extraction of energy from food. (Section 22.6; Exercise 22.28)
8. Explain how ATP plays a central role in the production and use of cellular energy. (Section 22.7; Exercise 22.34)
9. Explain the role of coenzymes in the common catabolic pathway. (Section 22.8; Exercise 22.46)

SOLUTIONS FOR THE END OF CHAPTER EXERCISES
NUTRITIONAL REQUIREMENTS (SECTION 22.1)
☑22.2 Macronutrients are nutrients needed by the body in gram quantities every day. Micronutrients are nutrients needed by the body in milligram or microgram quantities every day.

THE MACRONUTRIENTS (SECTION 22.2)
☑22.4 Carbohydrates are used for energy and materials for the synthesis of cell and tissue components. Lipids are a concentrated source of energy; they store fat-soluble vitamins and help carry them through the body, and provide essential fatty acids. Proteins are used to make new tissues, maintain and repair cells, synthesize enzymes, hormones, and other nitrogen-containing compounds of the body, and provide energy.

22.6
a. potato chips — carbohydrates, lipids, proteins
b. buttered toast — carbohydrates, lipids, proteins
c. plain toast with jam — carbohydrates, proteins
d. cheese sandwich — carbohydrates, lipids, proteins
e. a lean steak — proteins, lipids
f. a fried egg — lipids, proteins, carbohydrates

22.8 The daily values are 65 g of total fat; 20 g of saturated fat, 300 g of total carbohydrates, and 25 g of fiber for a 2000-Calorie diet.

MICRONUTRIENTS I: VITAMINS (SECTION 22.3)

☑22.10   a. tocopherol   fat-soluble   c. folic acid   water-soluble
         b. niacin   water-soluble   d. retinol   fat-soluble

22.12   Large doses of fat-soluble vitamins cause more concern than large doses of water-soluble vitamins because excess fat-soluble vitamins will stay in the body; however, large doses of water-soluble vitamins will be excreted.

22.14   a. scurvy   vitamin C (ascorbic acid)   c. pernicious anemia   vitamin B_{12} (cobalamin)
         b. beriberi   vitamin B_1 (thiamin)   d. pellagra   niacin

MICRONUTRIENTS II: MINERALS (SECTION 22.4)

☑22.16   Calcium is used for inorganic structural components of bones and teeth.
Phosphorus is used for inorganic structural components of bones and teeth.
Potassium is used for ions distributed throughout the body's various fluids.
Sulfur is used for inorganic structural components of bones and component of proteins.
Sodium is used for ions distributed throughout the body's various fluids.
Chloride is used for ions distributed throughout the body's various fluids.
Magnesium is used for ions distributed throughout the body's various fluids.

22.18   Trace minerals are components of vitamins, enzymes, hormones, and specialized proteins.

THE FLOW OF ENERGY IN THE BIOSPHERE (SECTION 22.5)

22.20   Hydrogen is the fuel for the fusion process occurring on the sun.

22.22 $6\ CO_2 + 6\ H_2O + \text{energy} \rightarrow C_6H_{12}O_6 + 6\ O_2$

☑22.24   Plants use the energy from the sun to produce glucose. Animals eat the plants that contain glucose. During cellular respiration, glucose and other storage forms of energy are converted into energy. Some of that energy is captured within the cells in the form of ATP.

METABOLISM AND AN OVERVIEW OF ENERGY PRODUCTION (SECTION 22.6)

☑22.26   Catabolism includes all breakdown processes. Anabolism includes all synthesis processes.

☑22.28   a. Formation of ATP   Stage III   c. Consumption of O_2 Stage III
         b. Digestion of fuel molecules Stage I   d. Generation of acetyl CoA   Stage II

22.30   The main purpose of the catabolic pathway is to produce ATP.

ATP: THE PRIMARY ENERGY CARRIER (SECTION 22.7)

22.32   ATP is involved in both anabolic and catabolic processes.

☑22.34   ATP is called the primary energy carrier because the catabolism of food results in the production of ATP which stores energy until it is delivered to cells for their use.

22.36 P_i stands for inorganic phosphate. PP_i stands for pyrophosphate.

Nutrition and Energy for Life 301

22.38 Phosphoenolpyruvate is called a high-energy compound because it releases 14.8 kcal per mole; however, glycerol 3-phosphate is not a high-energy compound because it only releases 2.2 kcal per mole.

22.40 The triphosphate end of ATP is the portion of the ATP molecule particularly responsible for its being described as a high-energy compound. The energy is stored in the bonds within the ATP.

22.42 Mitochondria serve as the site for most ATP synthesis in the cells.

22.44 Mitochondria have an outer membrane, an intermembrane space, and an inner membrane that forms cristae and the matrix. The enzymes needed for energy production are located within the inner membrane and inside the matrix.

IMPORTANT COENZYMES IN THE COMMON CATABOLIC PATHWAY (SECTION 22.8)

☑22.46 a. coenzyme A — This coenzyme is part of acetyl coenzyme A which is formed from all foods as they pass through Stage II of catabolism.
b. FAD — This coenzyme forms coenzyme $FADH_2$ by being reduced and carries hydrogen atoms from the citric acid cycle to the electron transport chain.
c. NAD^+ — This coenzyme forms coenzyme NADH by being reduced and carries hydrogen atoms from the citric acid cycle to the electron transport chain.

22.48 $^-OOC-CH_2-CH_2-COO^-$ is oxidized. FAD is reduced.

22.50 a. $HO-CH_2-COOH + NAD^+ \rightarrow O=CH-COOH + NADH + H^+$
b. $HO-CH_2CH_2-OH + FAD \rightarrow HO-CH=CH-OH + FADH_2$

22.52

FAD   FADH$_2$

CHEMISTRY AROUND US, CHEMISTRY AND YOUR HEALTH, AND OVER THE COUNTER

22.54 Vitamin B$_{12}$, vitamin D, calcium, iron, and zinc are often recommended supplements for vegetarians.

22.56 The amount of visceral fat can be lowered by decreasing the intake of foods high in saturated fat and increasing the amount of exercise.

ADDITIONAL ACTIVITIES

Section 22.1 Review:
(1) What do the prefixes *macro-* and *micro-* mean?
(2) What are the subclasses of macronutrients?
(3) What are the subclasses of micronutrients?

302 Chapter 22

(4) What are the four general categories of materials the human body must receive from diet?
(5) How are Reference Daily Intakes, Daily Reference Values, and Daily Values related?

Section 22.2 Review:
(1) Carbohydrates are the main dietary source of _____.
(2) What is the difference between simple and complex carbohydrates?
(3) Why must lipids be included in a healthy diet?
(4) What type of fat is considered the "wrong type" for healthy eating?
(5) Will a food that is a "complete protein" for a child always be a "complete protein" for an adult? Explain.

Section 22.3 Review:
(1) What is the main role of water-soluble vitamins in the body?
(2) Which water-soluble vitamin is an exception to (1)?
(3) Do the fat-soluble vitamins have the same role in the body as water-soluble vitamins? Explain.
(4) Can a person overdose on vitamins? Explain.

Section 22.4 Review:
(1) What is the difference between major minerals and trace minerals in terms of dietary requirements?
(2) What is the difference between major minerals and trace minerals in terms of functions in the body?
(3) If a patient is told to limit their sodium intake, what seasoning should they avoid?
(4) Does (3) contain elemental sodium? Explain.

Section 22.5 Review:
(1) Classify the following processes as endothermic or exothermic.
    (a) nuclear fusion
    (b) photosynthesis
    (c) cellular respiration
(2) Identify carbon dioxide as a reactant, product, or neither for the following processes.
    (a) nuclear fusion
    (b) photosynthesis
    (c) cellular respiration
(3) Repeat (2) for oxygen.
(4) Repeat (2) for water.

Section 22.6 Review:
(1) Metabolism can be divided into two subgroups as shown below. Label the subgroups with their scientific names.

(2) Which of the processes in (1) is exothermic?
(3) Which of the processes in (1) is endothermic?
(4) Using the construction/wrecking ball analogy, describe the three stages of the catabolism of food.

Section 22.7 Review:
(1) Where is the energy stored that is released during the oxidation of macronutrients?
(2) What are the components of ATP?
(3) How does ATP differ from the nucleotide containing adenine?
(4) What is the sign on ΔG for exothermic reactions?
(5) Complete the following diagram for the hydrolysis of ATP.

(6) Do the charges balance in each of the hydrolysis processes shown in (5)? Explain.
(7) Which of the processes in (5) produces more energy?
(8) What is one of the important processes that occurs in the mitochondria of a cell?

Section 22.8 Review:
(1) Label the following symbol for Coenzyme A.

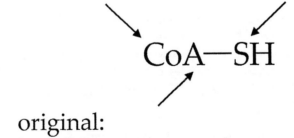

original:

now:
(2) Write a balanced equation for the reduction of oxygen to water by the addition of electrons and hydrogen ions.
(3) Write the equation for the decomposition of NADH into NAD$^+$, electrons, and hydrogen ions.
(4) Write the equation for the decomposition of FADH$_2$ into FAD, electrons, and hydrogen ions.
(5) Write the balanced equation for the reduction of oxygen to water by the addition of NADH. Include any additional hydrogen ions needed to balance the reaction.
(6) Write the balanced equation for the reduction of oxygen to water by the addition of FADH$_2$. Include any additional hydrogen ions needed to balance the reaction.
(7) If coenzymes are chemical group shuttles, what do the different coenzymes in this section shuttle?

Tying It All Together with a Laboratory Application:

The Calories in one serving of reduced-fat peanut butter, based on the food label on the next page, are (1) _____. Calories come from (2) _____, not from (3) _____, which include vitamins and minerals. The percentage of Calories from fat is (4) _____, carbohydrates is (5) _____, and protein is (6) _____. Reduced-fat peanut butter (7) _____ the recommended percentage of fat in the diet, which is no more than (8) _____. Reduced-fat peanut butter (9) _____ the recommended percentage of carbohydrates in the diet, which is about (10) _____. Reduced-fat peanut butter can still be included as part of a healthy diet, though.

Calories	?
Total Fat	12 g
Saturated Fat	2.5 g
Cholesterol	0 mg
Sodium	210 mg
Total Carbohydrate	15 g
Dietary Fiber	2 g
Sugars	4g
Protein	8 g
Vitamin A	0%
Vitamin C	0%
Calcium	0%
Iron	4%
Niacin	25%
Vitamin B$_6$	6%
Folic Acid	6%
Magnesium	15%
Zinc	6%
Copper	10%

Reduced-fat peanut butter contains (11) _____ g of complex carbohydrates, not including the fiber. It also contains (12) _____ g of unsaturated fats. If the ratio of unsaturated to saturated fats is higher for reduced-fat peanut butter, then the consistency of reduced-fat peanut butter may be more (13) _____ than regular peanut butter. The daily value percentage can be listed for all of the nutritional categories except sugars and proteins. The Daily Value for protein is called (14) _____, but it cannot be turned into a universal percentage because (15) _____.

Reduced-fat peanut butter might contain (16) _____ amino acids, which can only be obtained through diet; however, the labeling laws do not require that (16) amino acids be identified on product packaging. Reduced-fat peanut butter could be considered a source of (17) _____ protein for some people, if it contains all the (16) amino acids in the proper proportions needed by the body.

Vitamin A is a (18) _____-soluble vitamin, unlike Vitamin C which is a (19) _____-soluble vitamin. Neither vitamin A or vitamin C is contained in reduced-fat peanut butter. The magnesium and copper in reduced-fat peanut butter (20) _____ in metallic form, which is common for (21) _____. Magnesium is a (22) _____ mineral, while copper is a (23) _____ mineral.

The energy stored in peanut-butter originally came from (24) _____, where nuclear fusion occurs. The sunlight allows carbon dioxide and water to be converted through (25) _____ into glucose and oxygen. The glucose is then converted into other (26) _____ forms of energy.

When a person eats reduced-fat peanut butter, the body digests the peanut butter through (27) _____ processes that result in (28) _____ at the end of the common (27) pathway. This pathway utilizes a(n) (29) _____ called CoA to (30) _____ acyl groups.

SOLUTIONS FOR THE ADDITIONAL ACTIVITIES

Section 22.1 Review:
(1) *Macro-* means large. *Micro-* means very small.
(2) The subclasses of macronutrients are carbohydrates, proteins, and lipids.
(3) The subclasses of micronutrients are vitamins and minerals.
(4) The four general categories of materials the human body must receive from diet are macronutrients, micronutrients, water, and fiber.
(5) Reference Daily Intakes are the Daily Values for protein as well as 19 vitamins and minerals. Daily Reference Values are the Daily Values for other nutrients of public health importance.

Section 22.2 Review:
(1) energy
(2) Simple carbohydrates are monosaccharides and disaccharides. Complex carbohydrates are polysaccharides.
(3) Lipids are a necessary part of the diet as a source of energy, source and transportation for fat-soluble vitamins, and source of essential fatty acids.
(4) Saturated fat is considered the "wrong type" for healthy eating.
(5) A food that is a "complete protein" for a child will not always be a "complete protein" for an adult because a complete protein contains all of the essential amino acids in the proper proportions needed by the body. The daily needs of each essential amino acid will vary by individual.

Section 22.3 Review:
(1) The main role of water-soluble vitamins in the body is serving as coenzymes.
(2) Vitamin C does not function as a coenzyme. It participates in the synthesis of collagen for connective tissue.
(3) The fat-soluble vitamins do not have the same role in the body as water-soluble vitamins. They serve many different roles: participate in the synthesis of visual pigments, regulation of calcium and phosphorus metabolism, prevention of oxidation of vitamin A and fatty acids, synthesis of blood-clotting substances.
(4) A person can overdose on vitamins, especially fat-soluble vitamins. The water-soluble vitamins can be removed from the body by the kidneys; however, the fat-soluble vitamins are stored in fat reserves which are not filtered by the body.

Section 22.4 Review:
(1) Major minerals are found in the body in quantities of greater than 5 g. Minor minerals are found in the body in quantities smaller than 5 g. In general, a healthy diet includes more major minerals than trace minerals.
(2) Major minerals serve structural functions as well as being distributed throughout the body's fluids. Trace minerals are components of vitamins, enzymes, hormones, and specialized proteins.
(3) If a patient is told to limit their sodium intake, the patient should avoid table salt (sodium chloride).
(4) Table salt does not contain elemental sodium. It contains sodium ions.

Section 22.5 Review:
(1) (a) exothermic   (b) endothermic   (c) exothermic
(2) (a) neither   (b) reactant   (c) product

306 Chapter 22

(3) (a) neither   (b) product   (c) reactant
(4) (a) neither   (b) reactant   (c) product

Section 22.6 Review:

Metabolism

(4) Stage I of the catabolism of food is the "wrecking ball" phase. Large molecules are broken into smaller molecules (proteins → amino acids, carbohydrates → monosaccharides, lipids → fatty acids and glycerol). Stage II of the catabolism of food involves further breakdown of the small molecules into even simpler units, like the acetyl portion of acetyl CoA. These "refining" processes are unique for each type of material; however, the major product of these processes is the same. Stage III of the catabolism of food takes the refined product and "recycles" it through the citric acid cycle, the electron transport chain, and oxidative phosphorylation to produce high energy ATP.

Section 22.7 Review:

(1) The energy released during the oxidation of macronutrients is stored in some of the bonds of ATP.
(2) The components of ATP are ribose, adenine, and triphosphate.
(3) ATP differs from the nucleotide containing adenine only in the size of the phosphate group attached to the ribose. ATP contains a triphosphate group, but the nucleotide contains a monophosphate group.
(4) The sign on ΔG for exothermic reactions is negative.
(5)

(6) Yes, see calculations below.
   ATP + H$_2$O = 4- + 0 = 4-
   ADP + P$_i$ + H$^+$ = 3- + 2- + 1+ = 4-
   AMP + PP$_i$ + 2 H$^+$ = 2- + 4- + 2(1+) = 4-
   PP$_i$ + H$_2$O = 4- + 0 = 4-
   2 P$_i$ = 2(2-) = 4-

(7) The conversion of ATP to AMP produces more energy than the conversion of ATP to ADP, even if the pyrophosphate is not hydrolyzed.
(8) ATP synthesis occurs in the mitochondria of a cell.

Section 22.8 Review:

```
        coenzyme    key functional group
              ↘     ↙
              CoA—SH
                ↗
```

(1) original: transfers acetyl group
     now: transfers acyl groups
(2) $O_2 + 4\,e^- + 4\,H^+ \rightarrow 2\,H_2O$
(3) $NADH \rightarrow NAD^+ + 2\,e^- + H^+$
(4) $FADH_2 \rightarrow FAD + 2\,e^- + 2\,H^+$
(5) $O_2 + 2\,NADH + 2\,H^+ \rightarrow 2\,H_2O + 2\,NAD^+$
(6) $O_2 + 2\,FADH_2 \rightarrow 2\,H_2O + 2\,FAD$
(7) Coenzyme A shuttles acyl groups, while NADH and $FADH_2$ shuttle electrons and hydrogen ions.

Tying It All Together with a Laboratory Application:

(1) 190 Calories (fiber = Calorie-free)
(2) macronutrients
(3) micronutrients
(4) 56%
(5) 27%
(6) 17%
(7) exceeds
(8) 30%
(9) has less than
(10) 58%
(11) 9 g
(12) 9.5 g
(13) liquid
(14) Reference Daily Intake
(15) RDI varies based on age and gender
(16) essential
(17) complete
(18) fat
(19) water
(20) are not
(21) minerals
(22) major
(23) trace
(24) the sun
(25) photosynthesis
(26) storage
(27) catabolic
(28) ATP
(29) coenzyme
(30) transfer

SELF-TEST QUESTIONS

Multiple Choice

1. Nutrients required by the body in relatively large amounts are
   a. fiber   b. macronutrients   c. micronutrients   d. vitamins

2. The "D" in DV stands for
   a. daily   b. determined   c. dietary   d. deficiency

3. Which of the following is an example of a complex carbohydrate nutrient for humans?
   a. starch   b. cellulose   c. glucose   d. lactose

4. Which of the following has an established DV?
   a. carbohydrates   b. lipids   c. proteins   d. water

5. Which of the following is a water-soluble vitamin?
   a. vitamin A   b. vitamin B_2 c. vitamin E   d. vitamin D

6. Which of the following is a trace mineral
   a. phosphorus   b. sulfur   c. potassium   d. copper

7. It is recommended that no more than _____ percent of our calories be obtained from fats.
   a. 10   b. 30   c. 40   d. 50

8. Mitochondria are called "power stations of the cell" because
   a. most of the cellular energy is consumed there.
   b. mitochondria are rich in fat molecules.
   c. mitochondria exist at higher temperatures than other cell components.
   d. mitochondria are the major sites of ATP synthesis.

9. What product of the second stage is passed on to the third stage of the energy production process?
   a. CO_2 b. H atoms   c. acetyl CoA   d. H_2O

10. FAD often accompanies the enzyme-catalyzed
    a. oxidation of $-CH_2CH_2-$ to $-CH=CH-$
    b. oxidation of aldehydes
    c. oxidation of alcohol groups
    d. transfer of acetyl groups

Matching
Select the stage in the extraction of energy from food where the following molecules are produced.
11. majority of ATP
12. amino acids
13. acetyl CoA
14. carbon dioxide

a. stage I
b. stage II
c. stage III
d. more than one response is correct

Select the mineral that best matches each description on the left.
15. a major mineral found in the body in largest amounts
16. a vitamin component
17. a hormone component

a. cobalt
b. iodine
c. calcium
d. zinc

Select the coenzyme that best matches each description on the left.
18. serves as an electron acceptor in the oxidation of an alcohol
19. a derivative of the vitamin riboflavin
20. contains a sulfhydryl group (-SH)

a. coenzyme Q
b. coenzyme A
c. FAD
d. NAD^+

Select the process that best matches each description on the left.
21. energy-releasing solar process
22. conversion of CO_2 and H_2O to carbohydrates
23. oxidation of glucose to CO_2 and H_2O

a. cellular respiration
b. photosynthesis
c. nuclear fusion
d. cellular anabolism

True-False
24. Carbon dioxide is a source of carbon for the earth's organic compounds.
25. ΔG for the hydrolysis of ATP is a positive value.
26. Nine amino acids are listed as being essential.
27. The base present in ATP is dexoyribose.
28. An ATP molecule at pH 7.4 (body pH) has a -3 charge.
29. On a mass basis, water is the most abundant compound found in the human body.
30. Sucrose, a disaccharide, is correctly classified as a complex carbohydrate.
31. Most nutritional studies indicate that typical diets in the U.S. contain too high a percentage of complex carbohydrates.
32. Lipids are digested faster than carbohydrates or proteins.
33. Concern about vitamin overdoses focuses primarily on water-soluble vitamins.
34. Most fat-soluble vitamins are known to function in the body as coenzymes.
35. Vitamin K is important in the process of blood clotting.

Chapter 23: Carbohydrate Metabolism

CHAPTER OUTLINE
23.1 The Digestion of Carbohydrates
23.2 Blood Glucose
23.3 Glycolysis
23.4 The Fates of Pyruvate
23.5 The Citric Acid Cycle
23.6 The Electron Transport Chain
23.7 Oxidative Phosphorylation
23.8 The Complete Oxidation of Glucose
23.9 Glycogen Metabolism
23.10 Gluconeogenesis
23.11 The Hormonal Control of Carbohydrate Metabolism

LEARNING OBJECTIVES/ASSESSMENT
When you have completed your study of this chapter, you should be able to:
1. Identify the products of carbohydrate digestion. (Section 23.1; Exercise 23.2)
2. Explain the importance to the body of maintaining proper blood sugar levels. (Section 23.2; Exercise 23.6)
3. List the starting material and products of the glycolysis pathway. (Section 23.3; Exercise 23.10)
4. Describe how the glycolysis pathway is regulated in response to cellular needs. (Section 23.3; Exercise 23.14)
5. Name the three fates of pyruvate. (Section 23.4; Exercise 23.20)
6. Identify the two major functions of the citric acid cycle. (Section 23.5; Exercise 23.24)
7. Describe how the citric acid cycle is regulated in response to cellular energy needs. (Section 23.5; Exercise 23.32)
8. Explain the function of the electron transport chain and describe how electrons move down the chain. (Section 23.6; Exercises 23.34 and 23.36)
9. List the major feature of the chemiosmotic hypothesis. (Section 23.7; Exercise 23.46)
10. Calculate the amount of ATP produced by the complete oxidation of a mole of glucose. (Section 23.8; Exercise 23.48)
11. Explain the importance of the processes of glycogenesis and glycogenolysis. (Section 23.9; Exercise 23.50)
12. Describe gluconeogenesis and the operation of the Cori cycle. (Section 23.10; Exercise 23.58)
13. Describe how hormones regulate carbohydrate metabolism. (Section 23.11; Exercise 23.62)

SOLUTIONS FOR THE END OF CHAPTER EXERCISES
THE DIGESTION OF CARBOHYDRATES (SECTION 23.1)
☑23.2 a. starch — glucose
       b. lactose — glucose, galactose
       c. sucrose — fructose, glucose
       d. maltose — glucose

BLOOD GLUCOSE (SECTION 23.2)
23.4 Blood sugar level is the concentration of glucose in the blood.
Normal fasting level is the concentration of glucose in the blood after an 8-12 hour fast.

☑23.6 a. hypoglycemia — blood sugar level is below the normal fasting level
       b. hyperglycemia — blood sugar level is above the normal fasting level
       c. renal threshold — blood sugar level is above 108 mg/100 mL and the sugar is not completely reabsorbed by the kidneys
       d. glucosuria — blood sugar level is above the renal threshold and glucose appears in the urine

310 Chapter 23

23.8 Severe hypoglycemia can be very serious because it can cause convulsions and shock.

GLYCOLYSIS (SECTION 23.3)

☑23.10 The starting material for glycolysis is glucose. The product of glycolysis is pyruvate.

23.12 Two steps in glycolysis require ATP. Two steps produce ATP. The net production of ATP from one glycose molecule undergoing glycolysis is 2 molecules of ATP.

☑23.14 The third control point in glycolysis is the conversion of phosphoenolpyruvate to pyruvate catalyzed by pyruvate kinase. It is responsive to cellular needs because the enzyme is inhibited by higher concentrations of ATP.

23.16 A high concentration of glucose 6-phosphate inhibits glycolysis.

THE FATES OF PYRUVATE (SECTION 23.4)

23.18 Aerobic describes a process that requires oxygen to take place. Anaerobic describes a process that does not require oxygen to take place.

☑23.20 a. The fate of pyruvate in the body under aerobic conditions is conversion to acetyl CoA.
b. The fate of pyruvate in the body under anaerobic conditions is conversion to lactate.

23.22 The fate of the pyruvate molecule's carboxylate group during fermentation is conversion to a hydroxy group as hydrogen is added across the carbon-oxygen double bond. Eventually, carbon dioxide is produced.

THE CITRIC ACID CYCLE (SECTION 23.5)

☑23.24 The primary function of the citric acid cycle in ATP production is meeting the cellular needs for ATP by generating the reduced coenzymes NADH and FADH$_2$. The other vital role served by the citric acid cycle is the production of intermediates for biosynthesis.

23.26 a. The fuel needed by the cycle — acetyl CoA
b. The form in which carbon atoms leave the cycle — CO$_2$
c. The form in which hydrogen atoms and electrons leave the cycle — NADH, FADH$_2$

23.28 a. NADH 3 b. FADH$_2$ 1 c. GTP 1 d. CO$_2$ 2

23.30 succinyl CoA + GDP + P$_i$ → GTP + succinate + CoA-SH

☑23.32 a. The enzymes at the three control points of the citric acid cycle are citrate synthetase (step 1), isocitrate dehydrogenase (step 3), and the α-ketoglutarate dehydrogenase complex (step 4).
b. When the NADH levels are high, NADH serves as an inhibitor to slow the production of NADH and FADH$_2$. When ATP levels are high, the entry of acetyl CoA into the citric acid cycle is reduced.
c. Two of the control point enzymes are activated by ADP. This benefits the cell because not only can ADP can be used to synthesize ATP, but it is also in high concentration when ATP is in low concentration.

Carbohydrate Metabolism

THE ELECTRON TRANSPORT CHAIN (SECTION 23.6)

☑23.34   NADH and FADH₂ bring electrons to the electron transport chain.

☑23.36   The role of the cytochromes in the electron transport chain is to pass electrons along the chain to oxygen atoms and provide H⁺ ions to the oxygen in order to form water.

OXIDATIVE PHOSPHORYLATION (SECTION 23.7)

23.38   NADH and FADH₂ are oxidized during oxidative phosphorylation.
ADP is phosphorylated during oxidative phosphorylation.

23.40   1.5 ATP molecules are synthesized when one FADH₂ passes electrons to molecular oxygen.

23.42   When six acetyl CoA molecules enter the common catabolic pathway, 60 ATP molecules are formed by oxidation of acetyl CoA.

23.44 $\dfrac{686 \text{ kcal}}{1 \text{ mole glucose}} \left(\dfrac{1 \text{ mol ATP}}{7.3 \text{ kcal}} \right) = 94 \text{ mol ATP}$

☑23.46   ATP synthesis results from protons crossing the mitochondrial membranes.

THE COMPLETE OXIDATION OF GLYCOSE (SECTION 23.8)

☑23.48   a. Glycolysis produces 2 moles of ATP.
b. The citric acid cycle produces 2 moles of ATP.
c. The electron transport chain and oxidative phosphorylation produces 28 moles of ATP.

GLYCOGEN METABOLISM (SECTION 23.9)

☑23.50   Glycogenesis is the process by which glucose is converted to glycogen. Glycogenolysis is the process by which glycogen is broken back down into glucose. These are opposite processes.

23.52   The high-energy compound involved in the conversion of glucose to glycogen is uridine triphosphate (UTP).

23.54   Liver glycogen can be used to raise blood sugar levels because glycogenolysis can occur in the liver, but not in muscle tissue.

GLUCONEOGENESIS (SECTION 23.10)

23.56   The principal site for gluconeogenesis is the liver.

☑23.58   The sources of compounds that undergo gluconeogenesis are lactate, glycerol, and certain amino acids.

23.60   The Cori cycle would not operate if the body's muscles were completely oxidizing the glucose supplied because the body would not produce lactate (required for the Cori cycle).

THE HORMONAL CONTROL OF CARBOHYDRATE METABOLISM (SECTION 23.11)

☑23.62   a. Glucagon increases the blood sugar level by enhancing the breakdown of glycogen in the liver, while insulin reduces the blood sugar level by enhancing the absorption of glucose from the blood into cells of active tissue.

b. Glucagon inhibits glycogen formation, while insulin increases the rate of glycogen formation.

OVER THE COUNTER, CHEMISTRY AND YOUR HEALTH, AND CHEMISTRY AROUND US

23.64   Lactose intolerance is caused by a lactase deficiency.
The symptoms of lactose intolerance are abdominal distention, cramps, and diarrhea when lactose-containing foods (like milk or cheese) are eaten.

23.66   When engaging in strenuous exercise people breath heavily because the muscles need oxygen to convert pyruvate to acetyl CoA.

23.68   Carbohydrate loading is the practice of building glycogen stores in the muscles by eating a low-carbohydrate diet for 3-4 days with heavy exercise, followed by a high-carbohydrate diet for 1-2 days with light exercise.

23.70   Type I diabetes is insulin-dependent because the practically no insulin is produced by the β-cells of the pancreas. Type II diabetes, on the other hand, is non-insulin-dependent because the body secrets the normal amount of insulin, but the body does not respond to it properly.

23.72   The consequences of unchecked diabetes are hyperglycemia, glucosuria, lowered blood pH, blindness, kidney failure, and death.

ADDITIONAL ACTIVITIES

Section 23.1 Review:
(1) Describe the digestion of carbohydrates using the words monosaccharides, disaccharides, and polysaccharides.
(2) How do digested carbohydrates enter the bloodstream?

Section 23.2 Review:
Complete the following diagram with the vocabulary from this section.

blood sugar level
(regulated by: _____)

sugar not completely absorbed by the kidneys: _____

glucose appears in urine: _____

above: _____

| normal (fasting level) |

below: _____

Section 23.3 Review: Complete the following diagram for glycolysis.

Section 23.4 Review:
(1) What is meant by the term *fates* in the expression, "The Fates of Pyruvate"?
(2) What is common to all three fates of pyruvate from a coenzyme perspective?
(3) What is the fate of pyruvate when sufficient oxygen is present?
(4) What are the "fates" of the product from (3)?
(5) What are the fates of pyruvate when insufficient oxygen is present?
(6) Which of the processes in (5) are possible in the human body?

Section 23.5 Review:
(1) What are the other names for the citric acid cycle?
(2) Where does the citric acid cycle occur within the cell?
(3) How many main reactions are part of the citric acid cycle?
(4) Which coenzymes play an important role in the citric acid cycle?
(5) Which compound is the fuel for the citric acid cycle?
(6) Which high energy compound is produced by the citric acid cycle?
(7) What is the relationship between ATP and the allosteric enzyme citrate synthetase?
(8) Which other regulating enzyme behaves similarly to citrate synthetase?
(9) What inhibits the α-ketoglutarate dehydrogenase complex?

Section 23.6 Review:
(1) What compound is produced as NADH and FADH₂ participate in the electron transport chain?
(2) Why are the electron carriers in the electron transport chain in order of increasing electron affinity?
(3) What are the similarities among the cytochromes?

Section 23.7 Review:
(1) Which synthesis reaction stores some of the free energy released by the electron transport chain?
(2) Why is (1) called oxidative phosphorylation?
(3) What is the yield of ATP molecules for every molecule of acetyl CoA that enters the citric acid cycle?
(4) Describe the "flow of protons" according to the chemiosmotic hypothesis.

Section 23.8 Review:
(1) Which produces more ATP from glucose: aerobic catabolism or anaerobic catabolism?
(2) ATP production stores 34.1% of the energy released by the complete oxidation of glucose. Why is this considered efficient?

Section 23.9 Review:
(1) What is the storage form of glucose in the human body?
(2) What is the name of the process for converting glucose to its storage form from (1)?
(3) What is the energy source for the process in (2)?
(4) What are the two types of linkages in (1)?
(5) How are the two type of linkages in (4) formed?
(6) What is the name of the process for converting the storage form of glucose (1) back into glucose?
(7) Where does the process in (6) occur?
(8) Which of the linkages in (1) are broken first during the process in (6)?
(9) Muscle cells can complete two of the three reactions of (6). What process can be performed with the product from the second reaction of (6) instead of completing the final step of (6)?

Section 23.10 Review:
(1) What is the name of the process by which glucose is synthesized from noncarbohydrate materials?
(2) Where does the process from (1) occur?
(3) Common advice for athletes with sore muscles is to exercise in order to "get the lactic acid out of the muscles." Explain why exercise helps relieve sore muscles.

Section 23.11 Review:
(1) Which hormones are produced in the pancreas?
(2) Which cells in the pancreas produce each of the hormones in (1)?
(3) Which of the "glucose reactions" does each of the hormones in (1) stimulate?
(4) Epinephrine is also known as the "fight or flight" hormone. Explain how this hormone provides an opportunity to "fight or flight."

Tying It All Together with a Laboratory Application:
After fasting overnight, a sample of blood is taken from three students and checked for their blood sugar (or (1)_____) levels. The results of the test are:

Student	Blood Sugar Level (mg/dL)
A	65
B	115
C	72

Student A could be classified as (2) _____ because these results indicate a blood sugar level that is (3) _____ compared to the normal fasting level. Student B could be classified as (4) _____ because these results indicate a blood sugar level that is (5) _____ compared to the normal fasting level. Student C could be classified as (6) _____ because these results indicate a blood sugar level that is (7) _____ compared to the normal fasting level.

Student B also had a sample of urine tested for glucose. The sample came back positive for (8) _____ (glucosuria or insulin shock), which indicates the blood sugar level is above the (9) _____. Student B may have (10) _____.

The students then participated in an oral glucose tolerance test. They were given "glucola" (a solution with 75 g of glucose) and their blood was tested after 30 minutes, 1 hour, 2 hours, and 3 hours. After 30 minutes and 1 hour, the blood sugar level was elevated in all three students; however, after 2 hours, the blood sugar levels in Students A and C were significantly lower then after the 1 hour reading. The blood sugar level for Student B was still elevated at 2 hours. At the 3 hour mark, Students A and C have "normal" blood sugar level, but Student B does not.

The hormone (11) _____ is responsible for the lowering of the blood sugar level in Students A and C. It causes glucose to breakdown into (12) _____ in a process called (13) _____. Under aerobic conditions (12) is converted into (14) _____. Under anaerobic conditions (12) is converted into (15) _____ which is stored in the
(16) _____.

(14) is the fuel for the (17) _____. The coenzymes (18) _____ and _____ undergo (19) _____ during the (17). After the citric acid cycle, (20) _____ and _____ participate in the (21) _____ to produce water and regenerate (18). (21) releases a significant amount of energy that can be used to synthesize (22) _____.

Excess glucose can be stored in the form of (23) _____ through a process called (24) _____. When the glucose is needed, (23) can be broken down through a process called (25) _____. The hormone (26) _____ activates (25) in the liver. If the first two stages of (25) occur in the muscles rather than the liver, glucose cannot be formed because the muscles lack (27) _____; however, the intermediate (28) _____ can enter the (29) _____ pathway to produce energy. This process can also be stimulated in the muscles by the hormone (30) _____.

Student A's overnight fast lasted about 19 hours. Consequently, this student may have produced glucose from noncarbohydrate materials in a process called (31) _____. If (31) involved lactate from muscle tissue, this is one of the steps in the (32) _____ cycle.

SOLUTIONS FOR THE ADDITIONAL ACTIVITIES

Section 23.1 Review:
(1) Polysaccharides and disaccharides are hydrolyzed into monosaccharides during the digestion of carbohydrates.
(2) Digested carbohydrates enter the bloodstream through the lining of the small intestine.

Section 23.2 Review:

blood sugar level
(regulated by: **the liver**)

sugar not completely absorbed by the kidneys: **renal threshold**

glucose appears in urine: **glucosuria**

above: **hyperglycemia**

normal (fasting level)

below: **hypoglycemia**

Section 23.3 Review:

Section 23.4 Review:
(1) The term *fates* in the expression, "The Fates of Pyruvate" means what happens to the pyruvate molecule.
(2) All three fates of pyruvate regenerate NAD⁺ so that glycolysis can continue.
(3) When sufficient oxygen is present, pyruvate is converted to acetyl CoA.
(4) The acetyl CoA either undergoes the citric acid cycle to produce carbon dioxide or biosynthesis to produce compounds like the fatty acid components of triglycerides.
(5) Lactate fermentation or alcoholic fermentation are the fates of pyruvate when insufficient oxygen is present.
(6) Alcohol fermentation cannot occur in the human body; however, lactate fermentation can occur in the human body.

Section 23.5 Review:
(1) The citric acid cycle is also known as the tricarboxylic acid cycle and the Krebs cycle.
(2) The citric acid cycle occur within the mitochondria.
(3) Eight main reactions are part of the citric acid cycle.
(4) NAD⁺ and FAD are coenzymes that play an important role in the citric acid cycle.
(5) Acetyl CoA is the fuel for the citric acid cycle.
(6) GTP is a high energy compound produced by the citric acid cycle.
(7) ATP is an inhibitor for the allosteric enzyme citrate synthetase.
(8) Isocitrate dehydrogenase is another regulating enzyme that behaves similarly to citrate synthetase.
(9) The α-ketoglutarate dehydrogenase complex is inhibited by succinyl CoA, NADH, and ATP.

Section 23.6 Review:
(1) Water is produced as NADH and FADH₂ participate in the electron transport chain.
(2) The electron carriers in the electron transport chain are in order of increasing electron affinity because the electrons must flow through the chain towards the oxygen atom.
(3) Cytochromes are proteins that contain an iron group.

Section 23.7 Review:
(1) The synthesis of ATP from ADP and P$_i$ stores some of the free energy released by the electron transport chain.

(2) This process is called oxidative phosphorylation because the ADP reacts with inorganic phosphate as a result of the NADH and FADH$_2$ undergoing oxidation in the electron transport chain.
(3) Ten ATP molecules are produced for every molecule of acetyl CoA that enters the citric acid cycle.
(4) The "flow of protons" according to the chemiosmotic hypothesis is across the inner mitochondrial membrane into the intermembrane space as electrons flow through the electron transport chain, then back through the inner mitochondrial membrane through a channel formed by the enzyme F$_1$-ATPase as a result of proton concentration and electric potential differences on either side of the inner mitochondrial membrane.

Section 23.8 Review:
(1) Aerobic catabolism of glucose produces more ATP than anaerobic catabolism of glucose.
(2) Compared to other systems that use/store chemical energy, 34.1% is a high percentage; therefore, the process is considered efficient.

Section 23.9 Review:
(1) The storage form of glucose in the human body is glycogen.
(2) The process for converting glucose to glycogen is called glycogenesis.
(3) The energy source for glycogenesis is UTP.
(4) The two types of linkages in glycogen are α(1→4) and α(1→6).
(5) The α(1→4) linkages are formed as each glucose unit lengthens the chain. The α(1→6) linkages are formed by a branching enzyme.
(6) The process for converting glycogen back into glucose is glycogenolysis.
(7) Glycogenolysis occurs in the liver, kidneys, and intestinal cells.
(8) The α(1→4) linkages in glycogen are broken first during glycogenolysis.
(9) Muscle cells can complete two of the three reactions of glycogenolysis. Glucose 6-phosphate is the product of the second reaction of glycogenolysis. This compound can then enter the glycolysis pathway and produce energy.

Section 23.10 Review:
(1) Gluconeogenesis is the process by which glucose is synthesized from noncarbohydrate materials. The name can be divided into three parts "gluco-" refers to glucose, "neo-" refers to in a new, different, or modified way, and "-genesis" refers to production. This could be combined to define gluconeogenesis as the production of glucose in a modified way (not from carbohydrates).
(2) Gluconeogenesis occurs primarily in the liver.
(3) During exercise, blood goes to the muscles to deliver oxygen and retrieve waste products. Some of the lactate in the muscles is taken by the blood to the liver where the lactate can be converted back into pyruvate and then into glucose by gluconeogenesis. The glucose then reenters the bloodstream. If the exercise is aerobic and enough oxygen is available to the cells, then the glucose can undergo aerobic glycolysis to produce acetyl CoA.

Section 23.11 Review:
(1) Insulin and glucagon are produced by the pancreas.
(2) Insulin is produced by the β-cells of the pancreas. Glucagon is produced by the α-cells of the pancreas.
(3) Insulin stimulates glycolysis. Glucagon stimulates glycogenolysis.
(4) Epinephrine stimulates glycogen breakdown in the muscles and the liver. When glycogen is broken down in the muscles, glucose 6-phosphate is produced which undergoes glycolysis to produce energy. This energy is produced quickly and enables a person (or animal) with raised epinephrine levels either "fight or flight."

Chapter 23

Tying It All Together with a Laboratory Application:

(1) glucose	(10) diabetes	(19) reduction	(27) glucose 6-phosphatase
(2) hypoglycemic	(11) Insulin	(20) NADH, FADH$_2$	(28) glucose 6-phosphate
(3) low	(12) pyruvate	(21) electron transport chain	(29) glycolysis
(4) hyperglycemic	(13) glycolysis	(22) ATP	(30) epinephrine
(5) high	(14) acetyl CoA	(23) glycogen	(31) gluconeogenesis
(6) normal	(15) lactate	(24) glycogenesis	(32) Cori
(7) average	(16) muscles	(25) glycogenolysis	
(8) glucosuria	(17) citric acid cycle	(26) glucagon	
(9) renal threshold	(18) NAD$^+$, FAD		

SELF-TEST QUESTIONS

Multiple Choice

1. The digestion of carbohydrates produces glucose, fructose, and
   a. startch
   b. amylose
   c. lactose
   d. galactose

2. The central compound in carbohydrate metabolism is
   a. glycogen
   b. glucose
   c. fructose
   d. galactose

3. The glycolysis pathway is located within the _____ of cells.
   a. cytoplasm
   b. nucleus
   c. mitochondria
   d. ribosomes

4. The net production of ATP in glycolysis from one molecule of glucose is _____ ATP molecules.
   a. 0
   b. 1
   c. 2
   d. 4

5. The net production of NADH in glycolysis from one molecule of glucose is _____ NADH molecules.
   a. 0
   b. 1
   c. 2
   d. 4

6. Enzymes regulate the glycolysis pathway at _____ control points.
   a. 2
   b. 3
   c. 4
   d. 5

7. The muscle pain which follows prolonged and vigorous contraction of skeletal muscles is the result of
   a. lactate
   b. citrate
   c. NADH
   d. pyruvate

8. How many molecules of pyruvate are produced from the glycolysis of one molecule of glucose?
   a. 0
   b. 1
   c. 2
   d. 3

9. How many net molecules of ATP can be formed in the electron transport chain by oxidative phosphorylation from each molecule of cytoplasmic NADH produced during glycolysis?
   a. 1.0
   b. 1.5
   c. 2.0
   d. 2.5

10. The complete oxidation of glucose in the liver results in the formation of _____ molecules of ATP.
    a. 18
    b. 32
    c. 34
    d. 38

11. Which of the following exerts an effect on blood sugar levels opposite to that of insulin?
    a. cholesterol
    b. glucagon
    c. vasopressin
    d. aldosterone

12. What disease is commonly associated with glucosuria?
    a. diabetes mellitus   b. hepatitis   c. hypoglycemia   d. hyperinsulinism

13. The first step of the citric acid cycle involves the reaction of oxaloacetate with _____ to form citrate.
    a. coenzyme A   b. acetyl CoA   c. CO_2 d. malate

14. The high-energy compound formed in the citric acid cycle is
    a. CTP   b. ATP   c. UTP   d. GTP

15. One turn of the citric acid cycle produces _____ molecules of NADH.
    a. 1   b. 2   c. 3   d. 4

16. The citric acid cycle is inhibited by
    a. ATP   b. CO_2 c. NAD^+ d. FAD

17. One product of the electron transport chain is
    a. O_2 b. H_2O c. CO_2 d. NADH

18. How many sites in the electron transport chain can support the synthesis of ATP?
    a. 1   b. 2   c. 3   d. 4

Matching

Match each description on the left with the condition or disease on the right.
19. a high glucose level in the blood
20. a low glucose level in the blood
21. a high glucose level in the urine

    a. glucosuria
    b. hypoglycemia
    c. hyperglycemia
    d. galactosemia

Match each description on the left with a product on the right.
22. the product of alcoholic fermentation
23. produced by carbohydrate oxidation under aerobic conditions
24. produced by glycolysis under anaerobic conditions in the human body

    a. acetaldehyde
    b. lactate
    c. acetyl CoA
    d. ethanol

Match each characteristic on the left with the cycle or process on the right.
25. glucose is converted to glycogen
26. synthesis of glucose from non-carbohydrate sources
27. breakdown of glycogen to glucose

    a. glycogenesis
    b. gluconeogenesis
    c. glycogenolysis
    d. glycolysis

True-False
28. Fructose and galactose are not metabolized by humans.
29. The blood sugar level in a hypoglycemic individual is higher than the normal fasting level.
30. The glycolysis pathway is inhibited by ATP.
31. Part of the Cori cycle involves the conversion of pyruvate to glucose.
32. About 90% of gluconeogenesis takes place in the liver.
33. Oxidative phosphorylation is a process coupled with the electron transport chain.
34. The chemiosmotic hypothesis pertains to the flow of cytoplasmic NADH across the mitochondrial membrane.
35. The net production of ATP from one mole of glucose in brain cells is greater than that in liver cells.

Chapter 24: Lipid and Amino Acid Metabolism

CHAPTER OUTLINE
24.1 Blood Lipids
24.2 Fat Mobilization
24.3 Glycerol Metabolism
24.4 The Oxidation of Fatty Acids
24.5 The Energy from Fatty Acids
24.6 Ketone Bodies
24.7 Fatty Acid Synthesis
24.8 Amino Acid Metabolism
24.9 Amino Acid Catabolism: The Fate of the Nitrogen Atoms
24.10 Amino Acid Catabolism: The Fate of the Carbon Skeleton
24.11 Amino Acid Biosynthesis

LEARNING OBJECTIVES/ASSESSMENT
When you have completed your study of this chapter, you should be able to:
1. Describe the digestion, absorption, and distribution of lipids in the body. (Section 24.1; Exercises 24.2 and 24.4)
2. Explain what happens during fat mobilization. (Section 24.2; Exercise 24.8)
3. Identify the metabolic pathway by which glycerol is catabolized. (Section 24.3; Exercise 24.12)
4. Outline the steps of β-oxidation for fatty acids. (Section 24.4; Exercise 24.22)
5. Determine the amount of ATP produced by the complete catabolism of a fatty acid. (Section 24.5; Exercise 24.26)
6. Name the three ketone bodies and list the conditions that cause their overproduction. (Section 24.6; Exercise 24.28)
7. Describe the pathway for fatty acid synthesis. (Section 24.7; Exercise 24.34)
8. Describe the source and function of the body's amino acid pool. (Section 24.8; Exercise 24.38)
9. Write equations for transamination and deamination reactions. (Section 24.8; Exercise 24.44)
10. Explain the overall results of the urea cycle. (Section 24.9; Exercise 24.54)
11. Describe how amino acids can be used for energy production, the synthesis of triglycerides, and gluconeogenesis. (Section 24.10; Exercise 24.58)
12. State the relationship between intermediates of carbohydrate metabolism and the synthesis of nonessential amino acids. (Section 24.11; Exercise 24.64)

SOLUTIONS FOR THE END OF CHAPTER EXERCISES
BLOOD LIPIDS (SECTION 24.1)
☑24.2 The products of triglyceride digestion are glycerol, fatty acids, and monoglycerides.

☑24.4 Insoluble lipids form complexes with proteins to form lipoprotein aggregates (chylomicrons).

24.6 The major types of lipoproteins are chylomicrons, very low-density lipoproteins, low-density lipoproteins, and high-density lipoproteins.

FAT MOBILIZATION (SECTION 24.2)
☑24.8 Fat mobilization is the process of epinephrine stimulating the hydrolysis of triglycerides (in adipose tissue) into fatty acids and glycerol, which enter the bloodstream.

24.10 Resting muscles and liver cells use fatty acids to satisfy their energy needs.

GLYCEROL METABOLISM (SECTION 24.3)
☑24.12 The two fates of glycerol after it has been converted to an intermediate of glycolysis are to provide energy to cells or be converted to glucose.

THE OXIDATION OF FATTY ACIDS (SECTION 24.4)

24.14 A fatty acid is prepared for catabolism by a conversion into fatty acyl CoA. This activation occurs outside the mitochondria in the cytoplasm of the cell.

24.16 The oxidation of fatty acids is aided by the coenzymes FAD and NAD⁺.

24.18

$$CH_3(CH_2)_{16}-\overset{O}{\underset{\|}{C}}-OH + HS-CoA + ATP \longrightarrow CH_3(CH_2)_{16}-\overset{O}{\underset{\|}{C}}-S-CoA + H_2O + AMP + 2\,P_i$$

24.20 The oxidation of fatty acids is referred to as the fatty acid spiral rather than the fatty acid cycle because the process results in the production of acetyl CoA only. The fatty acid is not regenerated as it would be in a cycle.

☑24.22

a. $CH_3CH_2CH_2CH_2CH_2-\overset{O}{\underset{\|}{C}}-S-CoA + FAD \longrightarrow CH_3CH_2CH_2CH=CH-\overset{O}{\underset{\|}{C}}-S-CoA + FADH_2$

b. $CH_3CH_2CH_2CH=CH-\overset{O}{\underset{\|}{C}}-S-CoA + H_2O \longrightarrow CH_3CH_2CH_2\overset{OH}{\underset{|}{CH}}-CH_2-\overset{O}{\underset{\|}{C}}-S-CoA$

c. $CH_3CH_2CH_2\overset{OH}{\underset{|}{CH}}-CH_2-\overset{O}{\underset{\|}{C}}-S-CoA + NAD^+ \longrightarrow CH_3CH_2CH_2\overset{O}{\underset{\|}{C}}-CH_2-\overset{O}{\underset{\|}{C}}-S-CoA + NADH + H^+$

d. $CH_3CH_2CH_2\overset{O}{\underset{\|}{C}}-CH_2-\overset{O}{\underset{\|}{C}}-S-CoA + HS-CoA \longrightarrow CH_3CH_2CH_2\overset{O}{\underset{\|}{C}}-S-CoA + CH_3-\overset{O}{\underset{\|}{C}}-S-CoA$

THE ENERGY FROM FATTY ACIDS (SECTION 24.5)

24.24 For every molecule of acetyl CoA that enters the citric acid cycle, 10 molecules of ATP are produced.

☑24.26 As a 12-carbon fatty acid passes through the fatty acid spiral, 6 acetyl CoA, 5 FADH₂, and 5 NADH molecules are produced. Consequently, 78 molecules of ATP can be produced.

Oxidation product or step	ATP/unit	Total ATP
Activation step		-2
6 acetyl CoA	10	60
5 FADH₂	1.5	7.5
5 (NADH + H⁺)	2.5	12.5
Total		78

KETONE BODIES (SECTION 24.6)

☑24.28 Ketone bodies are the compounds synthesized within the liver from excess acetyl CoA. They are acetoacetate, β-hydroxybutyrate, and acetone.

24.30 Ketone bodies are formed in the liver. The brain, heart, and skeletal muscles can use these compounds to meet energy needs.

24.32 Ketosis is frequently accompanied by acidosis because two of the ketone bodies are acids. As the concentration of the ketone bodies rises in the blood, the pH of the blood decreases.

FATTY ACID SYNTHESIS (SECTION 24.7)

☑24.34 Most fatty acids within the body have an even number of carbon atoms because they are synthesized in two carbon units (acetyl CoA).

24.36 The liver can convert glucose to fatty acids via acetyl CoA; however, it does not have the enzyme that can catalyse the conversion of acetyl CoA to pryuvate which is needed for gluconeogenesis.

AMINO ACID METABOLISM (SECTION 24.8)

☑24.38 The amino acid pool is the body's available amino acids for maintenance. The pool is supplied by proteins eaten (and hydrolyzed), the body's degraded tissue, and the synthesis of amino acids in the liver.

24.40 Protein turnover is the dynamic process in which body proteins are continuously hydrolyzed and resynthesized.

24.42 Four important biomolecules other than proteins that are synthesized from amino acids are purine and pyrimidine bases of nucleic acids, heme structures for hemoglobin and myoglobin, choline and ethanolamine for phospholipids, and neurotransmitters such as acetylcholine and dopamine.

AMINO ACID CATABOLISM: THE FATE OF THE NITROGEN ATOMS (SECTION 24.9)

☑24.44
a. An amino group is removed from an amino acid and donated to an α-keto acid.   *transamination*
b. Ammonium ion is produced.   *deamination*
c. New amino acids are synthesized from other amino acids.   *transamination*
d. A keto acid is produced from an amino acid.   *deamination and transamination*

24.46

24.48

24.50 Urea is synthesized in the liver.

24.52 Carbamoyl phosphate enters the urea cycle by combining with ornithine.

☑24.54

$$NH_4^+ + HCO_3^- + 3\ ATP + H_2O + \begin{matrix} COO^- \\ | \\ CH_2 \\ | \\ CH-NH_3^+ \\ | \\ COO^- \end{matrix} \longrightarrow \begin{matrix} H_2N-C(=O)-NH_2 \\ + \\ \text{(fumarate)} \end{matrix} COO^- + 2\ ADP + 2\ P_i + AMP + PP_i + 5\ H^+$$

AMINO ACID CATABOLISM: THE FATE OF THE CARBON SKELETON (SECTION 24.10)

24.56 The amino acids isoleucine, phenylalanine, tryptophan, and tyrosine can be both glucogenic and ketogenic.

☑24.58 a. aspartate → oxaloacetate   glucogenic
         b. leucine → acetyl CoA   ketogenic
         c. tyrosine → acetoacetyl CoA ketogenic

AMINO ACID BIOSYNTHESIS (SECTION 24.11)

24.60 Essential amino acids are amino acids must be obtained from the diet because, unlike nonessential amino acids, they cannot be synthesized within the liver in the amounts needed by the body.

24.62 Two vital functions performed by transaminases are participating in the biosynthesis of nonessential amino acids and transferring the nitrogen atoms of all amino acids to α-keto acids when the disposal of nitrogen is necessary.

☑24.64 Two general sources of intermediates for the biosynthesis of amino acids are the glycolysis pathway and citric acid cycle.

CHEMISTRY AROUND US, CHEMISTRY AND YOUR HEALTH, AND OVER THE COUNTER

24.66 The essential roles served by cholesterol are aiding in the formation of cell membranes, the insulation of nerves, the synthesis of several hormones, and the digestion of food.

24.68 Vegetarians may have the problem of obtaining enough Calories to maintain a healthy body because the Caloric content of vegetables is much lower than that of meat. Consequently, vegetarians need to eat 5-6 times the weight of vegetables as they would have eaten in meat.

24.70 It is important to include an adequate amount of protein in my diet because otherwise I might experience protein-energy malnutrition. Proteins are needed as a source of nitrogen atoms for protein synthesis and the synthesis of other nitrogen containing compounds.

24.72 The treatment for PKU is restricting the amount of phenylalanine in the diet and supplementing tyrosine.

ADDITIONAL ACTIVITIES

Section 24.1 Review:
(1) Why are triglycerides and phosphoglycerides digested into component substances before they are absorbed into cells of the intestinal mucosa, if they are resynthesized once they are absorbed?
(2) What does the term *lipoprotein* imply?
(3) How do the densities of proteins and lipids compare?

Lipid and Amino Acid Metabolism

Section 24.2 Review:
(1) What part of the cell is responsible for fatty acid oxidation?
(2) Fatty acids cannot cross the blood-brain barrier; however, fatty acids can be used to supply energy to the brain during a fast. Explain.
(3) Which horomone stimulates the hydrolysis of triglycerides?
(4) What is formed when fatty acids combine with serum albumin?
(5) How is glycerol transported in the body?

Section 24.3 Review:
(1) Glycerol can be converted to dihydroxyacetone phosphate. What is the chemical process by which dihydroxyacetone phosphate can be converted into pyruvate?
(2) What could the pyruvate from (1) be used for in the body?

Section 24.4 Review:
(1) Is the process for converting a fatty acid to a fatty acyl CoA endothermic or exothermic? Explain.
(2) What part of the cell can a fatty acyl CoA enter that a fatty acid molecule cannot?
(3) What process can occur once the fatty acyl CoA enters the part of the cell identified in (2)?
(4) What is the name of the process when one fatty acyl CoA repeats the process in (3) until all of the carbon atoms from the original molecule are in molecules of acetyl CoA?

Section 24.5 Review:
(1) As food is digested and turned into ATP, is the food oxidized or reduced?
(2) If more energy is stored in fat, why are low-fat diets encouraged?

Section 24.6 Review:
(1) What happens if more acetyl CoA is produced by fatty acid oxidation than can be processed through the citric acid cycle?
(2) How does the body process ketone bodies at a low concentration in the blood?
(3) How does the body process ketone bodies at a high concentration in the blood?
(4) What are the consequences of prolonged ketoacidosis?

Section 24.7 Review:
(1) Does the phrase, "Use it or lose it," apply to the energy provided by excess fat or carbohydrates in the diet?
(2) Where does biosynthesis of fatty acids occur in the cell?
(3) Acetyl CoA is produced during the fatty acid spiral in the mitochondria. How is it involved in the biosynthesis of fatty acids?
(4) Which type of fatty acids cannot be synthesized in the body?
(5) What would a nutritionist call the fatty acids in (4) that are needed by the body?
(6) Acetyl CoA is one of the "fates of pyruvate." Is pyruvate one of the "fates of acetyl CoA" in humans?

Section 24.8 Review:
(1) Does the phrase, "Use it or lose it," apply to proteins?
(2) What are the sources for the amino acid pool?
(3) What are the uses of the amino acid pool?

Section 24.9 Review:
(1) What processes do transaminases catalyze?
(2) What undergoes oxidation during oxidative deamination?
(3) On the next page, identify and label the Stage 1 reaction(s).
(4) On the next page, identify and label the Stage 2 reaction(s).
(5) On the next page, identify and label the Stage 3 reaction(s).

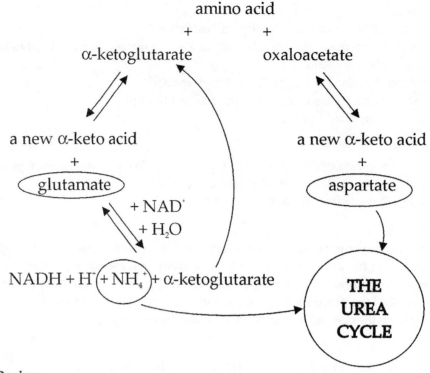

Section 24.10 Review:
(1) Can amino acids be used as a source of energy in the body?
(2) Can amino acids be converted into pyruvate?
(3) Can amino acids be converted into source material for gluconeogenesis?
(4) Can amino acids be converted into source material for ketone bodies and fatty acids?
(5) Give one reason why a person suffering from anorexia nervosa might have heart trouble.

Section 24.11 Review:
(1) Why is cirrhosis of the liver dangerous for the health of the entire body?
(2) What is the relationship between essential and nonessential amino acids?
(3) Which enzymes are a part of the relationship in (2)?

Tying It All Together with a Laboratory Application:
When a patient has (1) _____ breath, it suggests that (2) _____ is the macronutrient serving as a major energy source for the body. The cause of this condition could be a number of reasons, including that the patient might have (3) _____ or be fasting.

Urinalysis that reveals glucose and (4) _____ indicates that (3) is likely because both of these materials are above their respective (5) _____. The presence of glucose in the urine is called (6) _____ and the presence of (4) in the urine is called (7) _____. The blood condition in which (4) is revealed is called (8) _____. A patient who suffers from (1) breath, (7), and (8) is said to have (9) _____. This condition can lead to (10) _____ blood pH (or (11) _____) because two of the compounds associated with (4) are (12) _____.

(13) _____ can be used as a treatment for patients with (3) in order to use the glucose from their diet and retard the (14) _____ mechanism that was activated in order to meet the body's energy needs. The sequence of reactions for producing energy from lipids is called the (15) _____. Lipids store (16) _____ energy than carbohydrates of similar mass. Consequently, when excess carbohydrates or lipids are consumed, they undergo (17) _____ to produce fatty acids which can be converted to (18) _____ and stored for later use.

Blood work done within 6 hours of eating a lipid-rich meal will likely show elevated (19) _____ levels (compared to fasting). These (19) are distinguished by (20) _____ because of the (20) differences between lipids and proteins.

Excess proteins are not stored as (18); consequently, if the (21) _____ (from diet, protein degradation, and biosynthesis) becomes too large, then more (22) _____ will be formed and excreted in the urine. (23) _____ and _____ are the amino acids produced by transamination and (24) _____ that can be used to form (22) by participating in or supplying intermediates to the (25) _____ cycle.

The degradation and rebuilding of proteins is called (26) _____. Some of the amino acids required for this process are (27) _____ and must be supplied by the diet. Other can be synthesized in the body. Protein can also be used as a source of (28) _____ because the (29) _____ of amino acids, specifically the carbon skeletons, produces intermediates of the (30) _____.

SOLUTIONS FOR THE ADDITIONAL ACTIVITIES

Section 24.1 Review:
(1) Triglycerides and phosphoglycerides are too large to pass through cell walls. When these compounds are eaten they pass into the stomach where lipases and acid digest these compounds into their smaller components. These components are able to pass through the cell walls of the intestinal mucosa where they are used to resynthesize triglycerides and phosphoglycerides.
(2) The term *lipoprotein* implies the compound contains both lipid and protein components.
(3) Proteins are more dense than lipids.

Section 24.2 Review:
(1) The mitochondria are responsible for fatty acid oxidation.
(2) Glucose and other compounds can cross the blood-brain barrier; therefore, the fatty acids must be broken down and used to form compounds that can cross the barrier in order to supply energy to the brain during a fast.
(3) Epinephrine stimulates the hydrolysis of triglycerides.
(4) A lipoprotein forms when fatty acids combine with serum albumin.
(5) Glycerol is water soluble and it dissolves in the blood for transport in the body.

Section 24.3 Review:
(1) Glycolysis is the chemical process by which dihydroxyacetone phosphate can be converted into pyruvate.
(2) Pyruvate can be used for energy production or gluconeogenesis.

Section 24.4 Review:
(1) The process for converting a fatty acid to a fatty acyl CoA is endothermic because a molecule of ATP must be broken down into AMP and PP_i to release the energy required for this reaction.
(2) A fatty acyl CoA can enter the mitochondria inside the cell, but a fatty acid molecule cannot.
(3) Once the fatty acyl CoA enters the mitochondria, β-oxidation can occur.
(4) The fatty acid spiral is the process by which one fatty acyl CoA molecule undergoes repeated β-oxidation until all of the carbon atoms from the original molecule are in molecules of acetyl CoA.

Section 24.5 Review:
(1) Food is oxidized as it is digested and turned into ATP.
(2) Low-fat diets are encouraged for many reasons. The body will use energy from carbohydrate sources before it uses energy from fat sources. If excess carbohydrates are consumed, they can be stored in the form of fat. The American diet typically includes too much dietary fat. This fat is excessive because the body does not need as much energy as the fat provides; therefore, this fat is stored in the

328 Chapter 24

body. The body does need some fat-soluble vitamins; however, these can be obtained in adequate amounts from a low-fat diet.

Section 24.6 Review:
(1) Ketone bodies form in the liver if more acetyl CoA is produced by fatty acid oxidation than can be processed through the citric acid cycle.
(2) Ketone bodies at a low concentration in the blood can be oxidized to meet energy needs of cells.
(3) Ketone bodies at a high concentration in the blood resulting in a condition called ketonemia (ketones in the blood) as well as ketonuria (ketones in the urine) if the renal threshold for ketone bodies is exceeded. If the level is high enough, the acetone will be expelled through the lungs and "acetone breath" results. At this level, the condition is called ketosis.
(4) The consequences of prolonged ketoacidosis are debilitation, coma, and death.

Section 24.7 Review:
(1) No, the phrase, "Use it or lose it," does not apply to the energy provided by excess fat or carbohydrates in the diet. They are stored as fat.
(2) The biosynthesis of fatty acids occurs in the cytoplasm of the cell.
(3) Acetyl CoA reacts with oxaloacetate and water in the mitochondria to produce citrate that can be transported into the cytoplasm. In the cytoplasm, the citrate is converted back into acetyl CoA and oxaloacetate. The acetyl CoA is then used to synthesize fatty acids.
(4) Polyunsaturated fatty acids cannot be synthesized in the body.
(5) A nutritionist would call those polyunsaturated fatty acids essential fatty acids.
(6) No, pyruvate is not one of the "fates of acetyl CoA" in humans.

Section 24.8 Review:
(1) Yes, the phrase, "Use it or lose it," applies to proteins. They cannot be stored for later use.
(2) The sources for the amino acid pool are dietary proteins, biosynthesis in the liver, and degradation of body tissues.
(3) The uses of the amino acid pool include rebuilding body tissues, biosynthesis of nitrogen-containing compounds, and catabolism for energy and for excretion of excess nitrogen.

Section 24.9 Review:
(1) Transaminases catalyze reactions that involve the transfer of an amino group from amino acids to other amino acids or α-keto acids.
(2) During oxidative deamination, glutamate is oxidized.

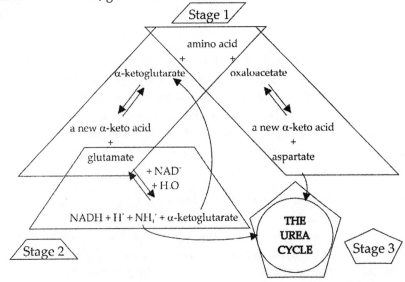

Section 24.10 Review:
(1) Yes, amino acids can be used as a source of energy in the body.
(2) Yes, amino acids can be converted into pyruvate.
(3) Yes, amino acids can be converted into source material for gluconeogenesis.
(4) Yes, amino acids can be converted into source material for ketone bodies and fatty acids.
(5) A person suffering from anorexia nervosa might have heart trouble because the heart is a muscle and the body will hydrolyze proteins from body tissues, including muscles, in order to supply amino acids for gluconeogenesis during starvation. The loss of protein from the heart may weaken it.

Section 24.11 Review:
(1) The liver is extremely active biochemically. It is a major site of biosynthesis. Liver damage, including cirrhosis, will decrease the ability of the liver to function.
(2) Essential amino acids or other compounds can be used to synthesize nonessential amino acids in the body.
(3) Transaminases are needed for the synthesis of nonessential amino acids.

Tying It All Together with a Laboratory Application:
(1) acetone
(2) fat
(3) diabetes
(4) ketone bodies
(5) renal thresholds
(6) glucosuria
(7) ketonuria
(8) ketonemia
(9) ketosis
(10) low
(11) ketoacidosis
(12) acids
(13) Insulin
(14) fat mobilization
(15) fatty acid spiral
(16) more
(17) biosynthesis
(18) body fat
(19) lipoprotein
(20) density
(21) amino acid pool
(22) urea
(23) Glutamate, aspartate
(24) deamination
(25) urea
(26) protein turnover
(27) essential
(28) energy
(29) catabolism
(30) citric acid cycle

SELF-TEST QUESTIONS
Multiple Choice

1. What reaction occurs during the digestion of triglycerides?
   a. hydrolysis   b. hydrogenation   c. hydration   d. oxidation

2. The products of triglyceride digestion are fatty acids, some monoglycerides, and
   a. triglycerides   b. diglycerides   c. glucose   d. glycerol

3. Lipoproteins which contain the greatest amount of protein are the
   a. chylomicrons
   b. low-density lipoproteins
   c. very-low density lipoproteins
   d. high-density lipoproteins

4. What reaction occurs during fat mobilization?
   a. reduction   b. hydrolysis   c. oxidation   d. hydration

5. One of the hormones involved in fat mobilization is
   a. aldosterone   b. vasopressin   c. insulin   d. epinephrine

6. The entry point for glycerol into the glycolysis pathway is
   a. dihydroxyacetone phosphate
   b. pyruvate
   c. acetyl CoA
   d. fructose 1,6-diphsophate

330 Chapter 24

7. In order for fatty acids to enter the mitochondria for degradation, they must first be converted to
   a. acetyl CoA   b. fatty acyl CoA   c. pyruvate   d. malonate

8. Which of the following processes takes place during the fatty acid spiral?
   a. addition of water to a double bond
   b. addition of H atoms to a double bond
   c. oxidation of an –OH group to a ketone
   d. more than one response is correct

9. During one run through the fatty acid spiral, which bond of the following fatty acid would be broken?

   $$CH_3-(CH_2)_4-CH_2\overset{d}{-}CH_2\overset{c}{-}CH_2\overset{b}{-}\underset{a}{\overset{O}{\overset{\|}{C}}}-OH$$

   a. bond a   b. bond b   c. bond c   d. bond d

10. Which of the following is a product of the fatty acid spiral?
    a. pyruvate
    b. $CO_2 + H_2O$
    c. acetyl CoA
    d. more than one response is correct

11. How many runs through the fatty acid spiral would be required to completely break down one molecule of a 12-carbon fatty acid?
    a. 1   b. 5   c. 6   d. 12

12. How many FADH₂ molecules are produced from one turn of the fatty acid spiral?
    a. 1   b. 2   c. 3   d. 4

13. How many ATP molecules ultimately result from one turn of the fatty acid spiral?
    a. 8   b. 12   c. 14   d. 17

14. The concentration of ketone bodies builds up when increased amounts of _____ are oxidized.
    a. amino acids   b. fatty acids   c. glucose   d. glycerol

15. Which of the following types of proteins tends to have the shortest half-life in the body?
    a. enzymes
    b. connective tissue proteins
    c. plasma proteins
    d. muscle proteins

16. The primary function of the urea cycle in the body is to
    a. produce ATP from ADP
    b. convert keto acids into amino acids
    c. convert amino acids into keto acids
    d. convert ammonia ions into urea

17. Which of the following substances is a product of a transamination reaction?
    a. amino acid
    b. keto acid
    c. ammonia
    d. more than one response is correct

18. The nitrogen of amino acids appears in the urine of mammals primarily as
    a. uric acid   b. urea   c. ammonia   d. N₂

19. Which of the following substances enters into the urea cycle?
    a. malonyl CoA
    b. acetyl CoA
    c. phosphocreatine
    d. carbamoyl phosphate

20. Which one of the following is an intermediate formed in the urea cycle?
    a. CO_2 b. acetyl CoA   c. pyruvate   d. aspartate

21. Deamination of an amino acid produces
    a. ammonium ions
    b. CO_2
    c. an α-keto acid
    d. more than one response is correct

22. The carbon skeletons of amino acids are ultimately catabolized through
    a. glycolysis
    b. the citric acid cycle
    c. glycogenolysis
    d. the urea cycle

23. A number of amino acids can be synthesized in the body from intermediates of
    a. the citric acid cycle
    b. oxidative phosphorylation
    c. the urea cycle
    d. the fatty acid spiral

True-False
24. Fat has a caloric value more than twice that of glycogen and starch.
25. Upon complete oxidation to CO_2 and H_2O, fatty acids produce more net energy than a carbohydrate containing the same number of carbon atoms.
26. The glycerol resulting from triglyceride hydrolysis can be converted to pyruvate.
27. The glycerol resulting from triglyceride hydrolysis can be converted to glucose.
28. Starvation can lead to ketosis.
29. Amino acids in excess of immediate body requirements are stored for later use.
30. Amino acids can be catabolized for energy production.
31. Urea formation is an energy-yielding process.
32. The carbon atoms of some amino acids can be used in fatty acid synthesis.
33. The carbon atoms of some amino acids can be used in glucose synthesis.
34. Biosynthesis of fatty acids occurs in the mitochondria of the cell.
35. The liver is the most important organ in fatty acid synthesis.

Chapter 25: Body Fluids

CHAPTER OUTLINE
25.1 A Comparison of Body Fluids
25.2 Oxygen and Carbon Dioxide Transport
25.3 Chemical Transport to the Cells
25.4 The Constituents of Urine
25.5 Fluid and Electrolyte Balance
25.6 Acid-Base Balance
25.7 Buffer Control of Blood pH
25.8 Respiration Control of Blood pH
25.9 Urinary Control of Blood pH
25.10 Acidosis and Alkalosis

LEARNING OBJECTIVES/ASSESSMENT
When you have completed your study of this chapter, you should be able to:
1. Compare the chemical compositions of plasma, interstitial fluid, and intracellular fluid. (Section 25.1; Exercise 25.2)
2. Explain how oxygen and carbon dioxide are transported within the bloodstream. (Section 25.2; Exercise 25.10)
3. Explain how materials move from the blood into the body cells and from the body cells into the blood. (Section 25.3; Exercise 25.14)
4. List the normal and abnormal constituents of urine. (Section 25.4; Exercise 25.18)
5. Discuss how proper fluid and electrolyte balance is maintained in the body. (Section 25.5; Exercise 25.22)
6. Explain how acid-base balance is maintained in the body. (Section 25.6; Exercise 25.26)
7. Explain how buffers work to control blood pH. (Section 25.7; Exercise 25.28)
8. Describe respiratory control of blood pH. (Section 25.8; Exercise 25.30)
9. Describe urinary control of blood pH. (Section 25.9; Exercise 25.34)
10. List the causes of acidosis and alkalosis. (Section 25.10; Exercise 25.38)

SOLUTIONS FOR THE END OF CHAPTER EXERCISES
A COMPARISON OF BODY FLUIDS (SECTION 25.1)

☑25.2 Plasma and interstitial fluid have nearly the same chemical composition.

25.4
a. protein — intracellular
b. K^+ — intracellular fluid
c. Na^+ — plasma, interstitial fluid
d. HPO_4^{2-} — intracellular fluid
e. HCO_3^- — plasma, interstitial fluid
f. Mg^{2+} — intracellular fluid
g. Cl^- — plasma, interstitial fluid

25.6
a. The principal anion in blood plasma is chloride.
b. The principal anion in interstitial fluid is chloride.
c. The principal anion in intracellular fluid is hydrogen phosphate.

OXYGEN AND CARBON DIOXIDE TRANSPORT (SECTION 25.2)
25.8
a. $HHb + O_2 \rightleftharpoons HbO_2^- + H^+$
b. $HHb + CO_2 \rightleftharpoons HHbCO_2$

☑25.10
a. O_2 dissolved gas in plasma — 2%
b. O_2 as oxyhemoglobin — 98%
c. CO_2 as carbaminohemoglobin — 25%
d. CO_2 as dissolved gas in plasma — 5%
e. CO_2 as bicarbonate ions — 70%

334 Chapter 25

25.12 Large capillary surface areas are critical to O_2 and CO_2 transport processes because the body needs to be able to diffuse large amounts of O_2 and CO_2 through the capillaries.

CHEMICAL TRANSPORT TO THE CELLS (SECTION 25.3)
☑25.14 a. At the venous end of a capillary osmotic pressure has the greater influence on the direction of fluid movement through capillary walls.
b. At the arterial end of a capillary blood pressure has the greater influence on the direction of fluid movement through capillary walls.

THE CONSTITUENTS OF URINE (SECTION 25.4)
25.16 The organic material normally excreted in the largest amount in urine is urea.

☑25.18 Abnormal constituents of urine are large amounts of glucose, protein, ketone bodies, pus, hemoglobin, red blood cells, and large amounts of bile pigments.

FLUID AND ELECTROLYTE BALANCE (SECTION 25.5)
25.20 Four routes by which water leaves the body are the kidneys (urine), the lungs (water vapor in expired air), the skin (diffusion and perspiration), and the intestines (feces).

☑25.22 The primary way fluid balance is maintained in the body is urine output.

25.24 During the operation of the thirst mechanism, as the body loses water, salivary secretions decrease, a dry feeling develops in the mouth, thirst is recognized, and water is drunk to relieve the condition.

ACID-BASE BALANCE (SECTION 25.6)
☑25.26 Three systems that cooperate to maintain blood pH in an appropriate narrow range are blood buffer, respiratory, and urinary.

BUFFER CONTROL OF BLOOD pH (SECTION 25.7)
☑25.28 Three major buffer systems of the blood are the bicarbonate system, the phosphate system, and the plasma protein system.

RESPIRATORY CONTROL OF BLOOD pH (SECTION 25.8)
☑25.30 The respiratory mechanism for controlling blood pH is as follows. CO_2 and H_2O are exhaled, more carbonic acid is removed from the blood and the blood pH rises. A reduced rate of breathing will lower the blood pH.

25.32 Hypoventilation is a symptom of high acid concentration in the body. It can result in respiratory acidosis.

URINARY CONTROL OF BLOOD pH (SECTION 25.9)
☑25.34 When kidneys function to control blood pH, the CO_2 in the blood is converted to bicarbonate ions in the blood.

25.36 As hydrogen ions are removed from the blood by the kidneys, sodium ions enter to bloodstream. This ionic shift maintains electron charge balance when the kidneys function to increase the pH of blood.

ACIDOSIS AND ALKALOSIS (SECTION 25.10)

☑25.38 The differences between the causes of respiratory acid-base imbalances and metabolic acid-base imbalances in the body are these: respiratory acid-base imbalances are caused by breathing too quickly and deeply or too slowly and shallowly, whereas metabolic acid-base imbalances are caused by various body processes that either produce or use acid.

25.40 If an individual takes an overdose of a depressant narcotic, they will probably breath more slowly and shallowly. This will likely lead to respiratory acidosis.

CHEMISTRY AROUND US, CHEMISTRY AND YOUR HEALTH, AND OVER THE COUNTER

25.42 Food can influence drugs by affecting the rate of absorption or changing the way the drug is metabolized.

25.44 Some classes of drugs are not safe to take with alcohol; a patient should always gollow the advice of the physician.

25.46 One pint of blood is typically collected during blood donation.

ADDITIONAL ACTIVITIES

Section 25.1 Review:
(1) Rank intracellular fluid, plasma, and interstitial fluid in order of increasing volumes in the body.
(2) What class of fluids contains both plasma and interstitial fluid?
(3) What are the major differences between the cations and anions inside intracellular fluid and the category from (2)?
(4) Rank intracellular fluid, plasma, and interstitial fluid in order of increasing protein content.

Section 25.2 Review:
(1) How is most oxygen transported through the bloodstream?
(2) How is most carbon dioxide transported through the bloodstream?
(3) How does oxygen enter the bloodstream?
(4) Why is chloride shift physiologically important?
(5) What enzyme is responsible for breaking carbonic acid into carbon dioxide and water?
(6) What enzyme is responsible for the formation of carbonic acid from carbon dioxide and water?

Section 25.3 Review:
(1) What are the various mechanisms of chemical transport in the body?
(2) Do capillary walls act more like osmotic membranes or dialyzing membranes? Explain.
(3) What factors govern the movement of water and dissolved materials through the capillary walls?
(4) How does the importance of the factors in (3) change as blood is pumped from the heart through the body and back to the heart?

Section 25.4 Review:
(1) Explain why urinalysis may be the first test performed on a patient entering the hospital.
(2) What is the major component of urine?
(3) Approximately how many grams of (2) should be excreted by an adult, if they also excrete 40 g of dissolved solids?
(4) If the entire volume of urine was due to the major component of (2), how many milliliters of urine should be excreted?

(5) Draw a pH scale for the "normal" urine pH range. Mark the neutral pH, the average pH for urine, the pH range for fruit/vegetable-rich diets, and the pH range for protein-rich diets.
(6) Is the average pH for urine acidic, basic, or neutral?

Section 25.5 Review:
(1) Is the relationship between water content and fat content in the body generally direct or inverse?
(2) Why are marathon runners generally given sports drinks at aid stations rather than just water?
(3) Fluid enters the body through food and drink. How does fluid leave the body?
(4) Which hormones influence urine output?

Section 25.6 Review:
(1) What is the [H^+] range for blood in the normal pH range?
(2) Is normal blood pH acidic, basic, or neutral?
(3) What are the three systems to maintain normal blood pH?

Section 25.7 Review:
(1) Two of the three ketone bodies are acids. Explain how the blood is able to maintain a healthy pH, even when a low concentration of ketone bodies are present.
(2) What happens to blood pH when a high concentration of ketone bodies are present in the blood?
(3) What is buffering according to biochemists?

Section 25.8 Review:
(1) During exercise, which is more likely to occur: hyperventilation or hypoventilation?
(2) During meditation, which is more likely to occur: hyperventilation or hypoventilation?

Section 25.9 Review:
(1) Which buffer systems are involved in the urinary system?
(2) Which buffer system is most responsible for keeping urine near pH 6?

Section 25.10 Review:
(1) When blood pH is normal and acid-base balance exists, which is in higher concentration: bicarbonate ions or carbonic acid?
(2) During respiratory alkalosis, which is in higher concentration: bicarbonate ions or carbonic acid?
(3) What is the cause of respiratory alkalosis?
(4) During respiratory acidosis, which is in higher concentration: bicarbonate ions or carbonic acid?
(5) What is the cause of respiratory acidosis?
(6) During metabolic acidosis, which is in higher concentration: bicarbonate ions or carbonic acid?
(7) What is the cause of metabolic acidosis?
(8) During metabolic alkalosis, which is in higher concentration: bicarbonate ions or carbonic acid?
(9) What is the cause of metabolic alkalosis?

Tying It All Together with a Laboratory Application:
(1) _____ and urine are body fluids examined in a hospital setting on a regular basis. The pH of (1) should be in the range of (2) _____, while the pH of urine should be in the range of (3) _____. The pH of (1) is regulated by (4) _____, respiratory, and urinary systems. The pH of urine is regulated by the (5) _____ buffer.

One of the main pathways of chemical transport within the body is the (6) _____. The main cation found in the (1) plasma is (7) _____ and the main anion is (8) _____. Nutrients are delivered to the cells because of (9) _____ and waste products are reclaimed from the cells because of (10) _____. The body disposes of these waste products in (11) _____, _____, _____, _____, and _____. These processes also involve (12) _____ and electrolytes. When the (12) level in the body is low, (13) _____ helps maintain electrolyte balance.

Hyperventilation, prolonged vomiting, and ingesting excess antacid can lead to (14) _____, which is a condition in which the pH of the blood is (15) _____. Hypoventilation, uncontrolled diabetes mellitus, severe diarrhea, and aspirin overdose can lead to (16) _____, which is a condition in which the pH of the blood is (17) _____. These conditions are dangerous because death can occur if the pH of (1) is below (18) _____ or above (19) _____. Respiratory causes of (14) and (16) change the concentration of (20) _____ in the (1).

SOLUTIONS FOR THE ADDITIONAL ACTIVITIES

Section 25.1 Review:
(1) volume in the body: plasma < interstitial fluid < intracellular fluid
(2) Extracellular fluids contain both plasma and interstitial fluid.
(3) Intracellular fluids contain mostly potassium cations and hydrogen phosphate anions. Extracellular fluids contain mostly sodium cations and chloride anions.
(4) protein content: interstitial fluid < plasma < intracellular fluid

Section 25.2 Review:
(1) Most oxygen is transported through the bloodstream bound to hemoglobin in the form of oxyhemoglobin.
(2) Most carbon dioxide is transported through the bloodstream as bicarbonate ions.
(3) Oxygen enter the bloodstream by diffusing from the alveoli of the lungs through the capillary wall into the red blood cells.
(4) The chloride shift is physiologically important because it maintains electrolyte balance and osmotic pressure relationships between red blood cells and plasma.
(5) Carbonic anhydrase is responsible for breaking carbonic acid into carbon dioxide and water.
(6) Carbonic anhydrase is responsible for the formation of carbonic acid from carbon dioxide and water.

Section 25.3 Review:
(1) Chemical transport in the body requires the substances to become part of the moving bloodstream. They may dissolve in the water-based plasma, chemically bond to cellular components, or form a suspension in the plasma.
(2) Capillary walls act more like dialyzing membranes than osmotic membranes because they allow materials other than the solvent (water) to pass through the walls.
(3) Blood pressure and osmotic pressure govern the movement of water and dissolved materials through the capillary walls.
(4) Blood pressure exceeds osmotic pressure as blood is pumped away from the heart; however, osmotic pressure exceeds blood pressure as blood is returned to the heart.

Section 25.4 Review:
(1) Urine is an excellent indicator of health. The components of urine can be used to diagnosis illness.
(2) The major component of urine is water.
(3) 960 g of water

$$\frac{4\%}{100\%} = \frac{40\ g}{x}$$

$$4\%\ x = (100\%)(40\ g)$$

$$x = 1000\ g\ \text{urine}$$

water = 96% of urine = (96%)(1000 g) = 960 g water

(4) 960 mL

$$960 \text{ g water} \left(\frac{1 \text{ mL}}{1 \text{ g}}\right) = 960 \text{ mL}$$

(5)
(6) The average pH for urine is acidic.

Section 25.5 Review:
(1) The relationship between water content and fat content in the body is inverse. High fat content corresponds to low water content and vice versa.
(2) Marathon runners generally given sports drinks at aid stations rather than just water because they are sweating profusely and need to replenish not only the fluids lost, but also the electrolyte balance in the body.
(3) Fluids leave the body through urine, feces, perspiration, diffusion through the skin, and exhaled air.
(4) Vasopressin helps reabsorb water in the kidneys. Aldosterone helps reabsorb sodium ions from the kidneys.

Section 25.6 Review:
(1) The [H⁺] range for blood in the normal pH range is 3.5×10^{-8} M - 4.5×10^{-8} M.

$$[H^+] = 10^{-pH}$$
$$10^{-7.35} = 4.5 \times 10^{-8} \text{ M}$$
$$10^{-7.45} = 3.5 \times 10^{-8} \text{ M}$$

(2) Normal blood pH is basic.
(3) The three systems to maintain normal blood pH are the buffer system, respiratory system, and urinary system.

Section 25.7 Review:
(1) The blood has three buffer systems that respond to the presence of ketone bodies in the blood and attempt to maintain a "normal" pH.
(2) If the concentration of ketone bodies in the blood is too high, then the buffer capacity of the blood could be overwhelmed and the pH of the blood will change.
(3) According to biochemists, buffering is the replacement of a stronger acid with a weaker acid.

Section 25.8 Review:
(1) During exercise, hyperventilation is more likely to occur than hypoventilation because the rate of breathing is increased. Some athletes develop patterns for breathing to ensure the oxygen in the air breathed into the lungs is exchanged for the carbon dioxide waste products. These patterns help athletes guard against hyperventilation.
(2) During meditation, hypoventilation is more likely to occur than hyperventilation because the rate of breathing is slowed. Most "meditation" instructors remind their students to breathe as they become increasingly relaxed. This generally prevents hypoventilation.

Section 25.9 Review:
(1) The bicarbonate buffer system and the phosphate buffer system are involved in the urinary system.
(2) The phosphate buffer system is most responsible for keeping urine near pH 6.

Section 25.10 Review:
(1) In the blood, the bicarbonate ion concentration is higher than the carbonic acid concentration at a normal pH that is in a state of acid-base balance.
(2) During respiratory alkalosis, bicarbonate ions are in higher concentration than carbonic acid.
(3) The cause of respiratory alkalosis is hyperventilation which results in the excessive loss of carbon dioxide.
(4) During respiratory acidosis, although the the bicarbonate ion concentration in the blood is higher than the carbonic acid concentration, the bicarbonate concentration is lower than the bicarbonate concentration in blood at normal pH.
(5) The cause of respiratory acidosis is hypoventilation which results in the accumulation of carbon dioxide.
(6) During metabolic acidosis, the bicarbonate ion concentration is higher than the carbonic acid concentration; however, the blood contains a lower bicarbonate ion concentration than blood at normal pH.
(7) Metabolic acidosis is caused by excessive H^+ ions in the blood.
(8) During metabolic alkalosis, the bicarbonate ion concentration is greater than the carbonic acid concentration. The bicarbonate ion concentration is also higher than in blood at a normal pH.
(9) Metabolic alkalosis is caused by a loss of acid because of vomiting or ingestion of alkaline substances.

Tying It All Together with a Laboratory Application:
(1) Blood
(2) 7.35-7.45
(3) 4.5-8.0
(4) buffer
(5) phosphate
(6) bloodstream
(7) Na^+
(8) Cl^-
(9) blood pressure
(10) osmotic pressure
(11) urine, feces, perspiration, diffusion from the skin, water vapor
(12) water
(13) aldosterone
(14) alkalosis
(15) raised
(16) acidosis
(17) lowered
(18) 6.8
(19) 7.8
(20) CO_2

SELF-TEST QUESTIONS
Multiple Choice

1. Which of the following fluids contain similar concentrations of Na^+ and Cl^- ions?
   a. blood plasma and interstitial fluid
   b. interstitial fluid and intracellular fluid
   c. blood plasma and intracellular fluid
   d. blood plasma, interstitial fluid, and intracellular fluid

2. A body fluid is analyzed and found to contain a low concentration of Na^+ and a high concentration of K^+. The fluid is most likely
   a. blood plasma
   b. interstitial fluid
   c. intracellular fluid
   d. all three have about the same concentrations

3. Which of the following reactions takes place in red blood cells at the lungs?
   a. $H_2CO_3 \rightarrow H_2O + CO_2$
   b. $H^+ + HbO_2^- \rightarrow HHb + O_2$
   c. $CO_2 + H_2O \rightarrow H_2CO_3$
   d. $H_2CO_3 \rightarrow H^+ + HCO_3^-$

4. Most oxygen is carried to various parts of the body, via the bloodstream, in the form of
   a. a dissolved gas
   b. bicarbonate ion
   c. oxyhemoglobin
   d. carbon dioxide

5. During respiration reactions at the lungs and cells, the function of the chloride shift is to
   a. eliminate toxic chlorine from the body
   b. maintain charge balance in red blood cells
   c. maintain pH balance in red blood cells
   d. activate the carbonic anhydrase enzyme in red blood cells

6. Which of the following would be considered to be an abnormal constituent in urine?
   a. ammonium ion
   b. protein
   c. creatinine
   d. bicarbonate salts

7. The vasopressin mechanism and the aldosterone mechanism both tend to regulate the
   a. fluid and electrolyte levels in the body
   b. rate of hemoglobin production in the body
   c. rate of glucose oxidation in the body
   d. CO_2 levels in the cells

8. Which of the following maintains a constant pH for the blood?
   a. respiration reactions associated with breathing
   b. kidney activity
   c. formation and excretion of perspiration
   d. more than one response is correct

9. Which buffer system is regulated in part by the kidneys and by the respiratory system?
   a. bicarbonate
   b. protein
   c. phosphate
   d. ammonium

10. The three major buffer systems of the blood are the plasma proteins, the bicarbonate buffer, and the
    a. succinate buffer
    b. lactate buffer
    c. ammonium buffer
    d. phosphate buffer

11. Exhaling CO_2 and H_2O
    a. raises blood pH
    b. increases the blood concentration of H_2CO_3
    c. lowers blood pH
    d. has no effect on H_2CO_3 concentration

12. Blood pH normally remains in the range
    a. 7.15-7.25
    b. 7.25-7.35
    c. 7.35-7.45
    d. 7.45-7.55

Matching
For each cause of blood acid-base imbalance listed on the left, select the resulting condition from the responses.
13. prolonged diarrhea
14. excessive intake of baking soda
15. hypoventilation
16. hyperventilation

a. respiratory acidosis
b. respiratory alkalosis
c. metabolic acidosis
d. metabolic alkalosis

For each condition listed on the left, identify the resulting abnormal urine constituent on the right.

17. hepatitis
18. starvation
19. diabetes mellitus

a. ketone bodies
b. glucose (in large amounts)
c. bile pigments
d. more than one constituent results

Select a correct name for each formula on the left.

20. HbO_2^-
21. HHb
22. $HHbCO_2$

a. carbaminohemoglobin
b. oxyhemoglobin
c. deoxyhemoglobin
d. carboxyhemoglobin

True-False

23. Plasma is classified as an interstitial fluid.
24. Osmotic pressure differences between plasma and interstitial fluid always tend to move fluid into the blood.
25. Osmotic pressure exceeds heart pressure at the venous end of the circulatory system.
26. An increased rate of breathing tends to lower blood pH.
27. The excretion of H^+ ions decreases urine pH.
28. The pH of blood is increased by the excretion of H^+ in urine.
29. Vomiting may give rise to metabolic acidosis.
30. The concentration of protein in plasma is much higher than the protein concentration in the interstitial fluid.
31. The body is 82-90% water.
32. Aldosterone is also called the antidiuretic hormone (ADH).
33. Death can result if blood pH falls below 6.8.
34. H_2CO_3 is a moderately strong acid.
35. Urine is buffered by the phosphate buffer.

ANSWERS TO CHAPTER SELF-TESTS

CHAPTER 1
1. C
2. C
3. A
4. A
5. D
6. B
7. B
8. B
9. B
10. C
11. D
12. A
13. C
14. C
15. A
16. D
17. B
18. A
19. C
20. B
21. C
22. E
23. F
24. T
25. T
26. F
27. T
28. F
29. F
30. T
31. T
32. F
33. F
34. F
35. T

CHAPTER 2
1. D
2. B
3. A
4. C
5. D
6. C
7. B
8. A
9. B
10. D
11. D
12. B
13. C
14. D
15. C
16. A
17. B
18. B
19. F
20. F
21. T
22. T
23. T
24. T
25. F

CHAPTER 3
1. A
2. D
3. A
4. D
5. C
6. A
7. B
8. B
9. B
10. C
11. D
12. C
13. C
14. A
15. D
16. C
17. D
18. T
19. T
20. T
21. T
22. T
23. B
24. A
25. C
26. B
27. B
28. C
29. A
30. D
31. A
32. C
33. E
34. A
35. B

CHAPTER 4
1. A
2. D
3. A
4. D
5. A
6. C
7. D
8. B
9. B
10. B
11. C
12. B
13. B
14. B
15. D
16. F
17. F
18. T
19. F
20. F
21. F
22. F
23. D
24. B
25. B
26. B
27. C
28. A
29. E
30. A
31. D
32. A
33. C
34. D
35. B

CHAPTER 5
1. B
2. A
3. D
4. D
5. B
6. C
7. B
8. D
9. B
10. B
11. F
12. T
13. T
14. F
15. F
16. T
17. T
18. F
19. A
20. E
21. B
22. D
23. E
24. C
25. A
26. D
27. B
28. A
29. C
30. A
31. A; Fe
32. A; H_2
33. B
34. B
35. A; Ag_2O

Self-Test Answers

CHAPTER 6

1.	A	8.	B	15.	F	22.	T	29.	C
2.	A	9.	B	16.	T	23.	F	30.	B
3.	B	10.	A	17.	F	24.	F	31.	D
4.	B	11.	C	18.	T	25.	T	32.	D
5.	D	12.	D	19.	T	26.	F	33.	C
6.	A	13.	A	20.	F	27.	F	34.	B
7.	A	14.	C	21.	T	28.	D	35.	A

CHAPTER 7

1.	B	8.	D	15.	F	22.	T	29.	T
2.	A	9.	B	16.	F	23.	F	30.	F
3.	D	10.	C	17.	T	24.	F	31.	C
4.	B	11.	D	18.	T	25.	T	32.	B
5.	A	12.	B	19.	T	26.	T	33.	B
6.	B	13.	D	20.	T	27.	F	34.	A
7.	A	14.	T	21.	F	28.	F	35.	D

CHAPTER 8

1.	D	8.	B	15.	F	22.	A	29.	A
2.	A	9.	B	16.	F	23.	B	30.	D
3.	D	10.	B	17.	T	24.	C	31.	B
4.	C	11.	T	18.	T	25.	C	32.	A
5.	A	12.	T	19.	F	26.	A	33.	A
6.	B	13.	F	20.	F	27.	B	34.	B
7.	C	14.	T	21.	C	28.	C	35.	C

CHAPTER 9

1.	B	9.	D	17.	F	25.	B	33.	C
2.	D	10.	B	18.	F	26.	C	34.	D
3.	D	11.	F	19.	T	27.	B	35.	B
4.	C	12.	T	20.	T	28.	A	36.	A
5.	C	13.	T	21.	F	29.	A	37.	A
6.	D	14.	T	22.	C	30.	B	38.	C
7.	A	15.	F	23.	D	31.	D	39.	B
8.	C	16.	T	24.	A	32.	A	40.	A

CHAPTER 10

1.	D	5.	D	9.	B	13.	F	17.	F
2.	A	6.	C	10.	D	14.	T	18.	C
3.	C	7.	D	11.	T	15.	T	19.	D
4.	D	8.	A	12.	F	16.	F	20.	B

CHAPTER 11

1. D	9. C	17. A	25. T	33. F
2. A	10. A	18. C	26. F	34. T
3. C	11. C	19. C	27. T	35. F
4. A	12. D	20. B	28. T	36. T
5. B	13. D	21. A	29. F	37. F
6. B	14. A	22. B	30. T	38. T
7. B	15. B	23. F	31. F	39. F
8. A	16. D	24. F	32. T	40. T

CHAPTER 12

1. B	7. B	13. C	19. B	25. T
2. C	8. C	14. A	20. T	26. F
3. D	9. D	15. D	21. T	27. T
4. D	10. C	16. B	22. F	28. T
5. C	11. C	17. C	23. T	29. F
6. C	12. D	18. A	24. F	30. T

CHAPTER 13

1. C	7. B	13. C	19. C	25. T
2. C	8. A	14. D	20. A	26. T
3. A	9. C	15. B	21. F	27. F
4. B	10. A	16. C	22. T	28. T
5. A	11. B	17. A	23. F	29. F
6. B	12. A	18. B	24. T	30. F

CHAPTER 14

1. A	6. A	11. C	16. D	21. F
2. B	7. B	12. D	17. A	22. T
3. B	8. B	13. B	18. B	23. T
4. B	9. A	14. D	19. T	24. F
5. A	10. B	15. C	20. F	25. T

CHAPTER 15

1. B	7. C	13. C	19. A	25. F
2. B	8. C	14. A	20. B	26. T
3. D	9. D	15. B	21. F	27. T
4. B	10. B	16. D	22. F	28. T
5. D	11. B	17. C	23. T	29. F
6. D	12. B	18. D	24. T	30. F

CHAPTER 16

1. A	7. C	13. A	19. F	25. T
2. D	8. D	14. D	20. F	26. T
3. B	9. A	15. B	21. F	27. T
4. B	10. B	16. C	22. T	28. T
5. A	11. D	17. A	23. F	29. T
6. C	12. C	18. T	24. T	30. F

CHAPTER 17

1. B
2. A
3. B
4. C
5. C
6. D
7. B
8. C
9. C
10. A
11. B
12. D
13. B
14. D
15. C
16. B
17. C
18. A
19. C
20. A
21. D
22. T
23. F
24. F
25. T
26. T
27. T
28. T
29. F
30. T

CHAPTER 18

1. B
2. A
3. C
4. B
5. A
6. C
7. C
8. A
9. C
10. A
11. C
12. D
13. B
14. D
15. D
16. C
17. A
18. C
19. A
20. D
21. B
22. F
23. T
24. T
25. T
26. T
27. F
28. F
29. F
30. T

CHAPTER 19

1. C
2. B
3. B
4. A
5. B
6. C
7. D
8. A
9. D
10. D
11. D
12. B
13. B
14. C
15. D
16. A
17. C
18. A
19. B
20. D
21. B
22. D
23. A
24. C
25. C
26. D
27. A
28. C
29. C
30. F
31. T
32. T
33. T
34. F
35. F

CHAPTER 20

1. D
2. A
3. D
4. D
5. D
6. A
7. A
8. C
9. D
10. C
11. B
12. B
13. D
14. C
15. A
16. B
17. C
18. A
19. B
20. C
21. D
22. A
23. B
24. D
25. C
26. T
27. F
28. F
29. F
30. T
31. T
32. F
33. T
34. T
35. F

CHAPTER 21

1. B
2. B
3. D
4. B
5. C
6. A
7. D
8. D
9. A
10. C
11. A
12. A
13. D
14. A
15. B
16. C
17. A
18. C
19. C
20. D
21. B
22. T
23. F
24. F
25. F
26. F
27. T
28. T
29. F
30. F
31. T
32. F
33. T
34. T
35. T

Self-Test Answers 347

CHAPTER 22
1. B
2. A
3. A
4. C
5. B
6. D
7. B
8. D
9. C
10. A
11. C
12. A
13. B
14. C
15. C
16. A
17. B
18. D
19. C
20. B
21. C
22. B
23. A
24. T
25. F
26. T
27. F
28. F
29. T
30. F
31. F
32. F
33. F
34. F
35. T

CHAPTER 23
1. D
2. B
3. A
4. C
5. C
6. B
7. A
8. C
9. D
10. B
11. B
12. A
13. B
14. D
15. C
16. A
17. B
18. C
19. C
20. B
21. A
22. D
23. C
24. B
25. A
26. B
27. C
28. F
29. F
30. T
31. T
32. T
33. T
34. F
35. F

CHAPTER 24
1. A
2. D
3. D
4. B
5. D
6. A
7. B
8. D
9. C
10. C
11. B
12. A
13. C
14. B
15. A
16. D
17. D
18. B
19. D
20. D
21. D
22. B
23. A
24. T
25. T
26. T
27. T
28. T
29. F
30. T
31. F
32. T
33. T
34. F
35. T

CHAPTER 25
1. A
2. C
3. A
4. C
5. B
6. B
7. A
8. D
9. A
10. D
11. A
12. C
13. C
14. D
15. A
16. B
17. C
18. A
19. D
20. B
21. C
22. A
23. F
24. T
25. T
26. F
27. T
28. T
29. F
30. T
31. F
32. F
33. T
34. F
35. T